Lecture Notes in Mathematics

Edited by A. Dold and B. Eckmann

1243

Non-Commutative Harmonic Analysis and Lie Groups

Proceedings of the International Conference
held in Marseille-Luminy, June 24–29, 1985

Edited by J. Carmona, P. Delorme and M. Vergne

Springer-Verlag
Berlin Heidelberg New York London Paris Tokyo

Editors

Jacques Carmona
Patrick Delorme
Université d'Aix Marseille II, Département de Mathématiques de Luminy
70, Route Léon Lachamp, 13288 Marseille Cedex 2, France

Michèle Vergne
CNRS Université de Paris VII, UER Mathématiques
2, Place Jussieu, 75221 Paris Cedex 05, France

and

M.I.T.
Cambridge, MA. 02139, USA

Mathematics Subject Classification (1980): 20G05, 22E50, 22E30

ISBN 3-540-17701-9 Springer-Verlag Berlin Heidelberg New York
ISBN 0-387-17701-9 Springer-Verlag New York Berlin Heidelberg

Library of Congress Cataloging-in-Publication Data. Non commutative harmonic analysis and Lie groups. (Lecture notes in mathematics; 1243) "Le sixième Colloque 'Analyse harmonique et groupes de Lie' s'est tenu à Marseille-Luminy du 24 au 29 juin 1985 dans le cadre du Centre international de rencontres mathématiques (C.I.R.M.)"–Pref. Text in English and French. Bibliography: p. 1. Harmonic analysis–Congresses. 2. Lie groups–Congresses. I. Carmona, Jacques, 1934-. II. Delorme, Patrick. III. Vergne, Michèle. IV. Colloque "Analyse harmonique et groupes de Lie" (6th: 1985: Luminy, Marseille, France) V. Series: Lecture notes in mathematics (Springer-Verlag), 1243. QA3.L28 no. 1243 510 s 87-9444 [QA403] [512'.55]
ISBN 0-387-17701-9 (U.S.)

Printed in Germany

Printing and binding: Druckhaus Beltz, Hemsbach/Bergstr.
2146/3140-543210

P R E F A C E

Le sixième Colloque "Analyse Harmonique et Groupes de Lie" s'est tenu à Marseille-Luminy du 24 au 29 Juin 1985 dans le cadre du Centre International de Rencontres Mathématiques (C.I.R.M.)

Le présent volume contient le texte des conférences présentées, à l'exception de celles dont le contenu a déjà fait ou fera l'objet d'une publication par ailleurs.

Outre les participants à cette rencontre, nous tenons à remercier la Faculté de Luminy et le C.I.R.M. qui ont rendu possible la tenue de ce Colloque et le secrétariat du Laboratoire de Mathématiques (L.A. 225 du C.N.R.S. qui a assuré la préparation du présent volume.

Jacques CARMONA
Patrick DELORME
Michèle VERGNE

TABLE OF CONTENTS

M. Andler, Relationships of divisibility between local L-functions associated to representations of complex reductive groups 1

M.W. Baldoni-Silva, A.W. Knapp, Irreducible unitary representations of some groups of real rank two 15

M.W. Baldoni-Silva, A.W. Knapp, Vogan's algorithm for computing composition series 37

D. Barbasch, Unipotent representations and unitarity 73

J.-L. Clerc, Bochner-Riesz means of H^p functions ($0<p<1$) on compact Lie groups 86

P. Delorme, Injection de modules sphériques pour les espaces symétriques réductifs dans certaines représentations induites 108

A. Joseph, On the cyclicity of vectors associated with Duflo involutions 144

T. Kawazoe, Atomic Hardy spaces on semisimple Lie groups 189

J. Orloff, Orbital integrals on symmetric spaces 198

D. Peterson, M. Vergne, Recurrence relations for Plancherel functions 240

J. Rohlfs, B. Speh, A cohomological method for the determination of limit multiplicities 262

W. Rossmann, Springer representations and coherent continuation representations of Weyl groups 273

S. Sano, Distributions sphériques invariantes sur l'espace symétrique semi-simple et son c-dual 283

Relationships of divisibility between local L-functions associated to representations of complex reductive groups

Martin Andler
Massachussetts Institute of Technology
CNRS-Université Paris 7

1. Introduction

Let G be a complex connected reductive algebraic Lie group, $\mathfrak{g}_0$ its real Lie algebra (i.e. the underlying real algebra of the complex Lie algebra), $\mathfrak{g} = (\mathfrak{g}_0)_{\mathbb{C}}$ the complexification of $\mathfrak{g}_0$, and K a maximal compact subgroup of G. The Langlands classification asserts that the set of equivalence classes of irreducible representations of G (precise definitions will be given later) is parametrized by the set of (conjugacy classes of) morphisms of $\mathbb{C}^*$ into the set of semi-simple elements of the connected component of the L-group ${}^LG^0$ of G.

If r is a finite representation of ${}^LG^0$, a morphism ϕ of $\mathbb{C}^*$ into ${}^LG^0$ defines a finite dimensional representation of $\mathbb{C}^*$ and hence an L-function in the sense of Artin. We thus have a triangle whose vertices are :

a. representations of G

b. morphisms of $\mathbb{C}^*$ into ${}^LG^0$

c. L-functions.

Unfortunately, little is known about links between representations and L-functions (a. and c.) without going through b.. The only case which is well understood is the case of $Gl(n,\mathbb{C})$ and r the standard representation of ${}^LG^0 = G$. The results there are due to Jacquet (<J>, see also <G-J>, <J-L>). As a motivation for the present paper, let us describe Jacquet's results.

He gives a method for computing the L-function associated to a representation π (recall that r is the trivial representation of ${}^LG^0$). It goes roughly in the following way.

Let $M(n,\mathbb{C})$ be the set of nxn matrices with complex coefficients, and $S_0(n,\mathbb{C})$ the set of functions on $M(n,\mathbb{C})$ of the form

$$\phi(z) = P(z_{ij},\bar{z}_{ij}) \exp(-2\pi \sum z_{ij}\bar{z}_{ij})$$

where P is a polynomial. By the Langlands classification, π is a well defined subquotient of some principal series representation ρ. Let f be a coefficient of ρ , and define

$$\zeta(\Phi,f,s) = \int_G \Phi(g)\, f(g)\, |\det g|^s\, dg .$$

The integral converges in some half-plane $Re(s) > Re(s_0)$, extends meromorphically to $\mathbb{C}$. When Φ varies over $S_0(n,\mathbb{C})$ and f over the set of coefficients of ρ , the $\zeta(\Phi,f,s)$ can be written

$$\zeta(\Phi,f,s) = P(\Phi,f,s)\, L'(\rho,s)$$

where P is a polynomial in s and $L'(\rho,s)$ is a meromorphic function which is uniquely defined up to constants. The same results hold if we let f vary over the set of coefficients of a subquotient π' of ρ; the corresponding L'-factor is denoted $L'(\pi',s)$.

The results of Jacquet which are relevant to us are the following :

1° If π is the representation of G associated to the morphism $\phi : \mathbb{C}^* \to {}^LG^0$, we have, up to multiplication by constants

$$L(\phi,s) = L'(\pi,s).$$

We can assume from now on that the equality holds.

2° Let ρ be a principal series representation and π its Langlands subquotient. We have the following equality :

$$L'(\pi,s) = L'(\rho,s).$$

(This should be thought of as a motivation for the choice of the Langlands quotient.)

We can now forget about the ' and write simply $L(\pi,s) = L(\phi,s)$ if π is the representation associated to ϕ . Results 1° and 2° imply that

(*) $\frac{L(\pi',s)}{L(\pi,s)}$ is an entire function if π' is *a* subquotient in a principal series ρ whose *Langlands* subquotient is π.

3° The ζ factors verify a functional equation

$$\zeta(\Phi^\wedge,f^\vee,1-s+\tfrac{1}{2}(n-1)) = \gamma(\rho,s)\, \zeta(\Phi,f,s)$$

where $\Phi^\wedge$ is the Fourier transform of Φ and $f^\vee$ is the distribution contragredient to f, and the factor $\gamma(\rho,s)$ is a meromorphic function. If we restrict f to being a coefficient of a subquotient π' of ρ the corresponding $\gamma(\pi,s)$ verifies

$$\gamma(\pi,s) = \gamma(\rho,s).$$

4° The meromorphic function $\varepsilon(\pi,s) = \gamma(\pi,s)\, L(\pi,s)/L(\pi^{V},1-s)$ is entire and its inverse is entire. (π^{V} the representation contragredient to π)

These facts imply that

(**) $\dfrac{L(\pi,s)}{L(\pi^{V},1-s)}$ is, up to multiplication by a function ψ such that ψ and $1/\psi$ are both entire, independent of the choice of the irreducible subquotient π of the principal series ρ.

The results in this paper are a generalisation of (*) and (**) to the case of a general reductive group and of a general representation r of the L-group. Going back to the general case, we use the notations of the beginning of the paper : G, ${}^{L}G^{0}$, etc. We prove the following theorem.

Theorem. Let π be the Langlands subquotient of a principal series representation ρ and let π' be another irreducible subquotient of ρ. Let r be a representation of the L-group ${}^{L}G^{0}$, and denote by $L(\pi,r,s)$, $L(\pi',r,s)$ the L-functions associated to the data. Then

1° $L(\pi',r,s)/L(\pi,r,s)$ is an entire function.

2° $\dfrac{L(\pi,r,s)}{L(\pi^{V},r,1-s)} = \dfrac{L(\pi',r,s)}{L(\pi'^{V},r,1-s)}\,\psi(s)$, where $\psi(s)$ and $1/\psi(s)$ are both entire.

Remarks. *It can be useful to think of the results above in a slightly different way :*

- In terms of divisibility in the ring of differentiable functions, 1° states that $L(\pi,r,s)$ *divides* $L(\pi',r,s)$ *and 2° states that* $\dfrac{L(\pi',r,s)}{L(\pi'^{V},r,1-s)} = \dfrac{L(\pi,r,s)}{L(\pi^{V},r,1-s)}$ *are associated.*

- In terms of the sets of poles and zeroes of the L-functions considered, 1° means that the set of poles of $L(\pi,r,s)$ *contains the set of poles of* $L(\pi',r,s)$ *and the set of zeroes of* $L(\pi,r,s)$ *is contained in the set of zeroes of* $L(\pi',r,s)$ *. 2° means that the set of zeroes and the set of poles of* $\dfrac{L(\pi,r,s)}{L(\pi^{V},r,1-s)}$ *are independent of the choice of the irreducible subquotient* π *of* ρ*.*

Analogous results hold in the p-adic case (<Ro>). They should also extend to the real case, but the proofs may well be harder, be it only because the combinatorial description of the composition series of a principal series representation of a real group is more complicated. One may ask also whether a converse of the theorem holds. Along those lines here are two questions that might be worth looking into :

1° Do the sets of poles and zeroes of $L(\pi,r,s)/L(\pi^{\vee},r,1-s)$ determine the infinitesimal character of π?

2° If π and π' have same infinitesimal character, does the property " $L(\pi',r,s)/L(\pi,r,s)$ is entire for all r " imply that π' belongs to the Jordan-Hölder series of the principal series representation ρ whose Langlands quotient is π ?

The theorem in this paper was stated to me as an oral conjecture by Laurent Clozel, to whom I am also indebted for numerous conversations . I would also like to thank Frédéric Bien, Hervé Jacquet and David Vogan for their suggestions.

2. General notations

Let us first give the framework and notations to state the Langlands correspondence. Following Borel (<Bo>), we consider a connected algebraic group G over $\mathbb{C}$, T a maximal torus of G, $X^*(T)$ the group of characters of T and $X_*(T)$ the group of 1-parameter subgroups of T, Φ the set of roots of G with respect to T, $\Phi^{\vee}$ the associated set of coroots. Of course $X^*(T)$ is a sublattice of the lattice of weights P of Φ , $X_*(T)$ being a sublattice of the lattice of weights $P^{\vee}$ of $\Phi^{\vee}$. We call $\Psi(G) = (X^*(T),\Phi,X_*(T),\Phi^{\vee})$ the <u>root datum</u> associated to G; there is a bijection between the set of connected algebraic groups defined over $\mathbb{C}$, and the set of root data.

The choice of a Borel subgroup B of G containing T is equivalent to that of a basis Δ of Φ. We thus get a bijection between the set of isomorphism classes of triples (G,B,T) and the set of <u>based root data</u> $\Psi_0(G) = (X^*(T),\Delta, X_*(T),\Delta^{\vee})$.

Consider the based root datum $\Psi_0 = \Psi_0(G) = (X^*(T),\Delta,X_*(T),\Delta^{\vee})$ associated to G. The <u>inverse system</u> $\Psi_0^{\vee}$ is by definition $(X_*(T),\Delta^{\vee},X^*(T),\Delta)$, and defines an algebraic connected group ${}^LG^0$ over $\mathbb{C}$, along with a torus ${}^LT^0$ and a Borel subgroup ${}^LB^0$, such that

$$\Psi_0({}^LG^0) = \Psi_0^{\vee} .$$

The <u>Weil group</u> $W_{\mathbb{C}}$ of $\mathbb{C}$ is simply $\mathbb{C}^*$, and the <u>L-group</u> ${}^LG$ of G is the direct product

$${}^LG = W_{\mathbb{C}} \times {}^LG^0 .$$

Examples. If G is semi-simple classical, the process of going from G to ${}^LG^0$ amounts to going from the simply connected to the adjoint group in the A_n and D_n series and to going from the simply connected B_n and C_n to the adjoint C_n and B_n respectively. If G is $Gl(n,\mathbb{C})$ then ${}^LG^0$ is also $Gl(n,\mathbb{C})$.

Systematically, we shall use the following notations. Let $\mathfrak{g}_0$ be the (complex) Lie algebra of G, $\mathfrak{g}_0$ the underlying real Lie algebra, and $\mathfrak{g}$ its complexification. Choose a Chevalley basis of $\underline{\mathfrak{g}}_0$, $(H_\alpha, X_\alpha)_{\alpha \in \Phi}$ where the H_α generate the Lie algebra $\underline{\mathfrak{t}}_0$ of T. The H_α and X_α generate a real form with respect to which we define a conjugation

$$X \to \overline{X} .$$

Consider the anti-automorphism $X \to {}^tX$ defined by :

$${}^tX_\alpha = X_{-\alpha}$$

$${}^tH_\alpha = H_\alpha .$$

Then the map θ :

$$\theta(X) = -{}^t\overline{X}$$

is a Cartan involution of $\mathfrak{g}_0$. We get a Cartan decomposition :

$$\mathfrak{g}_0 = \mathfrak{k}_0 + \mathfrak{p}_0 .$$

Since $\mathfrak{t}_0$ is stable under θ , we have

$$\mathfrak{t}_0 = \mathfrak{a}_0 + \mathfrak{m}_0 ,$$

where $\mathfrak{a}_0 = \mathfrak{t}_0 \cap \mathfrak{p}_0$ and $\mathfrak{m}_0 = \mathfrak{t}_0 \cap \mathfrak{k}_0$.

Put

$$\mathfrak{n}_0 = \sum \mathbb{C}.X_\alpha , \text{ the sum being on } \alpha \in \Phi^+.$$

Then $\mathfrak{g}_0 = \mathfrak{k}_0 + \mathfrak{a}_0 + \mathfrak{n}_0$ is an Iwasawa decomposition of $\mathfrak{g}_0$.

The Langlands decomposition establishes a one-to-one map between

1) the set of admissible irreducible $(\mathfrak{g},K)$ modules

2) the set of morphisms from $W_\mathbb{C}$ into ${}^LG^0$ with semi-simple image, up to conjugacy.

In the context of complex groups, the Langlands classification was known by the work of Zhelobenko. Since it will anyway be useful later, we might as well describe it now. We follow Duflo (<Du 1>).

There is a natural identification of $\mathfrak{g}$ with $\underline{\mathfrak{g}}_0 \times \overline{\mathfrak{g}}_0$, sending an element Z of $\mathfrak{g}_0$ to $(Z,\overline{Z}) \in \underline{\mathfrak{g}}_0 \times \underline{\mathfrak{g}}_0$. This map restricts to an identification of

$\mathfrak{t}$ with $\underline{t}_0 \times \underline{t}_0$. Using this isomorphism, we get an identification $\mathfrak{t}^* \sim \underline{t}_0^* \times \underline{t}_0^*$. We shall denote (p,q) $(p,q \in \underline{t}_0^*)$ the element of $\mathfrak{t}^*$ defined by

$$(p,q)(H) = p(H) + q(\overline{H})$$

for H in $\underline{t}_0$. On the other hand, since $\mathfrak{t}_0 = \mathfrak{a}_0 + \mathfrak{m}_0$, $\mathfrak{t}^*$ can also be identified to $\mathfrak{a}^* + \mathfrak{m}^*$. Now $\mathfrak{a}$ and $\mathfrak{m}$ are both isomorphic naturally to $\underline{t}_0$ so we get another identification of $\mathfrak{t}^*$ with $\underline{t}_0^* \times \underline{t}_0^*$. We will write $\mu+\lambda$ for the form

$$(\mu+\lambda)\ (H+H') = \mu(H) + \lambda(H')$$

for H in $\mathfrak{m}_0$ and H' in $\mathfrak{a}_0$.

The dictionary to go from one description to the other is the following :

$$(p,q) = \mu+\lambda \quad \text{if and only if} \quad \begin{cases} p = \frac{1}{2}(\mu+\lambda) \\ q = \frac{1}{2}(-\mu+\lambda). \end{cases}$$

Finally, let $\sigma \in \underline{t}_0^*$ be half the sum of positive roots, and $\rho = (\sigma,\sigma) = 0+2\sigma$.

At the level of groups, we have T = MA, where M = exp $\mathfrak{m}_0$ and A = exp $\mathfrak{a}_0$. A quasi-character of MA is determined by an element $(p,q) = \mu+\lambda$ of $\mathfrak{t}^*$ such that $\mu = p-q$ is an element of $X^*(T)$. The corresponding character of T is written

$$ma \to m^{\mu} a^{\lambda} .$$

Let M' be the normalizer of $\mathfrak{a}_0$ in K. the quotient M'/M = W acts on $\mathfrak{a}_0$ as the Weyl group of $(\mathfrak{g}_0, \mathfrak{a}_0)$ and on $\mathfrak{t}_0$ as the Weyl group of $(\underline{\mathfrak{g}}_0, \underline{t}_0)$. The Weyl group of $(\mathfrak{g},\mathfrak{t})$ is the product $W \times W$; it acts on $\mathfrak{t}^*$ as follows :

$$(w,w')(p,q) = (wp,w'q)$$

$$(w,w')(\mu+\lambda) = \frac{1}{2}(w\lambda+w'\lambda+w\mu-w'\mu) + \frac{1}{2}(w\lambda-w'\lambda+w\mu+w'\mu).$$

The reflection with respect to a root α is denoted s_α.

Recall that we have chosen a basis Δ of Φ . An element δ of $\underline{t}_0^*$ will be said to be dominant if

$$\delta(\alpha^{\vee}) \notin \{-1,-2,\ldots\}$$

for all $\alpha \in \Delta$. The set of dominant weights will be written P^+ , and we write $X^*(T)^+ = P^+ \cap X^*(T)$. An element $\delta \in X^*(T)^+$ is the highest weight of a finite dimensional representation E^{δ} of K.

3. Classification of irreducible admissible $(\mathfrak{g},K)$ modules

We will call $U = U(\mathfrak{g})$, $U(\mathfrak{g}_0)$, $U(\mathfrak{k})$... the enveloping algebras of $\mathfrak{g}, \mathfrak{g}_0, \ldots$

An admissible $(\mathfrak{g},K)$ module is a U-module V such that

$$V = \bigoplus_{\delta \in K^{\wedge}} V^{\delta}$$

where V^{δ} is the sum of all K-submodules of V which are isomorphic to E^{δ}. Consider an element $(\mu+\lambda)$ in $\mathfrak{t}^*$, with $\mu \in X^*(T)$. The space of the principal series representation $L^{\infty}(\mu,\lambda)$ is the set of C^{∞} functions on G which verify, for all g in G, m in M, a in A and n in N

$$\psi(gman) = m^{-\mu} a^{-\lambda-\rho} \psi(g).$$

Let $L(\mu,\lambda)$ be the set of K-finite vectors in $L^{\infty}(\mu,\lambda)$; it is a Harish Chandra module. It is easy to see that the K-type E^{μ_0} (μ_0 being the unique dominant element in the W-orbit of μ) has multiplicity one in $L(\mu,\lambda)$. Let us call $V(\mu,\lambda)$ the unique irreducible subquotient of $L(\mu,\lambda)$ containing E^{μ_0}. It is, by definition, the Langlands subquotient of $L(\mu,\lambda)$.

For reasons which will soon become clear, we will often use the (p,q) description of $\mathfrak{t}^*$. If $(p,q) = \mu+\lambda$, we shall write $L\langle p,q\rangle$ and $V\langle p,q\rangle$ instead of $L(\mu,\lambda)$ and $V(\mu,\lambda)$. The results of Zhelobenko are as follows.

* The set of representations $V(\mu,\lambda)$ exhaust the set of irreducible admissible $(\mathfrak{g},K)$ modules.

* The infinitesimal character of $V\langle p,q\rangle$ is $\chi_{(p,q)}$ where χ is the Harish Chandra homomorphism

* Two modules $V(\mu,\lambda)$ and $V(\mu',\lambda')$ are equivalent if and only if there exists an element w in W such that $\mu' = w\mu$ and $\lambda' = w\lambda$.

4. Restatement of the classification in terms of the L-group

The Langlands classification uses as parameters morphisms ϕ of $W_{\mathbb{C}}$ into ${}^LG$ such that the diagram

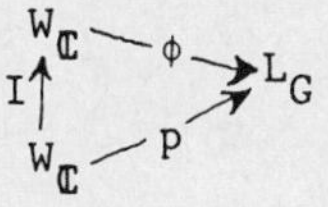

I is the identity of $W_{\mathbb{C}}$

p is the projection of ${}^LG$ on $W_{\mathbb{C}}$

is commutative. Since in our case, ${}^LG$ is the direct product ${}^LG^0 \times W_{\mathbb{C}}$, and $W_{\mathbb{C}}$ is $\mathbb{C}^*$, it is equivalent to consider morphisms

$$\phi : \mathbb{C}^* \longrightarrow {}^LG^0 .$$

We ask further that the image of ϕ be semi-simple.

The classification associates to such a morphism ϕ a representation

T_ϕ (remember that there is no L-indisinguishability in the complex case). Any admissible ($\mathfrak{g}$,K) module is of the form T_ϕ for some ϕ; ϕ and ϕ' define the same representation if and only if ϕ' is conjugate to ϕ under ${}^LG^0$.

Since the classification is known, we need only to establish a correspondence between the ϕ's and the (p,q) description of the preceeding paragraph. A conjugacy class of morphisms of $\mathbb{C}^*$ into ${}^LG^0$ with semi-simple image is determined if we assume that the image of ϕ is contained in the torus ${}^LT^0$. Such a ϕ can be written

$$\phi(z) = z^p \bar{z}^q$$

where p and q are elements of $X_*({}^LT^0) \otimes \mathbb{C}$ and $p-q \in X_*({}^LT^0)$. By definition of ${}^LT^0$, the sets $X_*({}^LT^0)$ and $X^*(T)$ coincide, so p and q are elements of $X^*(T) \otimes \mathbb{C}$, and $p-q \in X^*(T)$. Furthermore $X^*(T) \otimes \mathbb{C} = \underline{t}_0^*$. Finally, we can define T_ϕ ($\phi = \phi_{(p,q)}$) as

$$T_\phi = V\langle p,q\rangle.$$

5. Description of the composition series of the principal series representations

Since we are interested only in the description of the irreducible subquotients and not of their multiplicities, the results are fairly easy to state. They have been studied by several authors, and are closely related to the analogous results for Verma modules. (<Du 2>, <B-G>, <Hi>)

Let us fix a pair (p,q), p and q in $\underline{t}_0^*$ and p-q in $X^*(T)$. Consider the subroot system of Φ

$$\Phi_p = \Phi_q = \{\alpha \in \Phi,\ \langle p,\alpha^\vee\rangle\ \mathbb{Z}\}$$

and also

$$\Phi_p^0 = \{\alpha \in \Phi,\ \langle p,\alpha^\vee\rangle = 0\}$$

$$\Phi_q^0 = \{\alpha \in \Phi,\ \langle q,\alpha^\vee\rangle = 0\}$$

The Weyl groups W_p, W_q, W_p^0, W_q^0 of the root systems Φ_p, Φ_q, Φ_p^0, Φ_q^0 have the following properties :

$$W_p = W_q = \{w \in W,\ wp-p \text{ is a linear combination of roots}\}$$

$$W_p^0 = \{w \in W_p,\ wp = p\}$$

$$W_q^0 = \{w \in W_q,\ wq = q\}$$

Using the results of Zhelobenko mentioned above, the problem of the determination of the composition series of a principal series L<p,q> can be reduced to the following question : Assuming p and q to be <u>dominant</u> and given w in W_p, what are the elements w' of W_p such that V<p,w'q> is a subquotient of L<p,wq> ?

To formulate the answer, we have to define the Bruhat order in a Weyl group W. Let S be a set of generators of W (the reflections with respect to a set of simple roots); each element $w \in W$ has a length with respect to S, which we denote $\ell(w)$. Following Dixmier <Di> (surprisingly there are two (opposite) conventions for the Bruhat order), we first define $w \to w'$ if there exists a reflection s such that $w' = sw$ and $\ell(w') = \ell(w) + 1$. The Bruhat order is

$w \geq w'$ if and only if $w = w'$ or there exists $w_0 = w, w_1, w_2, \ldots w_n = w'$ such that $w_0 \to w_1 \to w_2 \to \ldots \to w_n$.

<u>Theorem (Hirai)</u> (sees also <B-G>, <Du 2>). The irreducible module V<p,w'q> is a subquotient of L<p,wq> if and only if $w' \geq w$ for the Bruhat order in W_p.

<u>Remark</u>. The set of irreducible subquotients of L<p,wq> is parametrized by a subset of $W_p^0 \backslash W / W_q^0$. It is possible to choose a set of representatives of $W_p^0 \backslash W / W_q^0 = W'_{pq}$ verifying

$$x \in W'_{pq} \text{ iff } \begin{array}{l} \ell(s_\alpha x) > \ell(x) \text{ for all } \alpha \text{ simple in } \Phi^0_{p_0} \\ \ell(x s_\alpha) > \ell(x) \text{ for all } \alpha \text{ simple in } \Phi^0_q \end{array} .$$

6. <u>L-functions</u>

Let us first recall the definition of an abelian L-function in the complex case (see Tate <Ta>). Consider a quasi-character χ of $\mathbb{C}^* = W_{\mathbb{C}}$. It can be written

$$\chi(z) = z^p \bar{z}^q$$

for some pair (p,q) of complex numbers with p-q $\mathbb{Z}$. Let us put $\sigma = \max(p,q)$. We can write

$$\chi(z) = (z\,\bar{z})^\sigma\ \bar{z}^{-(p-q)} \quad \text{if } p \geq q$$

$$\chi(z) = (z\,\bar{z})^\sigma\ \bar{z}^{-(q-p)} \quad \text{if } p \leq q.$$

The L-function associated to χ is, by definition the meromorphic function

$$L(\chi,s) = 2\,(2\pi)^{-(s+\sigma)}\Gamma(s+\sigma).$$

Consider now a finite dimensional representation θ of $\mathbb{C}^*$ in $\mathbb{C}^n$ with semi-simple image. It can be decomposed

$$\theta = \bigoplus \chi_i \quad \text{(i ranging from 1 to n).}$$

By definition, the L-function associated to θ is

$$L(\theta,s) = \prod L(\chi_i,s).$$

Let T be an irreducible admissible representation of G. By the Langlands classification, it is associated to a morphism ϕ of $\mathbb{C}^*$ into ${}^LG^0$ with semi-simple image. Let r be a rational representation of ${}^LG^0$. Define the L-function associated to T as the L-function $L(r_\circ\phi,s)$ defined above :

$$L(T_\phi,s) = L(r_\circ\phi,s).$$

Explicit computations of the L-factors

Let p and q be elements of $\underline{t}_0^*$ with $p-q \in X^*(T)$. This determines (§4) a morphism ϕ of $\mathbb{C}^*$ into ${}^LG^0$, and therefore a representation $T_\phi = V\langle p,q\rangle$ of G. Let r be a finite dimensional rational representation of ${}^LG^0$. It is entirely determined by its highest weight ν, which is an element of $X^*({}^LT^0)^+$. (Recall that we have been given a set of simple roots in Φ). Let us call r_ν the representation of ${}^LG^0$ with highest weight ν, and $P(r_\nu)$ its set of weights.

We need to compute the 1-dimensional representations of $\mathbb{C}^*$ in which $r_{\nu}\circ\phi$ breaks up. It is easy to see that

$$r_\nu\circ\phi(z) = \bigoplus z^{\langle p,\mu\rangle}\bar{z}^{\langle q,\mu\rangle} ,$$

the sum being extended to the set $\mu\in P(r_\nu)$. We must take multiplicities into account; but that also is easy, the multiplicity of $z^{\langle p,\mu\rangle}\bar{z}^{\langle q,\mu\rangle}$ being equal to the multiplicity of μ in r_ν. The notation $\langle p,\mu\rangle$, $\langle q,\mu\rangle$ is easily understood : $P(r_\nu)$ is a subset of $X^*({}^LT^0) = X_*(T)$, while p and q are in $X^*(T)\times\mathbb{C}$; the symbol $\langle\ ,\ \rangle$ simply means the duality between $X_*(T)$ and $X^*(T)$. Using the formulas for abelian L-functions, we get the following formula for $L(T_\phi,r_\nu,s) = L(V\langle p,q\rangle,r_\nu,s) = L(p,q,r_\nu,s)$:

$$L(p,q,r_\nu,s) = 2^{\dim r_\nu}\prod_\mu (2\pi)^{-(s+\max(\langle p,\mu\rangle,\langle q,\mu\rangle))}\,\Gamma(s+\max(\langle p,\mu\rangle,\langle q,\mu\rangle))$$

each of the factors being repeated with its multiplicity.

7. Proof of the first part of the theorem

The notations are those of parts 5 and 6. We assume p and q dominant in $\mathfrak{t}_0^*$ (a set of simple roots in Φ_p being given), and take $w \geqq w'$ in W_p for the Bruhat order. We need to prove that

$$\frac{L(p,w'q,r_\nu,s)}{L(p,wq,r_\nu,s)}$$

is an entire function.

By induction on the length, it is enough to prove it if $w' = s_\alpha w$ with $\ell(w') = \ell(w)+1$, and α is a positive root in Φ_p. The Lemme 7.7.2 in <Di> implies that in our case we have

$$s_\alpha wq = wq - n\alpha \text{ , and } n \geqq 0.$$

Consider the subgroup $W_p^{(\alpha)}$ generated by s_α. It acts on $P(r_\nu)$ and preserves multiplicities. We can therefore write

$$\frac{L(p,s_\alpha wq,r_\nu,s)}{L(p,wq,r_\nu,s)} = (*) \prod_{P(r_\nu)/W_p^{(\alpha)}} (F_\mu(s))^{m(\mu)}$$

where $F_\mu(s)$ is :

$$F_\mu(s) = \frac{\Gamma(s+\max(\langle\mu,p\rangle,\langle\mu,s_\alpha wq\rangle))\ \Gamma(s+\max(\langle s_\alpha\mu,p\rangle,\langle s_\alpha\mu,s_\alpha wq\rangle))}{\Gamma(s+\max(\langle\mu,p\rangle,\langle\mu,wq\rangle))\ \Gamma(s+\max(\langle s_\alpha\mu,p\rangle,\langle s_\alpha\mu,wq\rangle))}$$

($m(\mu)$ is the multiplicity of μ)

and (*) is a product of exponentials of s.

Put
$$A = \langle\mu,p\rangle \qquad A_\alpha = \langle s_\alpha\mu,p\rangle$$
$$B = \langle\mu,wq\rangle \qquad B_\alpha = \langle s_\alpha\mu,wq\rangle \ .$$

We have

$$F_\mu(s) = \frac{\Gamma(s+\max(A,B_\alpha))\ \Gamma(s+\max(A_\alpha,B))}{\Gamma(s+\max(A,B))\ \Gamma(s+\max(A_\alpha,B_\alpha))}$$

Since
$$A - A_\alpha = \langle p,\alpha^\vee\rangle\ \langle\mu,\alpha\rangle$$
$$B - B_\alpha = \langle wq,\alpha^\vee\rangle\ \langle\mu,\alpha\rangle$$

and $\langle p,\alpha^\vee\rangle \geqq 0$ (p dominant) and $\langle wq,\alpha^\vee\rangle \geqq 0$ ($\ell(s_\alpha w) > \ell(w)$), we can choose μ in its orbit under $W_p^{(\alpha)}$ so that

$$A \geqq A_\alpha$$
$$B \geqq B_\alpha \ .$$

Besides, α is an element of Φ_p and w is in W_p , so that $A - A_\alpha$ and $B - B_\alpha$ are integers. We also know that $p - q$ is a weight, so that $A - B$ is an integer.

Therefore the ordering on $\mathbb{Z}$ induces an order on the set A, A_α, B, B_α. Let us write these elements in increasing order :

$$C_1 \leq C_2 \leq C_3 \leq C_4 .$$

The factor $F_\mu(s)$ is then of the form :

$$\frac{\Gamma(s+C_3)\ \Gamma(s+C_4)}{\Gamma(s+C_i)\ \Gamma(s+C_4)}$$

with i = 2 or i = 3

Using finally the equality

$$\Gamma(s+n+1) = (s+n)(s+n-1)\ldots s\ \Gamma(s)$$

valid for n a non negative integer, we obtain the fact that $F_\mu(s)$ is a polynomial.

8. Proof of the second part of the theorem

It is well known that the contragredient representation to $V\langle p,q\rangle$ is $V\langle -p,-q\rangle$. We assume again that p and q have been chosen dominant, and that w and w' are in the Weyl group W_p. We want to prove that

$$\frac{L(p,wq,r_\nu,s)}{L(-p,-wq,r_\nu,1-s)} = \frac{L(p,w'q,r_\nu,s)}{L(-p,-w'q,r_\nu,1-s)}\ \Psi(s)$$

or, equivalently, that

$$\frac{L(p,wq,r_\nu,s)}{L(p,w'q,r_\nu,s)} = \frac{L(-p,-wq,r_\nu,1-s)}{L(-p,-w'q,r_\nu,1-s)}\ \Psi(s)$$

where $\Psi(s)$ denotes an entire function such that its inverse is also entire. Of course it is enough to prove the equality for $w' = s_\alpha w$, making the computations slightly easier. Going back to the formula for the L function, we need to prove the equality

$$\prod_{P(r_\nu)} \frac{\Gamma(s+\max(\langle p,\mu\rangle,\langle wq,\mu\rangle))}{\Gamma(s+\max(\langle p,\mu\rangle,\langle w'q,\mu\rangle))} = \prod_{P(r_\nu)} \frac{\Gamma(1-s+\max(-\langle p,\mu\rangle,-\langle wq,\mu\rangle))}{\Gamma(1-s+\max(-\langle p,\mu\rangle,-\langle w'q,\mu\rangle))}\ \Psi(s) .$$

To simplify the notations, we will write = instead of equal up to multiplication by a function $\psi(s)$ ... Using the same method for grouping the terms in the product as in the last paragraph, and changing q to wq, we need to prove that for any μ in $P(r_\nu)$, the following "equality" holds :

$$\frac{\Gamma(s+\max(\langle p,\mu\rangle,\langle q,\mu\rangle))\ \Gamma(s+\max(\langle p,s_\alpha\mu\rangle,\langle q,s_\alpha\mu\rangle))}{\Gamma(s+\max(\langle p,\mu\rangle,\langle s_\alpha q,\mu\rangle))\ \Gamma(s+\max(\langle p,s_\alpha\mu\rangle,\langle s_\alpha q,s_\alpha\mu\rangle))}$$

$$= \frac{\Gamma(1-s+\max(-\langle p,\mu\rangle,-\langle q,\mu\rangle))\ \Gamma(1-s+\max(-\langle p,s_\alpha\mu\rangle,-\langle q,s_\alpha\mu\rangle))}{\Gamma(1-s+\max(-\langle p,\mu\rangle,-\langle s_\alpha q,\mu\rangle))\ \Gamma(1-s+\max(-\langle p,s_\alpha\mu\rangle,-\langle s_\alpha q,s_\alpha\mu\rangle))}$$

Put $A = \langle p,\mu\rangle \qquad A_\alpha = \langle p, s_\alpha\mu\rangle$

$B = \langle q,\mu\rangle \qquad B_\alpha = \langle q, s_\alpha\mu\rangle$.

The set A, A_α, B, B_α is totally ordered, as in the last paragraph, and we can choose μ in the set $\{\mu, s_\alpha\mu\}$ so that $A \geq A_\alpha$. We have to prove the equality

$$\frac{\Gamma(s+\max(A,B))\ \Gamma(s+\max(A_\alpha,B_\alpha))}{\Gamma(s+\max(A,B_\alpha))\ \Gamma(s+\max(A_\alpha,B))} = \frac{\Gamma(1-s-\min(A,B))\ \Gamma(1-s-\min(A_\alpha,B_\alpha))}{\Gamma(1-s-\min(A,B_\alpha))\ \Gamma(1-s-\min(A_\alpha,B))} .$$

Put the set A, A_α, B, B_α in increasing order :

$$C_1 \geq C_2 \geq C_3 \geq C_4 .$$

We clearly have the following facts :

$\{\max(A,B), \max(A_\alpha,B_\alpha)\} = \{C_4, C_i\}$ for $i = 2$ or 3 .

$\{\max(A,B_\alpha), \max(A_\alpha,B)\} = \{C_4, C_j\}$ for $j = 2$ or 3 .

$\{\min(A,B), \min(A_\alpha,B_\alpha)\} = \{C_1, C_{i'}\}$ for $i' = 2$ or 3, $i' \neq i$.

$\{\min(A,B_\alpha), \min(A_\alpha,B)\} = \{C_1, C_{j'}\}$ for $j' = 2$ or 3, $j' \neq j$.

There are only two possibilities : either $i = j$, and then necessarily $i' = j'$, or $i \neq j$, and then $i = j'$ and $i' = j$.

Finally, the equality we want to prove is, according to these two cases :

<u>First case</u> $\qquad 1 = 1$.

<u>Second case</u> $\qquad \dfrac{\Gamma(s+C_i)}{\Gamma(s+C_j)} = \dfrac{\Gamma(1-s-C_j)}{\Gamma(1-s-C_i)}$.

Leaving the proof of the first case to the reader, we need only remark that the second case is a consequence of the duplication formula for the Gamma function :

$$\Gamma(z)\,\Gamma(1-z) = \frac{\pi}{\sin(\pi z)} .$$

We obtain $\qquad \sin\pi(s+C_i) = \sin\pi(s+C_j)$,

equality which is true up to a constant since $C_i - C_j$ is an integer.

<u>Remark</u>. Going through the computations in a slightly more thorough way, one sees that the function $\psi(s)$ in assertion 2° of the theorem is <u>equal</u> to the sign $(-1)^{(C_i-C_j)}$. One wanders whether this sign has any special meaning.

Bibliography

<B-G> J. Bernstein I. Gelfand Tensor product of finite and infinite dimensional representations of semi-simple Lie algebras, Compositio Math.41(80

<Bo> A. Borel Automorphic L-functions in Proceedings of Symposia in Pure Mathematics vol.33 (1979), part 2, pp. 27-61.

<Di> J. Dixmier Algèbres enveloppantes Gauthier-Villars, Paris, 1974.

<Du 1> M. Duflo Représentations irréductibles des groupes semi-simples complexes, in Lecture Notes in Mathematics 497, Springer Verlag, Heidelberg, 1975.

<Du 2 > M. Duflo Sur la classification des idéaux primitifs dans l'algèbre enveloppante d'une algèbre de Lie semi-simple, Annals of Math. 105 (1977) pp. 107-120.

<G-J> R. Godement H. Jacquet Zeta functions of simple algebras, Lecture Notes in Mathematics 260, Springer Verlag, Heidelberg, 1972.

<Hi> T. Hirai Structure of induced representations and characters of irreducible representations of complex semi-simple Lie groups, in Lecture Notes in Mathematics 266, Springer Verlag, Heidelberg, 1972.

<J> H. Jacquet Principal L-functions of the linear group in Proceedings of Symposia in Pure Mathematics vol.33 (1979), part 2, pp. 63-86.

<J-L> H. Jacquet R. Langlands Automorphic forms on GL(2), Lecture Notes in Mathematics 114, Springer Verlag, Heidelberg, 1970.

<La> R. Langlands, On the classification of irreducible representations of real reductive groups, to appear.

<Ro> F. Rodier, Décomposition de la série principale des groupes réductifs p-adiques, in Lecture Notes in Mathematics 880, Springer Verlag, Heidelberg, 1981.

IRREDUCIBLE UNITARY REPRESENTATIONS OF SOME GROUPS OF REAL RANK TWO

M. W. Baldoni-Silva* and A. W. Knapp**

We have been collecting data on the unitary duals of various linear connected semisimple Lie groups in an effort to find out whether it is reasonable to have a simply-stated explicit classification for all such groups. For groups of real rank one, Baldoni-Silva and Barbasch [2] obtained an explicit classification, and our paper [5], when specialized to these groups, shows how that classification can be stated simply.

Our concern here is with simple groups G of real rank two. We prefer to think of these as divided into two classes, those with $\text{rank } G = \text{rank } K$ (for K maximal compact) and those with $\text{rank } G > \text{rank } K$. Within each class, some of the groups appear to us as "regular cases," some are variants of regular cases, and some are exceptions.

For $\text{rank } G = \text{rank } K$, the regular cases are the "single-line" cases: $SU(n,2)$, $\widetilde{SO}(2n,2)$, $SO^*(10)$, and $E_{6(-14)}$. The groups $Sp(n,2)$ may be viewed as variants of $SU(n,2)$, while the various $\widetilde{SO}(2n+1,2)$ are variants of $\widetilde{SO}(2n,2)$. The group $G_2^{\mathbb{R}}$ is exceptional. We gave a classification for $SU(n,2)$ in [4]. Angelopoulos [1] announced a classification for $\widetilde{SO}(2n,2)$ and $\widetilde{SO}(2n+1,2)$, but we are unable to relate his results to Langlands parameters nor do we have enough details to check his results. Thus we have recently obtained our own classification (unpublished) for these groups. Our methods appear to handle $Sp(n,2)$ as well, but they are insufficient for

---

* Partially supported by National Science Foundation Grant DMS 85-01793.

** Supported by National Science Foundation Grants DMS 80-01854 and DMS 85-01793.

$SO^*(10)$ and $E_{6(-14)}$. The classifications for all these groups have a qualitative similarity to them, but a detailed statement of the classification requires (at present) treatment of one class at a time. In short, we find these classifications discouraging.

The situation is nicer for rank $G >$ rank K. The regular cases are those with one conjugacy class of Cartan subgroups and with A_2 as restricted root diagram: $SL(3,\mathbb{C})$, $SL(3,\mathbb{H})$, and $E_{6(-26)}$ (which we regard as $SL(3,\mathbb{O})$ with $\mathbb{O} = \{\text{Cayley numbers}\}$). The group $SL(3,\mathbb{R})$ is a variant of $SL(3,\mathbb{C})$, while $Sp(2,\mathbb{C})$ and $G_2^{\mathbb{C}}$ are exceptional. Classifications were done by Tsuchikawa [18] for $SL(3,\mathbb{C})$, Vogan [23] for $SL(3,\mathbb{H})$, Vahutinskii [19] for $SL(3,\mathbb{R})$, and Duflo [6] for $Sp(2,\mathbb{C})$ and $G_2^{\mathbb{C}}$. Our objective in this paper will be to complete the classifications for the real rank two groups with rank $G >$ rank K by doing $E_{6(-26)}$. In doing so, we shall work with an abstract group with one conjugacy class of Cartan subgroups and with A_2 as restricted root diagram. This refusal to use explicit knowledge of $E_{6(-26)}$ is in line with our desire to have a simple final classification.

Turning to the precise statement of our result, we begin with notation and background. We let G be linear connected simple, K be maximal compact, and $G = KAN$ be an Iwasawa decomposition. For any subgroup we denote the Lie algebra by the corresponding lower-case German letter. We assume that G has just one conjugacy class of Cartan subgroups, that $\dim A = 2$, and that the restricted roots (the roots of $(\mathfrak{g},\mathfrak{a})$) form a root system of type A_2. Let $M = Z_K(A)$ be the (compact) centralizer of A in K, so that $P = MAN$ is a minimal parabolic subgroup of G. For σ an irreducible (finite-dimensional) representation of M and ν in $(\mathfrak{a}')^{\mathbb{C}}$, the representation $U(P,\sigma,\nu)$ given by normalized induction as

$$U(P,\sigma,\nu) = \operatorname{ind}_P^G(\sigma \otimes e^{\nu} \otimes 1)$$

is a member of the nonunitary principal series. If ν has the additional property that $\operatorname{Re}\nu$ is in the closed positive Weyl chamber of $\mathfrak{a}'$ relative to N, then it follows from Langlands [16], Miličić [17], and Knapp [9] for this kind of G that $U(P,\sigma,\nu)$ has a unique irreducible quotient $J(P,\sigma,\nu)$, which is known as the Langlands quotient.

The representations $J(P,\sigma,\nu)$ exhaust the candidates for irreducible unitary representations, and the classification problem is to decide which of them are infinitesimally unitary. From [14] and [12], it is known that $J(P,\sigma,\nu)$ admits a nonzero invariant Hermitian form (on its K-finite vectors) if and only if the following formal symmetry condition holds: there exists w in the normalizer $N_K(A)$ such that w^2 is in M, $w\sigma \cong \sigma$, and $w\nu = -\bar{\nu}$. This form lifts to a form $\langle\cdot,\cdot\rangle$ on $U(P,\sigma,\nu)$ that is given by an explicit intertwining operator $\mathcal{A}$ on $L^2(K)$: $\langle f,g\rangle = (\mathcal{A}f,g)_{L^2(K)}$. Moreover, J is infinitesimally unitary if and only if $\langle\cdot,\cdot\rangle$ is semidefinite, if and only if $\mathcal{A}$ is semidefinite. By a theorem of Vogan (see Theorem 16.10 of [11]), it is enough to decide the unitarizability for ν real.

Let us denote the simple restricted roots by $\alpha_R' = e_1 - e_2$ and $\alpha_R'' = e_2 - e_3$. The Weyl group $W(A:G) = N_K(A)/M$ is the symmetric group on three letters, and the only Weyl group element of order two sending a real element $\nu \neq 0$ in the closed positive Weyl chamber into its negative is the reflection s_{α_R} in the sum $\alpha_R = \alpha_R' + \alpha_R'' = e_1 - e_3$. The set of ν's to study is therefore the one-dimensional set $\nu = c\alpha_R$ with $c \geq 0$. Let w_{12}, w_{23}, and w_{13} be representatives in K of the Weyl group elements $s_{\alpha_R'}$, $s_{\alpha_R''}$, and s_{α_R}. The formal symmetry condition $w_{13}\sigma \cong \sigma$ imposes a certain nontrivial condition on σ that we consider later.

The group M is compact and connected, and thus σ is given by the theory of the highest weight. Let $\mathfrak{b} \subseteq \mathfrak{m}$ be a maximal abelian subspace, so that $\mathfrak{b} \oplus \mathfrak{a}$ is a Cartan subalgebra of $\mathfrak{g}$. Let Δ_M^+

be a positive system for the root system $\Delta_M = \Delta(\mathfrak{m}^{\mathbb{C}}, \mathfrak{b}^{\mathbb{C}})$, let λ be the highest weight of σ, and let δ_M be half the sum of the members of Δ_M^+, so that $\lambda_0 = \lambda + \delta_M$ is the infinitesimal character of σ.

Let $\Delta = \Delta(\mathfrak{g}^{\mathbb{C}}, (\mathfrak{b} \oplus \mathfrak{a})^{\mathbb{C}})$ be the root system of $\mathfrak{g}$; we can regard Δ_M as the set of members of Δ that vanish on $\mathfrak{a}$. We introduce a positive system Δ^+ for Δ containing Δ_M^+ so that λ is Δ^+ dominant and $i\mathfrak{b}$ comes before $\mathfrak{a}$. (For example, we can use the lexicographic order obtained by adjoining an orthogonal basis of $\mathfrak{a}'$ at the end of an orthogonal basis of $i\mathfrak{b}'$ that starts with λ.)

Let $L = L(\sigma)$ be the analytic subgroup of G containing $\mathfrak{b} \oplus \mathfrak{a}$ and corresponding to the set Δ^L of all roots β in Δ with $\langle \lambda, \beta \rangle = 0$. Since λ is dominant, Δ^L is generated by Δ^+ simple roots. The group L is another real rank two group, though not necessarily simple, and it has an Iwasawa decomposition $L = (K \cap L)A(N \cap L)$. Let ρ_L be half the sum of its positive restricted roots, counting multiplicities; ρ_L is a positive multiple of $\alpha_R = e_1 - e_3$.

<u>Main Theorem</u>. Let G be linear connected simple of real rank two with just one conjugacy class of Cartan subgroups and with restricted root diagram of type A_2. Let σ be an irreducible representation of M such that $w_{13}\sigma \cong \sigma$, let $L = L(\sigma)$, and let L_{ss} be the semisimple part of L. Then L_{ss} has real rank one or two. Moreover, for ν real in the closed positive Weyl chamber, $J(P, \sigma, \nu)$ is infinitesimally unitary if and only if $\nu = c\rho_L$ with

$$\begin{cases} 0 \leq c \leq 1 & \text{if } L_{ss} \text{ has real rank one} \\ 0 \leq c \leq \tfrac{1}{2} \text{ or } c = 1 & \text{if } L_{ss} \text{ has real rank two.} \end{cases} \tag{0.1}$$

<u>Remarks</u>. The classification of real groups shows that L_{ss} is locally $SL(2)$ over $\mathbb{C}$, $\mathbb{H}$, or $\mathbb{O}$ if L_{ss} has real rank one; alternatively L_{ss} is locally $SO(n,1)$ for $n = 3, 5,$ or 9 if L_{ss} has real rank one. If L_{ss} has real rank two, L_{ss} is locally

$SL(3)$ over $\mathbb{C}$, $\mathbb{H}$, or $\mathbb{O}$. The statement of the theorem is that the unitarity for the series for σ in G is the same as the unitarity for the series for 1 in L_{ss}, which is given by the simple formula (0.1).

1. Structure of L and the roots

We proceed with L as in the introduction but temporarily do not assume $w_{13}\sigma \cong \sigma$. The positive restricted roots are $\alpha_R' = e_1 - e_2$, $\alpha_R'' = e_2 - e_3$, and $\alpha_R = \alpha_R' + \alpha_R'' = e_1 - e_3$.

Let us bring to bear some results from [9]. Since there is just one conjugacy class of Cartan subgroups, there are no real roots in Δ. Then Lemma 2.2 of [9] says that all restricted roots are "even," in the sense of that paper. Moreover, all the restricted roots are "useful," as one sees from §4 of [8]. For any root β in Δ, we decompose β as $\beta = \beta_I + \beta_R$, its parts on $i\mathfrak{b}$ and $\mathfrak{a}$, respectively. If β is complex, Lemma 2.5 of [9] says that

$$|\beta|^2 = 2|\beta_I|^2 = 2|\beta_R|^2. \tag{1.1}$$

Therefore all complex roots have the same length.

For each of the simple restricted roots β_R $(= \alpha_R'$ or $\alpha_R'')$, it is possible by Proposition 3.1 of [9] to choose β_I in $i\mathfrak{b}'$ so that $\beta_I + \beta_R$ is in Δ, so that the reflection s_{β_I} preserves Δ_M^+, and so that the linear extension of the map

$$\alpha_R' \to \alpha_I', \quad \alpha_R'' \to \alpha_I''$$

to $\mathfrak{a}'$ is an isometry of $\mathfrak{a}'$ into $i\mathfrak{b}'$. Put $\alpha_I = \alpha_I' + \alpha_I''$. Then it follows that $\alpha_I + \alpha_R$ is in Δ. Theorem 3.7 of [9] says that we obtain an action of $W(A:G)$ on $i\mathfrak{b}'$ from this correspondence—with $s_{\alpha_R'}$ acting by $s_{\alpha_I'}$ and so forth.

The group M is connected. In fact, $M = M^{\#}$ in the notation of

[9] (see §1 of that paper), and Lemma 2.1 of [15] shows that $M^{\#} = M_0$. Therefore σ is determined by its highest weight. If the highest weight of σ is λ and if w is a representative in K of a member s of $W(A:G)$, then Proposition 4.7 of [9] says that $w\sigma$ has highest weight $s\lambda$, with $s\lambda$ defined from the previous paragraph. Therefore

$$w_{12}\sigma \cong \sigma \quad \text{if and only if} \quad \langle \lambda, \alpha_I' \rangle = 0 \tag{1.2a}$$

$$w_{23}\sigma \cong \sigma \quad \text{if and only if} \quad \langle \lambda, \alpha_I'' \rangle = 0 \tag{1.2b}$$

$$w_{13}\sigma \cong \sigma \quad \text{if and only if} \quad \langle \lambda, \alpha_I \rangle = 0 . \tag{1.2c}$$

Lemma 1.1. Suppose $\beta = \beta_I + \alpha_R$ is in Δ and $\langle \lambda, \beta \rangle = 0$. Then $\langle \lambda, \alpha_I \rangle = 0$. Similar results hold for α_R' and α_R''.

Proof. Without loss of generality, we may take β to be positive. By (1.1) we may assume β_I is not a multiple of α_I. Then we have

$$\langle \beta, \alpha_I + \alpha_R \rangle = \langle \beta_I, \alpha_I \rangle + |\alpha_R|^2 > 0 \tag{1.3}$$

by (1.1) and the converse of the Schwarz inequality, and similarly

$$\langle \beta, \alpha_I - \alpha_R \rangle = \langle \beta_I, \alpha_I \rangle - |\alpha_R|^2 < 0 .$$

Therefore $\beta_I + \alpha_I$ and $\beta_I - \alpha_I$ are both roots. From (1.3) it follows that $2\langle \beta, \alpha_I + \alpha_R \rangle / |\alpha_I + \alpha_R|^2 = 1$, and from (1.1) we can then conclude that $\langle \beta_I, \alpha_I \rangle = 0$. Hence $s_{\alpha_I}(\beta_I + \alpha_I) = \beta_I - \alpha_I$, and the defining property of α_I forces $\beta_I + \alpha_I$ and $\beta_I - \alpha_I$ to have the same sign.

Meanwhile, $\beta = \beta_I + \alpha_R$ is positive, and our choice of Δ^+ makes $\beta_I - \alpha_R$ positive. Thus $2\beta_I$ is the sum of positive roots, and it follows that $\beta_I + \alpha_I$ and $\beta_I - \alpha_I$ are both positive. Finally β is equal to

$$(\beta_I - \alpha_I) + (\alpha_I + \alpha_R) = (\beta_I + \alpha_I) + (-\alpha_I + \alpha_R), \tag{1.4}$$

and either $\alpha_I + \alpha_R$ or $-\alpha_I + \alpha_R$ will be positive, by our choice of Δ^+. Hence one of the expressions in (1.4) exhibits β as the sum of positive roots. Since $\langle \lambda, \beta \rangle = 0$ and λ is Δ^+ dominant, we conclude $\langle \lambda, \alpha_I \rangle = 0$.

Proposition 1.2. Under the assumption $w_{13}\sigma \cong \sigma$, L_{ss} has real rank 1 or 2 and contains the roots $\pm \alpha_I \pm \alpha_R$. Moreover, the following conditions are equivalent:

(a) L_{ss} has real rank 2

(b) L_{ss} contains the roots $\pm \alpha_I' \pm \alpha_R'$ and $\pm \alpha_I'' \pm \alpha_R''$

(c) The whole Weyl group $W(A:G)$ fixes the class of σ.

Proof. In any case, the subspace $\mathfrak{a}$ is an Iwasawa $\mathfrak{a}$ for L. Thus the real rank of L_{ss} is equal to the dimension of the span of the restricted roots that contribute to L. Since we are assuming $w_{13}\sigma \cong \sigma$, (1.2c) shows that $\pm \alpha_R \pm \alpha_I$ contribute to L; hence $\pm \alpha_R$ are restricted roots for L. Thus the real rank of L_{ss} is 1 or 2.

With these considerations in mind, we see from (1.2) that (b) and (c) are equivalent and imply (a). On the other hand, (a) implies (b) by Lemma 1.1. This proves the proposition.

We need to relate our positive system Δ^+ to the positive systems in various other papers, seeing that they are the same. What we need to see is that the infinitesimal character $\lambda_0 = \lambda + \delta_M$ of σ is Δ^+ dominant and that any positive system for Δ that takes $i\mathfrak{b}$ before $\mathfrak{a}$ and makes $\lambda + \delta_M$ dominant automatically makes λ dominant.

Lemma 1.3. Let $(\Delta^+)'$ be any positive system for Δ that takes $i\mathfrak{b}$ before $\mathfrak{a}$ and contains Δ_M^+. If β is a $(\Delta^+)'$ simple root that is complex, then $\langle \delta_M, \beta \rangle = 0$.

Proof. We shall pair the members ϵ of Δ_M^+ having $\langle \epsilon, \beta \rangle < 0$ with the members ϵ' of Δ_M^+ having $\langle \epsilon', \beta \rangle > 0$, the pairing being $\epsilon' = \epsilon + 2\beta_I$ (where $\beta = \beta_I + \beta_R$) and satisfying $\langle \epsilon + \epsilon', \beta \rangle = 0$.

Notice from (1.1) that $\langle \beta, \theta\beta \rangle = 0$. Let $\langle \epsilon, \beta \rangle < 0$, and put

$\gamma = \epsilon + \beta$. Then $\theta\gamma$ and β are complex $(\Delta^+)'$ positive roots (necessarily of the same length) with

$$\langle \theta\gamma, \beta \rangle = \langle \epsilon + \theta\beta, \beta \rangle = \langle \epsilon, \beta \rangle < 0 .$$

Hence $\theta\gamma + \beta = \epsilon + \theta\beta + \beta = \epsilon + 2\beta_I$ is a positive root in Δ_M of the same length as β.

Conversely if ϵ' in Δ_M^+ has $\langle \epsilon', \beta \rangle > 0$, then $\gamma = \epsilon' - \beta$ is in $(\Delta^+)'$ (since β is simple). Our choice of $i\mathfrak{b}$ before $\mathfrak{a}$ makes $\theta\gamma$ be positive, too, and

$$\langle \theta\gamma, \beta \rangle = \langle \epsilon' - \theta\beta, \beta \rangle = \langle \epsilon', \beta \rangle > 0 .$$

Since β is simple, $\theta\gamma - \beta = \epsilon' - \theta\beta - \beta = \epsilon' - 2\beta_I$ is a positive root in Δ_M. The roots $\theta\gamma$ and β are complex and must be of the same length; hence $\epsilon' - 2\beta_I$ has that same length.

Consequently we have a pairing $\epsilon \longleftrightarrow \epsilon'$ by addition or subtraction of $2\beta_I$, and what we have just seen implies $|\epsilon| = |\beta| = |\epsilon'|$. Thus $2\langle \epsilon, \beta \rangle / |\beta|^2 = -1$ and $2\langle \epsilon', \beta \rangle / |\beta|^2 = +1$, and it follows that $\epsilon + \epsilon'$ is orthogonal to β. Summing on ϵ completes the proof of the lemma.

<u>Proposition 1.4</u>. The infinitesimal character $\lambda_0 = \lambda + \delta_M$ is Δ^+ dominant. Conversely if $(\Delta^+)'$ is any positive system for Δ containing Δ_M^+, taking $i\mathfrak{b}$ before $\mathfrak{a}$, and making $\lambda_0 = \lambda + \delta_M$ dominant, then λ is $(\Delta^+)'$ dominant.

<u>Proof</u>. A Δ^+ simple root β is either imaginary or complex. If it is imaginary, then $2\langle \delta_M, \beta \rangle / |\beta|^2 = 1$, while if it is complex, Lemma 1.3 gives $2\langle \delta_M, \beta \rangle / |\beta|^2 = 0$. Thus δ_M is Δ^+ dominant, and hence so is $\lambda + \delta_M$.

Conversely let β be $(\Delta^+)'$ simple. If β is imaginary, then $2\langle \lambda + \delta_M, \beta \rangle / |\beta|^2 \geq 1$ and hence $2\langle \lambda, \beta \rangle / |\beta|^2 \geq 0$. If β is complex, then $2\langle \lambda + \delta_M, \beta \rangle / |\beta|^2 = 2\langle \lambda, \beta \rangle / |\beta|^2$ by Lemma 1.3. Hence λ is $(\Delta^+)'$ dominant.

Let G^{13} be the centralizer in G of the subspace $\ker \alpha_R$ of $\mathfrak{a}$, and define G^{12} and G^{23} from α_R' and α_R'' similarly. These groups are mutually conjugate by representatives of members of $W(A:G)$. The Lie algebra $(\mathfrak{g}^{13})^{\mathbb{C}}$ is the sum of $(\mathfrak{a} \oplus \mathfrak{b})^{\mathbb{C}}$ and all root spaces for roots whose $\mathfrak{a}$ part is a multiple of α_R. Let Δ^{13} denote this subsystem of roots.

The semisimple part G^{13}_{ss} of G^{13} is then a group of real rank one with no real roots, and it is consequently locally isomorphic to the product of some $SO(2n+1,1)$ with a compact group. Similar remarks apply to G^{12} and G^{23}.

Suppose henceforth that $w_{13}\sigma \cong \sigma$. Let L^{13} be the centralizer in L of the subspace $\ker \alpha_R$ of $\mathfrak{a}$. If Δ^L denotes the root system of L, then the root system of L^{13} is $\Delta^L \cap \Delta^{13}$, the set of roots in Δ^{13} orthogonal to λ. From Theorem 1.1a of [5] and the identification of our positive system Δ^+ in Proposition 1.4, we obtain the following result.

<u>Proposition 1.5</u>. The standard intertwining operator of G^{13} for the nonunitary principal series of G^{13} with M parameter σ and A parameter $c\rho_{L^{13}}$ (i.e., the operator that defines the invariant Hermitian form on the Langlands quotient) is semidefinite and nonsingular for $0 \leq c < 1$ and is not semidefinite for $c > 1$.

If L_{ss} has real rank one, then $\pm\alpha_R$ are its restricted roots, by Proposition 1.2. Therefore

$$\rho_L = \rho_{L^{13}} \quad \text{if} \quad \text{real rank}(L_{ss}) = 1. \tag{1.5a}$$

Suppose L_{ss} has real rank two. Then the restricted root system of L is of type A_2. In the inclusion $W(A:L) \subseteq W(A:G)$ we thus have equality, and we can therefore take our representatives w_{12}, w_{23}, and w_{13} of Weyl group elements to be in L, hence in L_{ss}. Then we can define L^{12} and L^{23} to be the centralizers in L of the subspaces $\ker \alpha_R'$ and $\ker \alpha_R''$ of $\mathfrak{a}$, and the three subgroups L^{13}, L^{12}, and L^{23} are conjugate via representatives in L of members of $W(A:G)$. So α_R, α_R', and α_R'' all have the same multiplicity as restricted roots in L, and it follows that

$$\rho_L = 2\rho_{L^{13}} \quad \text{if real rank}(L_{ss}) = 2. \tag{1.5b}$$

2. Use of intertwining operators

We continue with the notation of the introduction, now assuming $w_{13}\sigma \cong \sigma$. In this section we shall prove the part of the Main Theorem that deals with $c\rho_L$ for $0 \leq c < 1$ and also the part for $c = 1$ when L_{ss} has real rank one.

The tool will be the intertwining operators of [13], except that we write them consistently with an action by G on the left in the induced space. For w in $N_K(A)$, the operator $A(w,\sigma,\nu)$ is given initially by

$$A(w,\sigma,\nu)f(x) = \int_{\theta N \cap w^{-1}Nw} f(xw\bar{n})\, d\bar{n}.$$

It is continued meromorphically and then normalized suitably. We require that the normalizing factor have no poles or zeros for $\operatorname{Re}\nu$ in the open positive Weyl chamber, and we let $\mathfrak{a}(w,\sigma,\nu)$ be the normalized operator. The operator that defines the Hermitian form of interest is

$$\sigma(w_{13})\mathfrak{a}(w_{13},\sigma,\nu). \tag{2.1}$$

Here $\sigma(w_{13})$ is defined by means of Lemma 18 of [13].

We can take w_{13} to be in L_{ss}, and we can take w_{12} to be in L_{ss} if L_{ss} has real rank two (see the end of §1). Then

$$w_{23} = w_{12}^{-1}w_{13}w_{12}$$

is a representative of $s_{\alpha_R''}$.

<u>Lemma 2.1</u>. For $w_{13}\nu = -\nu$ and ν real, the operator (2.1) satisfies

$$\sigma(w_{13})\mathcal{Q}(w_{13},\sigma,\nu) = \mathcal{Q}(w_{12}^{-1},\sigma,\nu)^*[(\sigma(w_{13})\mathcal{Q}(w_{23},w_{12}^{-1}\sigma,w_{12}^{-1}\nu)]\mathcal{Q}(w_{12}^{-1},\sigma,\nu)\,. \tag{2.3}$$

Moreover, the operator in brackets on the right side may be regarded as

$$w_{12}^{-1}\sigma(w_{23})\mathcal{Q}(w_{23},w_{12}^{-1}\sigma,w_{12}^{-1}\nu)\,. \tag{2.4}$$

<u>Proof</u>. From properties in [13], the operator (2.1) equals

$$\{\sigma(w_{13})\mathcal{Q}(w_{12},w_{23}w_{12}^{-1}\sigma,w_{23}w_{12}^{-1}\nu)\}\mathcal{Q}(w_{23},w_{12}^{-1}\sigma,w_{12}^{-1}\nu)\mathcal{Q}(w_{12}^{-1},\sigma,\nu)\,,$$

and the operator in braces equals

$$\mathcal{Q}(w_{12},\sigma(w_{13})(w_{23}w_{12}^{-1}\sigma)\sigma(w_{13})^{-1},w_{23}w_{12}^{-1}\nu)\sigma(w_{13})\,. \tag{2.5}$$

Here

$$\begin{aligned}\sigma(w_{13})(w_{23}w_{12}^{-1}\sigma)(m)\sigma(w_{13})^{-1} &= \sigma(w_{13})\sigma(w_{12}w_{23}^{-1}mw_{23}w_{12}^{-1})\sigma(w_{13})^{-1} \\ &= \sigma(w_{13}w_{12}w_{23}^{-1}mw_{23}w_{12}^{-1}w_{13}^{-1}) \\ &= \sigma(w_{12}mw_{12}^{-1}) \qquad\qquad \text{by (2.2)} \\ &= w_{12}^{-1}\sigma(m)\,.\end{aligned}$$

Moreover $w_{13}\nu = -\nu$ implies $w_{12}w_{23}w_{12}^{-1}\nu = -\nu$, and thus $w_{23}w_{12}^{-1}\nu = -w_{12}^{-1}\nu$. Thus (2.5) is

$$\begin{aligned}&= \mathcal{Q}(w_{12},w_{12}^{-1}\sigma,-w_{12}^{-1}\nu)\sigma(w_{13}) \\ &= \mathcal{Q}(w_{12}^{-1},\sigma,\nu)^*\sigma(w_{13}) \qquad\qquad \text{by [13].}\end{aligned}$$

This proves (2.3). To prove that the operator in brackets may be regarded as (2.4), we must verify that $w_{12}^{-1}\sigma(w_{23})$ intertwines $w_{12}^{-1}\sigma$ and $w_{23}^{-1}w_{12}^{-1}\sigma$ and that its square is equal to $w_{12}^{-1}\sigma(w_{23}^2)$, according to Lemma 18 of [13]. We have

$$w_{12}^{-1}\sigma(w_{23})w_{12}^{-1}\sigma(m)w_{12}^{-1}\sigma(w_{23}^{-1}) = \sigma(w_{13})\sigma(w_{12}mw_{12}^{-1})\sigma(w_{13})^{-1} \qquad \text{by (2.2)}$$

$$= \sigma(w_{13}w_{12}mw_{12}^{-1}w_{13}^{-1}) = \sigma(w_{12}w_{23}mw_{23}^{-1}w_{12}^{-1}) \qquad \text{by (2.2)}$$

$$= w_{23}^{-1}w_{12}^{-1}\sigma(m)$$

and

$$[w_{12}^{-1}\sigma(w_{23})]^2 = \sigma(w_{13})^2 = \sigma(w_{13}^2) = \sigma(w_{12}w_{23}^2w_{12}^{-1}) = w_{12}^{-1}\sigma(w_{23}^2) ,$$

and thus the lemma follows.

Now let us return to the Main Theorem. The operator in brackets in (2.3) is an operator for a simple reflection in G, and [13] and §5 of [12] show that this operator is semidefinite if and only if the same operator for G^{23} is semidefinite. Let us write the operator for G^{23} as in (2.4) and use the conjugation $G^{13} = w_{12}G^{23}w_{12}^{-1}$. This conjugation carries the induced space for G^{23}, $w_{12}^{-1}\sigma$, and $w_{12}^{-1}\nu$ to the induced space for G^{13}, σ, and ν, and the operator (2.4) is carried to the operator $\sigma(w_{13})G(w_{13},\sigma,\nu)$ defined in G^{13}. By Proposition 1.5 this operator is semidefinite for $\nu = c\rho_{L^{13}}$ with $0 \leq c \leq 1$. Thus the operator in brackets in (2.3) is semidefinite for $\nu = c\rho_{L^{13}}$ with $0 \leq c \leq 1$. Since (2.3) shows that (2.1) is just B^*CB, with C the operator in brackets, the operator (2.1) is semidefinite for $\nu = c\rho_{L^{13}}$ with $0 \leq c \leq 1$. Taking into account (1.5), we see that $J(P,\sigma,\nu)$ is infinitesimally unitary for $\nu = c\rho_L$ with

$$\begin{cases} 0 \leq c \leq 1 & \text{if } L_{ss} \text{ has real rank one} \\ 0 \leq c \leq \tfrac{1}{2} & \text{if } L_{ss} \text{ has real rank two.} \end{cases}$$

Now suppose L_{ss} has real rank 2. Then $W(A:G)$ has representatives in L, by Proposition 1.2, and it fixes the class of σ. From [13] and §5 of [12], we know that the nonsingularity of the right-hand operator $G(w_{12}^{-1},\sigma,\nu)$ in (2.3) is the same as for that operator in G^{12}, which depends only on the projection of ν in the direction of α_R'. Write $\nu = c\rho_L = 2c\rho_{L^{13}}$. Since ν is a multiple of α_R and

$$\frac{\langle \alpha_R, \alpha_R' \rangle}{|\alpha_R'|^2} = \frac{1}{2} ,$$

we can replace ν in our operator for G^{12} by $c\rho_{L^{12}}$. Proposition 1.5 says that our operator in G^{12} is nonsingular out to $\rho_{L^{12}}$, hence for $0 \leq c < 1$.

Thus for $\nu = c\rho_L$ with $0 \leq c < 1$, the operator (2.1) is of the form B^*CB with B nonsingular. Hence (2.1) is semidefinite only if C is semidefinite. But C is not semidefinite beyond $\rho_{L^{13}} = \frac{1}{2}\rho_L$, i.e., for $c > \frac{1}{2}$. Therefore $J(P,\sigma,\nu)$ is not infinitesimally unitary for $\frac{1}{2} < c < 1$.

3. Use of derived functor modules

We continue with the notation of the introduction, and we assume $w_{13}\sigma \cong \sigma$. Our goal in this section is to prove that $J(P,\sigma,\rho_L)$ is infinitesimally unitary if L_{ss} has real rank two; this result will be stated as Proposition 3.4. (We know already that $J(P,\sigma,\rho_L)$ is infinitesimally unitary if L_{ss} has real rank one, and that case will not concern us in this section.)

The proof is rather similar to the proof of unitarity of some isolated representations that occur in [5]. It uses Zuckerman's derived functor modules $A_{\mathfrak{q}}(\mu)$, as explained in Vogan and Zuckerman [24], but with the parameter μ outside the usual range. (See also Enright and Wallach [7].) A big theorem due to Vogan [22] establishes unitarity for such representations under suitable conditions. The supplementary arguments in §12 of [5] are due in part to Vogan as well, and we give the proofs only when they differ from those in §12 of [5].

Our subspace $\mathfrak{b}$ (see §1) is maximal abelian in $\mathfrak{k}$, and we let Δ_K be the set of roots $\Delta_K = \Delta(\mathfrak{k}^{\mathbb{C}}, \mathfrak{b}^{\mathbb{C}})$. One knows (see, e.g., [10]) that the members of Δ_K are the restrictions to $\mathfrak{b}^{\mathbb{C}}$ of all members of Δ except the noncompact roots of Δ_M (of which there are none in our case). We take Δ_K^+ to be the restrictions to $\mathfrak{b}^{\mathbb{C}}$ of the positive members of Δ; this is a positive system since we have taken $i\mathfrak{b}$ before $\mathfrak{a}$ in defining Δ^+. Let δ_K be half the sum of the members of Δ_K^+.

For ν in the closed positive Weyl chamber, $U(P,\sigma,\nu)$ in our situation has a unique minimal K-type Λ in the sense of Vogan [20], i.e., a subrepresentation τ_Λ of $U(P,\sigma,\nu)|_K$ with highest weight Λ for which $|\Lambda + 2\delta_K|^2$ is a minimum, and Theorem 1 of [10] and the present Proposition 1.4 together say that the minimal K-type is given by

$$\Lambda = \lambda = \lambda_0 + \delta - 2\delta_K. \tag{3.1}$$

Moreover, τ_Λ occurs with multiplicity one in $U(P,\sigma,\nu)$ and lies in $J(P,\sigma,\nu)$.

Proposition 3.1. In the situation of §1, suppose that $J(P,\sigma,\nu)$ and $J(P,\sigma',\nu')$ each admit a nonzero Hermitian form and that they have the same real infinitesimal character and same minimal K-type. Then $J(P,\sigma,\nu)$ and $J(P,\sigma',\nu')$ are infinitesimally equivalent.

Proof. The theory of [20] shows that the assumption of a minimal K-type in common implies we may take $\sigma = \sigma'$. (This is clear from (3.1) only after a little work, since one must first adjust the positive systems suitably.) Since our representations admit invariant Hermitian forms, ν and ν' must both be mapped into their negatives by w_{13}, hence must both be multiples of $\alpha_R = e_1 - e_3$. Since the infinitesimal characters are equal, we conclude $|\nu| = |\nu'|$ and then $\nu = \nu'$. This proves the proposition.

We shall use Proposition 3.1 to match $J(P,\sigma,\rho_L)$ with a suitable $A_{\mathfrak{q}}(\mu)$. We shall not need the detailed construction of $A_{\mathfrak{q}}(\mu)$, only its existence and properties in Theorem 3.2 below. We have already defined Δ^L and we let $\Delta(\mathfrak{u})$ be the set of positive roots (of Δ^+) outside Δ^L. Let $\delta(\mathfrak{u})$ be half their sum. In the Vogan theory, the symbol stands for $\mathfrak{l}^{\mathbb{C}} \oplus \mathfrak{u}$.

There being no noncompact roots in Δ_M, we let $\Delta(\mathfrak{p}^{\mathbb{C}})$ be the set of restrictions to $\mathfrak{b}^{\mathbb{C}}$ of the complex roots of Δ, and we let $\Delta(\mathfrak{u} \cap \mathfrak{p}^{\mathbb{C}})$ be the set of restrictions to $\mathfrak{b}^{\mathbb{C}}$ of the complex roots of $\Delta(\mathfrak{u})$. Let $\delta(\mathfrak{u} \cap \mathfrak{p}^{\mathbb{C}})$ be half the sum of the members of $\Delta(\mathfrak{u} \cap \mathfrak{p}^{\mathbb{C}})$. If $\delta_K(\mathfrak{u})$ denotes half the sum of the members of Δ_K^+ that arise by restriction from $\Delta(\mathfrak{u})$, then we have

$$\delta(\mathfrak{u}) = \delta_K(\mathfrak{u}) + \delta(\mathfrak{u} \cap \mathfrak{p}^{\mathbb{C}}) . \tag{3.2}$$

<u>Theorem 3.2</u> (Vogan [22]). Suppose G, Δ^L, and L are as in the introduction; here L is the analytic subgroup of G with Lie algebra

$$\mathfrak{l} = \mathfrak{g} \cap (\mathfrak{b} + \mathfrak{a} + \sum_{\beta \in \Delta^L} \mathfrak{g}_\beta) .$$

If μ in $i\mathfrak{b}'$ is the differential of a unitary (one-dimensional) character of L such that

$$\langle \mu + \delta(\mathfrak{u}), \beta \rangle \geq 0 \quad \text{for all } \beta \text{ in } \Delta(\mathfrak{u}) , \tag{3.3}$$

and if

$$\Lambda' = \mu + 2\delta(\mathfrak{u} \cap \mathfrak{p}^{\mathbb{C}}) ,$$

then there exists an admissible representation $A_{\mathfrak{q}}(\mu)$ of $\mathfrak{g}$ with infinitesimal character $\mu + \delta$ such that

(a) the K-types have multiplicities given by the following version of Blattner's formula:

$$\text{mult } \tau_{\Lambda''} = \sum_{s \in W_K} (\det s) \mathcal{P}(s(\Lambda'' + \delta_K) - (\Lambda' + \delta_K)) ,$$

where W_K is the Weyl group of Δ_K and $\mathcal{P}$ is the partition function relative to $\Delta(\mathfrak{u} \cap \mathfrak{p}^{\mathbb{C}})$, and

(b) the representation $A_{\mathfrak{q}}(\mu)$ admits a positive definite invariant inner product.

Proof. This is derived from [22] and [21] in the same way that Theorem 12.2 is proved in [5].

Proposition 3.3. Under the assumptions of Theorem 3.2, suppose that Λ' is Δ_K^+ dominant. Then $A_{\mathfrak{q}}(\mu)$ is nonzero and the K-type $\tau_{\Lambda'}$ occurs with multiplicity one. If in addition $\langle \Lambda'+2\delta_K, \beta\rangle \geq 0$ for all β in $\Delta(\mathfrak{u})$, then $\tau_{\Lambda'}$ is the unique minimal K-type of $A_{\mathfrak{q}}(\mu)$.

Proof. The argument is the same as for Proposition 12.3 of [5].

Proposition 3.4. In the setting of the introduction, let Λ be defined by (3.1). If L_{ss} has real rank two, then $J(P,\sigma,\rho_L)$ is infinitesimally unitary.

Proof. We shall use Δ^L from the introduction, together with the corresponding $\mathfrak{u}$ built from Δ^+, as data for Theorem 3.2. Put

$$\mu = \Lambda - 2\delta(\mathfrak{u} \cap \mathfrak{p}^{\mathbb{C}}). \tag{3.4}$$

First we exponentiate μ; this is a little tricky.

There is no loss in generality in assuming that $G^{\mathbb{C}}$ is simply connected. Vogan and Zuckerman [24] show that $2\delta(\mathfrak{u} \cap \mathfrak{p}^{\mathbb{C}})$ is integral, i.e., exponentiates to $(\mathfrak{b} \oplus \mathfrak{a})^{\mathbb{C}}$. To see that $\Lambda = \lambda$ is integral, let us note first that $2\langle \lambda, \beta\rangle / |\beta|^2$ is an integer if β is in Δ_M^+, since λ is the highest weight of σ. Suppose β is a complex root: $\beta = \beta_I + \beta_R$. The $\mathfrak{a}$ part β_R is one of $\pm\alpha_R$, $\pm\alpha_R'$, and $\pm\alpha_R''$. Since the proof will be the same in all cases, let us suppose $\beta_R = \alpha_R'$. Then it follows from the first part of the proof of Lemma 1.1 that $\beta_I + \alpha_I'$ and $\beta_I - \alpha_I'$ are both roots. Hence

$$\beta = (\beta_I - \alpha_I') + (\alpha_I' + \alpha_R')$$

and

$$\frac{2\langle\lambda,\beta\rangle}{|\beta|^2} = \frac{2\langle\lambda,\beta_I - \alpha_I'\rangle}{|\beta_I - \alpha_I'|^2} \tag{3.5}$$

by Proposition 1.2. (Here we use that L_{ss} has real rank two.) Since λ is integral for members of Δ_M, (3.5) shows λ is integral for Δ^+. Since $G^{\mathbb{C}}$ is simply connected, we conclude e^{μ} is a well defined one-dimensional quasicharacter of $(\mathfrak{b} \oplus \mathfrak{a})^{\mathbb{C}}$.

Next we show that μ is orthogonal to the members of Δ^L. We know that $\Lambda = \lambda$ is orthogonal to Δ^L, by definition of Δ^L, and we have to check the orthogonality of $2\delta(\mathfrak{u} \cap \mathfrak{p}^{\mathbb{C}})$ with Δ^L. It is clear that $2\delta(\mathfrak{u})$ is orthogonal to Δ^L, and (3.2) says it is enough to prove that $2\delta_K(\mathfrak{u})$ is orthogonal to Δ^L. Now $2\delta_K(\mathfrak{u})$ vanishes (term-by-term) on $\mathfrak{a}$ and therefore sees only the restrictions of members of Δ^L to $\mathfrak{b}$. We have observed that these restrictions are exactly the members of Δ_K^L (since $\mathfrak{m}$ has no noncompact roots), and thus we are to prove that $2\delta_K(\mathfrak{u})$ is orthogonal to Δ_K^L. But this is clear since the roots contributing to $2\delta_K(\mathfrak{u})$ are permuted by reflections in members of Δ_K^L.

Therefore the Theorem of the Highest Weight supplies an irreducible finite-dimensional representation of $L^{\mathbb{C}}$ with highest weight μ. Naturally this representation is one-dimensional. Since μ vanishes on $\mathfrak{a}$, the restriction of e^{μ} of this representation to L is unitary.

To apply Theorem 3.2, we need to verify (3.3). We have

$$\begin{aligned}
\mu + \delta(\mathfrak{u}) &= \Lambda - 2\delta(\mathfrak{u} \cap \mathfrak{p}^{\mathbb{C}}) + \delta(\mathfrak{u}) \\
&= \lambda_0 + \delta - 2\delta_K - 2\delta(\mathfrak{u} \cap \mathfrak{p}^{\mathbb{C}}) + \delta(\mathfrak{u}) && \text{by (3.1)} \\
&= \lambda_0 + \delta - 2\delta_K + 2\delta_K(\mathfrak{u}) - \delta(\mathfrak{u}) && \text{by (3.2)} \\
&= \lambda_0 + \delta(\mathfrak{l}^{\mathbb{C}}) - 2\delta_K(\mathfrak{l}^{\mathbb{C}}) && \text{in obvious notation.}
\end{aligned}$$

The right side is the sum of λ_0 and a real combination of members of

Δ^L, and any member γ of Δ^L satisfies $\sum_{w\in W(\Delta^L)} w\gamma = 0$, where $W(\Delta^L)$ is the Weyl group of the root system Δ^L. Since μ and $\delta(\mathfrak{u})$ are invariant under $W(\Delta^L)$, we obtain

$$\mu + \delta(\mathfrak{u}) = \sum_{w\in W(\Delta^L)} w\lambda_0 .$$

Thus β in $\Delta(\mathfrak{u})$ implies

$$\langle \mu + \delta(\mathfrak{u}), \beta\rangle = \sum_{w\in W(\Delta^L)} \langle \lambda_0, w^{-1}\beta\rangle \geq 0 ,$$

since $w^{-1}\beta$ is in $\Delta(\mathfrak{u})$ and $\langle \lambda_0, w^{-1}\beta\rangle$ is thus ≥ 0 (Proposition 1.4).

Thus Theorem 3.2 applies. The form Λ' in the theorem is our Λ, by (3.4). The theorem says that $A_{\mathfrak{q}}(\mu)$ has infinitesimal character $\mu + \delta$ and is unitary. By (3.1), $\Lambda + 2\delta_K$ is $\lambda_0 + \delta$, which is Δ^+ dominant by Proposition 1.4. Thus Proposition 3.3 applies, showing that $A_{\mathfrak{q}}(\mu)$ is nonzero and that Λ is the unique minimal K-type of $A_{\mathfrak{q}}(\mu)$.

Now $J(P,\sigma,\rho_L)$ has minimal K-type Λ and infinitesimal character $\lambda_0 + \rho_L$. By Proposition 3.1 the proof of Proposition 3.4 will be complete if we show that $\lambda_0 + \rho_L$ is conjugate to $\mu + \delta$ by the Weyl group of Δ. Here

$$\mu + \delta = \mu + \delta(\mathfrak{u}) + \delta(\mathfrak{l}^{\mathbb{C}}) .$$

The first two terms on the right side are fixed by $W(\Delta^L)$. Applying a member of $W(\Delta^L)$ that results in a positive system for Δ^L that takes $\mathfrak{a}$ before $i\mathfrak{b}$, we see from (3.1) and (3.4) that $\mu + \delta$ is conjugate to

$$\mu + \delta(\mathfrak{u}) + \delta_M(\mathfrak{l}^{\mathbb{C}}) + \rho_L = \lambda_0 + \delta - 2\delta_K - 2\delta(\mathfrak{u}\cap\mathfrak{p}^{\mathbb{C}}) + \delta(\mathfrak{u}) + \delta_M(\mathfrak{l}^{\mathbb{C}}) + \rho_L . \quad (3.6)$$

Now Lemma 3 of [10], applied to $\mathfrak{l}^{\mathbb{C}}$, says that

$$\delta(\mathfrak{l}^{\mathbb{C}}) - 2\delta_K(\mathfrak{l}^{\mathbb{C}}) = \delta_M(\mathfrak{l}^{\mathbb{C}}) - 2\delta_{M\cap K}(\mathfrak{l}^{\mathbb{C}}) .$$

Since $M\cap L$ is compact, the right side is just $-\delta_M(\mathfrak{l}^{\mathbb{C}})$. Thus

$$\delta_M(\mathfrak{l}^{\mathbb{C}}) = 2\delta_K(\mathfrak{l}^{\mathbb{C}}) - \delta(\mathfrak{l}^{\mathbb{C}}) ,$$

and (3.6) becomes

$$= \lambda_0 + \delta - 2\delta_K - 2\delta(\mathfrak{u} \cap \mathfrak{p}^{\mathbb{C}}) + \delta(\mathfrak{u}) + 2\delta_K(\mathfrak{l}^{\mathbb{C}}) - \delta(\mathfrak{l}^{\mathbb{C}}) + \rho_L$$

$$= \lambda_0 + \delta - 2\delta_K(\mathfrak{u}) - 2\delta(\mathfrak{u} \cap \mathfrak{p}^{\mathbb{C}}) + \delta(\mathfrak{u}) - \delta(\mathfrak{l}^{\mathbb{C}}) + \rho_L$$

$$= \lambda_0 + \delta - 2\delta(\mathfrak{u}) + \delta(\mathfrak{u}) - \delta(\mathfrak{l}^{\mathbb{C}}) + \rho_L$$

$$= \lambda_0 + \rho_L .$$

This proves the required conjugacy and completes the proof of the proposition.

4. Cut-off for unitarity

To complete the proof of the Main Theorem, we are to show that there is no unitarity beyond ρ_L. Our argument uses the techniques of [3], as amplified in §§1-2 of [4] and §3 of [5]: We produce a K-type on which the signature of the invariant Hermitian form rules out unitarity. This kind of argument involves a certain calculation that is briefly indicated in [3] and [4] and will be written in more detail later. Accordingly, in the present paper, we merely identify what is to be calculated, what the result is, and what its effect is on unitarity.

In the setting of the introduction, we assume that $w_{13}\sigma \cong \sigma$. Thus (1.2) gives $\langle \lambda, \alpha_I \rangle = 0$, and hence (3.1) gives $\langle \Lambda, \alpha_I \rangle = 0$ for the minimal K-type Λ. The representation σ of M occurs in $\tau_\Lambda|_M$, and we let B be a nonzero M map from the space on which τ_Λ operates to the space on which σ operates. If v_Λ is a nonzero highest weight vector for τ_Λ, then

$$f_0(k) = B(\tau_\Lambda(k)^{-1} v_\Lambda)$$

is a member of the induced space (in the compact picture). We normalize our invariant Hermitian form $\langle\cdot,\cdot\rangle$ so that $\langle f_0, f_0\rangle = 1$.

Let Λ' be the result of making $\Lambda + \alpha_I$ dominant for Δ_K^+ by applying a member of the Weyl group of Δ_K, and let $P_{\Lambda'}$ be the projection of the induced space to the $\tau_{\Lambda'}$ subspace. Let α be the root $\alpha_I + \alpha_R$, and let E_α be a root vector in $\mathfrak{g}^{\mathbb{C}}$ for α. Then we have the following result.

Proposition 4.1. In the setting of the introduction when $w_{13}\sigma \cong \sigma$, the function

$$f_1 = P_{\Lambda'} U(P,\sigma,c\rho_L, E_\alpha - \theta E_\alpha) f_0$$

is a nonzero member of the induced space, and $\langle f_1, f_1\rangle$ is a positive multiple of $1 - c^2$.

Then it follows that the form $\langle\cdot,\cdot\rangle$ is not semidefinite on the sum of its τ_Λ and $\tau_{\Lambda'}$ subspaces, and hence $U(P,\sigma,c\rho_L)$ is not infinitesimally unitary for $c > 1$.

References

[1] Angelopoulos, E., Sur les représentations unitaires irréductibles de $\overline{SO_0}(p,2)$, C. R. Acad. Sci. Paris Ser. I 292 (1981), 469-471.

[2] Baldoni Silva, M. W., and D. Barbasch, The unitary spectrum for real rank one groups, Invent. Math. 72 (1983), 27-55.

[3] Baldoni-Silva, M. W., and A. W. Knapp, Indefinite intertwining operators, Proc. Nat. Acad. Sci. USA 81 (1984), 1272-1275.

[4] Baldoni-Silva, M. W., and A. W. Knapp, Indefinite intertwining operators II, preprint, 1984.

[5] Baldoni-Silva, M. W., and A. W. Knapp, Unitary representations induced from maximal parabolic subgroups, preprint, 1985.

[6] Duflo, M., Représentations unitaires irréductibles des groupes simples complexes de rang deux, Bull. Soc. Math. France 107 (1979), 55-96.

[7] Enright, T. J., and N. R. Wallach, Notes on homological algebra and representations of Lie algebras, Duke Math. J. 47 (1980), 1-15.

[8] Knapp, A. W., Weyl group of a cuspidal parabolic, Ann. Sci. Ecole Norm. Sup. 8 (1975), 275-294.

[9] Knapp, A. W., Commutativity of intertwining operators for semisimple groups, Compositio Math. 46 (1982), 33-84.

[10] Knapp, A. W., Minimal K-type formula, "Non Commutative Harmonic Analysis and Lie Groups," Springer-Verlag Lecture Notes in Math. 1020 (1983), 107-118.

[11] Knapp, A. W., Representation Theory of Semisimple Groups: An Overview Based on Examples, Princeton University Press, to appear.

[12] Knapp, A. W., and B. Speh, Status of classification of irreducible unitary representations, "Harmonic Analysis," Springer-Verlag Lecture Notes in Math. 908 (1982), 1-38.

[13] Knapp, A. W., and E. M. Stein, Intertwining operators for semisimple groups, Ann. of Math. 93 (1971), 489-578.

[14] Knapp, A. W., and G. Zuckerman, Classification theorems for representations of semisimple Lie groups, "Non-Commutative Harmonic Analysis," Springer-Verlag Lecture Notes in Math. 587 (1977), 138-159.

[15] Knapp, A. W., and G. J. Zuckerman, Classification of irreducible tempered representations of semisimple groups, Ann. of Math. 116 (1982), 389-501. (See also 119 (1984), 639.)

[16] Langlands, R. P., On the classification of irreducible representations of real algebraic groups, mimeographed notes, Institute for Advanced Study, 1973.

[17] Miličić, D., Asymptotic behaviour of matrix coefficients of the discrete series, Duke Math. J. 44 (1977), 59-88.

[18] Tsuchikawa, M., On the representations of SL(3,C), III, Proc. Japan Acad. 44 (1968), 130-132.

[19] Vahutinskii, I. J., Unitary representations of GL(3,R), Math. Sbornik 75 (117) (1968), 303-320 (Russian).

[20] Vogan, D. A., The algebraic structure of the representation of semisimple groups I, Ann. of Math. 109 (1979), 1-60.

[21] Vogan, D. A., Representations of Real Reductive Lie Groups, Birkhäuser, Boston, 1981.

[22] Vogan, D. A., Unitarizability of certain series of representations, Ann. of Math. 120 (1984), 141-187.

[23] Vogan, D. A., The unitary dual of GL(n) over an archimedean field, preprint, 1985.

[24] Vogan, D. A., and G. J. Zuckerman, Unitary representations with non-zero cohomology, Compositio Math. 53 (1984), 51-90.

Dipartimento di Matematica
Università degli Studi di Trento
38050 Povo (TN), Italy

Department of Mathematics
Cornell University
Ithaca, New York 14853, U.S.A.

VOGAN'S ALGORITHM FOR COMPUTING COMPOSITION SERIES

M. W. Baldoni-Silva* and A. W. Knapp**

Building on joint work [8] with B. Speh, D. A. Vogan [9,10] has obtained an algorithm for computing composition series of the standard induced representations of a semisimple Lie group G. The algorithm takes a particularly simple form in the case that the representations are induced from a maximal parabolic subgroup.

On the one hand, this algorithm does not seem to be widely known, possibly because the emphasis in the Vogan papers is somewhat different. And on the other hand, the algorithm is particularly well suited to deciding some irreducibility questions that arise in our paper [1], which settles the contribution to the unitary dual of a linear connected G by Langlands quotients obtained from maximal parabolic subgroups. Thus it seems expedient to provide an exposition of Vogan's algorithm in the context of the examples needed for [1].

This approach to our irreducibility questions was suggested to us by Vogan, and it overlaps with the joint work of Barbasch and Vogan [3], which uses the algorithm to decide irreducibility in "critical cases" for classical groups. (In fact, the specific results that we give here for classical groups are some of the critical cases for Barbasch and Vogan, although our approach may be more direct.) Calculations with the algorithm appear in other papers as well, for example in [2].

The present paper is organized as follows. Section 1 collects some results from [8] and [9] and sketches the steps of the algorithm. Section 2 establishes a theorem that gives an efficient starting point

---

* Partially supported by National Science Foundation Grant DMS 85-01793.

** Supported by National Science Foundation Grants DMS 80-01854 and DMS 85-01793.

for use of the algorithm. And Sections 3-7 show how the algorithm works in $SO(2n-1,2)$, $Sp(3,\mathbb{R})$, $SO^*(10)$, $SO^*(2n)$, and some exceptional groups of type E, respectively. The worked examples are exactly the ones needed for [1]. We proceed only in the generality needed for our examples——induction from a maximal parabolic subgroup, G linear and connected, rank G = rank K, integral infinitesimal character. Readers interested in relaxing any of these assumptions should consult [8,9,10], particularly pp. 293-294 of [8], §6 of [9], and all of [10].

We are indebted to Vogan for several valuable conversations on this topic and to Barbasch and Vogan for sharing with us the details of their work announced as [3].

1. Notation and algorithm

Let G be a linear connected semisimple group with a simply-connected complexification, let θ be a Cartan involution, let K be the corresponding maximal compact subgroup, and assume that rank G = rank K. Let TA be a θ-stable Cartan subgroup (with $T \subseteq K$ and A equal to a vector group), and let MA be the corresponding Levi factor of the associated cuspidal parabolic subgroups.

Let $P = MAN$ be one of these parabolic subgroups. If σ is a discrete series representation of M or nondegenerate limit of discrete series and if ν is a real-valued linear functional on the Lie algebra $\mathfrak{a}$ of A, we let $U(P,\sigma,\nu)$ be the standard induced representation given by unitary induction as

$$U(P,\sigma,\nu) = \operatorname{ind}_P^G(\sigma \otimes e^\nu \otimes 1) .$$

It is known that $U(P,\sigma,\nu)$ has a finite Jordan-Holder series, and Vogan's algorithm addresses the problem of finding the irreducible subquotients and their multiplicities. (We shall be addressing this problem only when $\dim A = 1$ and $U(P,\sigma,\nu)$ has integral infinitesimal character.)

The algorithm takes place in three stages. In the first two stages one considers the same problem for regular integral infinitesimal character. In the first stage we identify some $U(P,\sigma_k,\nu_k)$ with the same infinitesimal character for which the composition series is known. In the second stage we go through a succession of wall-crossing operations, following what happens to the decomposition. At the third stage we pass to our original parameters by means of Zuckerman's ψ functor [11].

Let us introduce more convenient notation. Let $\Delta(\mathfrak{m}^{\mathbb{C}},\mathfrak{t}^{\mathbb{C}})$ be the set of roots of M, and let M_0 be the identity component of M. Pick any irreducible constituent of $\sigma|_{M_0}$, and let $(\lambda, \Delta^+(\mathfrak{m}^{\mathbb{C}},\mathfrak{t}^{\mathbb{C}}))$ be its Harish-Chandra parameter. Corresponding to each real root β in $\Delta(\mathfrak{g}^{\mathbb{C}},(\mathfrak{t}\oplus\mathfrak{a})^{\mathbb{C}})$ is an element γ_β of G that is the image of $\begin{pmatrix}-1 & 0\\ 0 & -1\end{pmatrix}$ in $SL(2,\mathbb{R})$ under the homomorphism of $SL(2,\mathbb{R})$ into G built from β. Let $F(T)$ be the finite abelian group generated by the elements γ_β. The restriction of σ to the M-central subgroup $F(T)$ determines a character χ, and $(\lambda, \Delta^+(\mathfrak{m}^{\mathbb{C}},\mathfrak{t}^{\mathbb{C}}))$ and χ together determine σ. Introduce a positive system $\Delta^+(\mathfrak{g}^{\mathbb{C}},(\mathfrak{t}\oplus\mathfrak{a})^{\mathbb{C}})$ for the roots of G containing $\Delta^+(\mathfrak{m}^{\mathbb{C}},\mathfrak{t}^{\mathbb{C}})$ such that $\lambda+\nu$ is dominant. (Such a system always exists; if $\lambda+\nu$ is regular, it is unique.)

Adding to $\lambda+\nu$ a suitable parameter μ that is dominant integral for $\Delta^+(\mathfrak{g}^{\mathbb{C}},(\mathfrak{t}\oplus\mathfrak{a})^{\mathbb{C}})$ and adjusting χ compatibly, we are led to a standard induced representation $U(P,\sigma_0,\nu_0)$ whose infinitesimal character $\gamma_0 = \lambda+\nu+\mu$ is regular ([5], Appendix B). Moreover, the Zuckerman ψ functor [11] satisfies

$$U(P,\sigma,\nu) \cong \psi_{\lambda+\nu}^{\lambda+\nu+\mu}\, U(P,\sigma_0,\nu_0)\,.$$

We shall be working with several Cartan subgroups, and we need to match carefully the corresponding root systems. Thus let us fix a compact Cartan subgroup B of G contained in K, and let $\mathfrak{b}$ be its Lie algebra. We may assume that all other Cartan subgroups of interest are obtained by Cayley transform relative to a succession of roots. In particular, for $\mathfrak{a}$ as above, we can write $\mathfrak{a} \leftrightarrow \{\ldots\}$, where $\{\ldots\}$ is an ordered set of roots of $(\mathfrak{g}^{\mathbb{C}}, \mathfrak{b}^{\mathbb{C}})$ with respect to which we have formed Cayley transforms. Then there will be no confusion if we refer parameters like $\gamma_0 = \lambda + \nu + \mu$ to our compact Cartan subalgebra without introducing new notation for them.

It is important to note that γ_0, $\mathfrak{a} \leftrightarrow \{\ldots\}$, and the adjusted χ determine the global character of $U(P, \sigma_0, \nu_0)$ completely. In fact, we build $\mathfrak{t} \oplus \mathfrak{a}$, transform γ_0 to it, decompose $\gamma_0 = \lambda_0 + \nu_0$, form σ_0, introduce any N' making ν_0 dominant, and induce $\sigma_0 \otimes e^{\nu_0} \otimes 1$ to obtain a representation $U(P', \sigma_0, \nu_0)$. This representation will have the same global character as $U(P, \sigma_0, \nu_0)$. We write $\pi(\gamma_0, \mathfrak{a} \leftrightarrow \{\ldots\})$ for this global character, dropping the adjusted χ from the notation. (Vogan's notation differs slightly from this: He uses $\Theta(\cdot)$ for the character, suppressing the explicit mention of $\mathfrak{a}$ and instead carrying it in the domain of γ_0.) In our notation, note that $\pi(\gamma_0, \mathfrak{a} \leftrightarrow \emptyset)$ is a discrete series character of G.

Under our assumption that γ_0 is regular, it follows essentially from [4] that the representation $U(P', \sigma_0, \nu_0)$ has a unique irreducible quotient $J(P', \sigma_0, \nu_0)$. We write $\overline{\pi}(\gamma_0, \mathfrak{a} \leftrightarrow \{\ldots\})$ for the global character of $J(P', \sigma_0, \nu_0)$. If N'' is a second choice of nilpotent subgroup such that ν_0 is dominant, then $J(P'', \sigma_0, \nu_0) \cong J(P', \sigma_0, \nu_0)$ and consequently the global character $\overline{\pi}(\gamma_0, \mathfrak{a} \leftrightarrow \{\ldots\})$ is independent of the choice of N'.

Now let us work with a character $\pi(\gamma, \mathfrak{a} \leftrightarrow \{\ldots\})$ or $\overline{\pi}(\gamma, \mathfrak{a} \leftrightarrow \{\ldots\})$ such that γ is regular and integral. Let α be a real root (relative to the specified choice of $\mathfrak{a}$), and let χ be the suppressed character of $F(T)$ for this representation. We say that α is a <u>cotangent case</u> for this representation if

$$\chi(\gamma_\alpha) = (-1)^{2\langle \rho_\alpha, \alpha\rangle/|\alpha|^2}$$

or is a <u>tangent case</u> if

$$\chi(\gamma_\alpha) = -(-1)^{2\langle \rho_\alpha, \alpha\rangle/|\alpha|^2} .$$

Here ρ_α is half the sum of the roots having positive inner product with α. In either event, $n = 2\langle\gamma,\alpha\rangle/|\alpha|^2$ is an integer. We say that α <u>satisfies the parity condition</u> for the representation if either

α is a cotangent case and n is even

or

α is a tangent case and n is odd.

If $\mathfrak{g}$ is simple, it can be shown that the parity condition can fail for integral γ only if α is long and $\mathfrak{g} \cong \mathfrak{sp}(n,\mathbb{R})$ for some n.

One starting place for the algorithm is the following character identity due to Schmid [6,7]. See p. 271 of Speh-Vogan [8]; recall we are assuming G is connected.

<u>Theorem 1.1</u> (Schmid's identity). Let γ be regular integral, and let α be a simple noncompact root (for the system $\Delta^+(\mathfrak{g}^{\mathbb{C}},\mathfrak{b}^{\mathbb{C}})$ that makes γ dominant). If α satisfies the parity condition for $\pi(\gamma, \mathfrak{a} \leftrightarrow \{\alpha\})$, then

$$\pi(\gamma, \mathfrak{a} \leftrightarrow \{\alpha\}) = \overline{\pi}(\gamma, \mathfrak{a} \leftrightarrow \{\alpha\}) + \overline{\pi}(\gamma, \mathfrak{a} \leftrightarrow \emptyset) + \overline{\pi}(s_\alpha\gamma, \mathfrak{a} \leftrightarrow \emptyset) ,$$

where s_α denotes reflection in α.

When the parity condition is not satisfied, the corresponding identity is as follows. (See Proposition 6.1 of Speh-Vogan [8].)

<u>Theorem 1.2</u>. Let γ be regular integral, and let α be a simple noncompact root. If α does not satisfy the parity condition for $\pi(\gamma, \mathfrak{a} \leftrightarrow \{\alpha\})$, then

$$\pi(\gamma, a \leftrightarrow \{\alpha\}) = \overline{\pi}(\gamma, a \leftrightarrow \{\alpha\}),$$

i.e., $\pi(\gamma, a \leftrightarrow \{\alpha\})$ is irreducible.

Next we recall the wall-crossing functors and the τ-invariant. Fix γ regular integral, let α be a simple root in the system that makes γ dominant, let Λ_α be the fundamental weight corresponding to α, and put $n = 2\langle\gamma, \alpha\rangle/|\alpha|^2$. In terms of Zuckerman's ψ and φ functors, define

$$\psi_\alpha = \psi_{\gamma - n\alpha}^{\gamma} \quad \text{and} \quad \varphi_\alpha = \varphi_{\gamma}^{\gamma - n\alpha}.$$

The <u>wall-crossing functor</u> is given by

$$s_\alpha \Theta = \varphi_\alpha \psi_\alpha \Theta - \Theta$$

for any virtual character Θ whose infinitesimal character is γ; s_α acts on the local expression for a global character by reflection in α (see Appendix C of [5]). We say that α is in the <u>τ-invariant</u> of $\overline{\pi}(\gamma, a \leftrightarrow \{\ldots\})$ if $\psi_\alpha \overline{\pi}(\gamma, a \leftrightarrow \{\ldots\}) = 0$.

For purposes of calculation, we shall want to regard the τ-invariant as a subset of integers, say of $\{1, \ldots, \ell\}$, where ℓ is the rank of $\mathfrak{g}$. To do so, we note that the only parameters of interest will be the Cartan subgroups and the various $w\gamma$ with γ fixed and w in the Weyl group of $\mathfrak{g}^{\mathbb{C}}$. The positive root systems for $w\gamma$ and γ are canonically identified via w, and the root systems for the different Cartan subgroups are all identified by our system of Cayley transforms. Thus we can number the roots in a single Dynkin diagram for $\mathfrak{g}^{\mathbb{C}}$ and obtain consistent numberings of the Dynkin diagrams of all the positive root systems we shall consider. In this way the τ-invariant of a character $\overline{\pi}$ can be regarded as a subset of $\{1, \ldots, \ell\}$. (This point will be clearer in the examples.)

If we replace γ by $w\gamma$ and if α is a simple root for the system that makes γ dominant, then $w\alpha$ is simple for the system making $w\gamma$ dominant. Moreover, as observed in [11], the functor $\psi_\alpha = \psi_{\gamma - n\alpha}^{\gamma}$ is the same as the functor $\psi_{w\alpha} = \psi_{w\gamma - nw\alpha}^{w\gamma}$. Similar remarks apply to φ, and thus s_α is the same functor as $s_{w\alpha}$. In keeping with the notation of the previous paragraph, we can thus denote our wall-crossing functors unambiguously by $s_1, \ldots, s_\ell$.

The observations about ψ_α in the previous paragraph make it clear that the τ-invariant is an invariant of the Langlands quotient in question. In particular, we can detect inequivalence of two Langlands quotients by seeing that their τ-invariants are different.

The τ-invariant controls what happens at the third stage of the algorithm when we pass back to our original parameters by means of the ψ functor $\psi_{\lambda+\nu}^{\lambda+\nu+\mu}$. Recall that $\lambda+\nu+\mu$ is dominant for the system $\Delta^+(\mathfrak{g}^{\mathbb{C}}, (\mathfrak{t}\oplus\mathfrak{a})^{\mathbb{C}})$. From this positive system we obtain a set of <u>singular roots</u>, namely those simple roots orthogonal to $\lambda+\nu$. As usual, we may canonically identify the singular roots with a subset of $\{1,\ldots,\ell\}$.

<u>Theorem 1.3</u> ([8], Theorems 5.15, 6.16, 6.18). Let Θ be an irreducible character with regular integral infinitesimal character γ conjugate to $\lambda+\nu+\mu$. Then $\psi_{\lambda+\nu}^{\lambda+\nu+\mu}\Theta$ is irreducible or 0. It is 0 if and only if the set of singular roots for $\lambda+\nu$ has nonempty intersection with the τ-invariant of Θ.

Computation of τ-invariants is a routine matter because of the next theorem.

<u>Theorem 1.4</u> ([8], Theorem 6.16). Let γ be regular integral, let $\overline{\pi}(\gamma, \mathfrak{a} \leftrightarrow \{\ldots\})$ be given, and let $\Delta^+(\mathfrak{g}^{\mathbb{C}}, (\mathfrak{t}\oplus\mathfrak{a})^{\mathbb{C}})$ be the corresponding positive system. Then a simple root α for this system is in the τ-invariant of $\overline{\pi}(\gamma, \mathfrak{a} \leftrightarrow \{\ldots\})$ if and only if α satisfies one of the following:

(a) α is imaginary and $\mathfrak{m}$-compact

(b) α is complex and $\theta\alpha$ is negative

(c) α is real and satisfies the parity condition for $\overline{\pi}(\gamma, \mathfrak{a} \leftrightarrow \{\ldots\})$.

Now we turn to computation of the effect of the wall-crossing functors. Theorem 1.5 will show the effect on a full induced character. Then we consider the constituents. If $\psi_\alpha \Theta = 0$, then it follows from the definition that $s_\alpha \Theta = -\Theta$. Thus we have only to know the effect of s_α on Θ in the case that α is not in the τ-invariant; this we write down in Theorem 1.6.

Theorem 1.5 ([8], Corollary 5.12). Let γ be regular integral, let $\pi(\gamma, \mathfrak{a} \leftrightarrow \{\ldots\})$ be given, and let $\Delta^+(\mathfrak{g}^{\mathbb{C}}, (\mathfrak{t} \oplus \mathfrak{a})^{\mathbb{C}})$ be the corresponding positive system. If α is a complex simple root for this system, then

$$s_\alpha \pi(\gamma, \mathfrak{a} \leftrightarrow \{\ldots\}) = \pi(s_\alpha \gamma, \mathfrak{a} \leftrightarrow \{\ldots\}) .$$

Remark. Here α complex and simple makes the positive roots vanishing on $\mathfrak{a}$ be the same for γ and $s_\alpha \gamma$. Hence this theorem is indeed implied by Corollary 5.12 and the sentence before Lemma 5.8 in [8].

Theorem 1.6 ([8], Theorem 6.16, and [9], Theorem 4.12). Let γ be regular integral, let $\bar{\pi}(\gamma, \mathfrak{a} \leftrightarrow \{\ldots\})$ be given, and let $\Delta^+(\mathfrak{g}^{\mathbb{C}}, (\mathfrak{t} \oplus \mathfrak{a})^{\mathbb{C}})$ be the corresponding positive system. Suppose that α is a simple root for this system that is not in the τ-invariant of $\bar{\pi}(\gamma, \mathfrak{a} \leftrightarrow \{\ldots\})$. Then the wall-crossing functor s_α satisfies

$$s_\alpha \bar{\pi}(\gamma, \mathfrak{a} \leftrightarrow \{\ldots\}) = \bar{\pi}(\gamma, \mathfrak{a} \leftrightarrow \{\ldots\}) + U_\alpha(\bar{\pi}(\gamma, \mathfrak{a} \leftrightarrow \{\ldots\})) ,$$

where $U_\alpha(\bar{\pi}(\gamma, \mathfrak{a} \leftrightarrow \{\ldots\}))$ is a sum of true characters as follows:

(a) If α is imaginary and $\mathfrak{m}$-noncompact, then

$$U_\alpha(\bar{\pi}(\gamma, \mathfrak{a} \leftrightarrow \{\ldots\})) = \begin{cases} \bar{\pi}(\gamma, \mathfrak{a} \leftrightarrow \{\ldots, \alpha\}) + \Theta_0 & \text{if } s_\alpha \notin W(M:T) \\ \bar{\pi}_1(\gamma, \mathfrak{a} \leftrightarrow \{\ldots, \alpha\}) + \bar{\pi}_2(\gamma, \mathfrak{a} \leftrightarrow \{\ldots, \alpha\}) + \Theta_0 & \\ & \text{if } s_\alpha \in W(M:T) \end{cases}$$

with $\bar{\pi}_1$ and $\bar{\pi}_2$ differing in how χ is defined (see p. 264 of [8]).

(b) If α is complex and $\theta\alpha$ is positive, then

$$U_\alpha(\overline{\pi}(\gamma, \mathfrak{a} \leftrightarrow \{\ldots\})) = \overline{\pi}(s_\alpha\gamma, \mathfrak{a} \leftrightarrow \{\ldots\}) + \Theta_0$$

with $\mathfrak{a}$ unchanged.

(c) If α is real and does not satisfy a parity condition for $\overline{\pi}(\gamma, \mathfrak{a} \leftrightarrow \{\ldots\})$, then

$$U_\alpha(\overline{\pi}(\gamma, \mathfrak{a} \leftrightarrow \{\ldots\})) = \Theta_0 .$$

Moreover, in all cases Θ_0 is a finite sum of irreducible characters, each of which has (the index corresponding to) α in its τ-invariant and each of which occurs in $\pi(\gamma, \mathfrak{a} \leftrightarrow \{\ldots\})$.

Remark. The main terms of U_α, as well, have α in their τ-invariant, by Lemma 3.11b of [9].

Finally there is a reciprocity theorem that is helpful in computing Θ_0.

Theorem 1.7 ([9], Theorem 4.14). Let γ be regular integral, and suppose Θ and Θ' are irreducible characters with infinitesimal character γ. If i is in $\tau(\Theta)$ but j is not in $\tau(\Theta)$, and if j is in $\tau(\Theta')$ but i is not in $\tau(\Theta')$ (and if indices i and j do not span a group G_2), then the multiplicity of Θ' in $U_j(\Theta)$ equals the multiplicity of Θ in $U_i(\Theta')$, and this common multiplicity is at most one.

Now we can state the algorithm roughly. We begin with the regular integral parameter γ_0 constructed earlier. Since $\dim \mathfrak{a} = 1$, we can write the character of $U(P', \sigma_0, \nu_0)$ as $\pi(\gamma_0, \mathfrak{a} \leftrightarrow \{\alpha\})$ for some α. By a succession of reflections in complex roots, we pass from γ_0 to $\gamma_1, \gamma_2, \ldots, \gamma_k$ to a point where we know how $\pi(\gamma_k, \mathfrak{a} \leftrightarrow \{\alpha\})$ decomposes. (For example, if α is simple in the system for γ_k, then either a Schmid identity (Theorem 1.1) or Theorem 1.2 will be available. In Section 2 we shall establish a more efficient starting point.) We write the decomposition of $\pi(\gamma_k, \mathfrak{a} \leftrightarrow \{\alpha\})$ and apply to the whole identity the reflection that takes us to $\pi(\gamma_{k-1}, \mathfrak{a} \leftrightarrow \{\alpha\})$, computing the individual terms by Theorems 1.6 and 1.7. (It is not clear whether these tools will be sufficient in general. However, they will suffice in our examples, and the additional tools in [10] will suffice in general.) Then we reflect again to pass to $\pi(\gamma_{k-2}, \mathfrak{a} \leftrightarrow \{\alpha\})$, and so on, until we have a decomposition of $\pi(\gamma_0, \mathfrak{a} \leftrightarrow \{\alpha\})$. Finally we use Theorems 1.3 and 1.4 to pass to our original parameters. If only one nonzero term survives, our original representation $U(P,\sigma,\nu)$ was irreducible.

Setting matters up with the initial reflections requires some further explanation. We illustrate matters for an example with $SO^*(10)$. The numbering of the Dynkin diagram will be

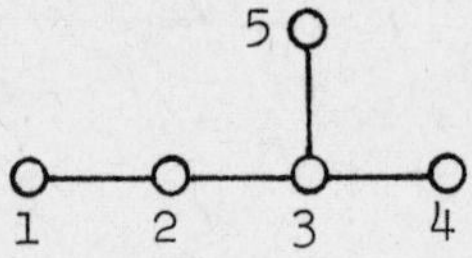

Our given data will be a positive system for $\mathfrak{g}^{\mathbb{C}}$ containing $\Delta^+(\mathfrak{m}^{\mathbb{C}}, \mathfrak{t}^{\mathbb{C}})$ such that the root defining $\mathfrak{a}$ is simple. This root α will be root 2. The white dots are the compact roots (when referred to $\mathfrak{b}^{\mathbb{C}}$), while the black dots are noncompact. We specify λ by attaching $2\langle\lambda,\beta\rangle/|\beta|^2$ to each simple root β. Then the diagram is

λ :

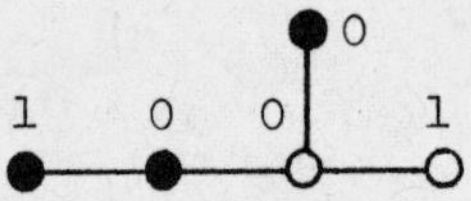

α = ②

and we investigate reducibility/irreducibility at $\nu = \alpha$. The diagram for $\lambda+\nu$ is

$\lambda+\nu$:

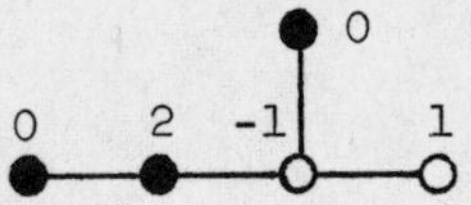

α = ② $\qquad \Delta_2^+$

We apply a succession of reflections in roots nonorthogonal to α in an effort to make $\lambda+\nu$ dominant. If $\Delta_1^+ = s_3\Delta_2^+$, we obtain

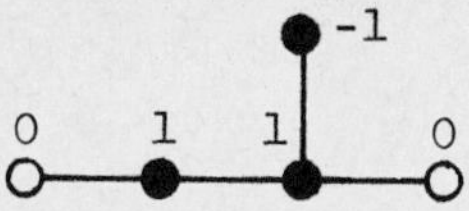

$\alpha =$ ② + ③ $\qquad \Delta_1^+$

Finally $\Delta_0^+ = s_5\Delta_1^+$ gives us

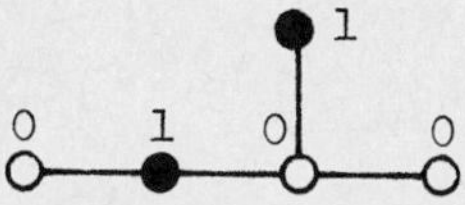

$\alpha =$ ② + ③ + ⑤ $\qquad \Delta_0^+$

This Δ_0^+ is a system compatible with $\Delta^+(\mathfrak{m}^{\mathbb{C}}, \mathfrak{t}^{\mathbb{C}})$ that makes $\lambda+\nu$ dominant. We can take it as the system in which γ_0 is to be dominant. The set of singular roots is $\{1,3,4\}$.

There is no need to carry along an explicit value of γ_0; having Δ_0^+ and the expression for α will be enough. Then we can define $\gamma_1 = s_5\gamma_0$, which will be dominant for Δ_1^+, and $\gamma_2 = s_3\gamma_1$, which will be dominant for Δ_2^+. Since α is simple for Δ_2^+ and satisfies the parity condition (our group is not $Sp(n,\mathbb{R})$), $\pi(\gamma_2, \alpha \leftrightarrow \{\alpha\})$ is given by a Schmid identity. So we have a starting place for the algorithm.

2. An inductive application

When we set up matters as at the end of Section 1 and then proceed with the wall crossings, we typically find that the first few wall-crossing steps are independent of our example. What is happening is that the first few steps take place in a common real rank one example. The theorem below formalizes this process and its result. Because of this theorem, we shall find that the set-up at the end of Section 1 should be done in such a way as to minimize the number of steps that are outside a real rank one subgroup.

<u>Theorem 2.1</u>. Let γ be a regular integral infinitesimal character dominant for Δ^+. Suppose that α is a $\mathfrak{g}$-noncompact root that is the sum of all the simple roots in a single-line real rank one subgroup with positive system $\Delta_0^+ \subseteq \Delta^+$, and suppose α satisfies the parity condition for $\pi(\gamma, a \leftrightarrow \{\alpha\})$. Let n be the number of simple roots in Δ_0^+, and suppose $n \geq 2$. Let ϵ_1 and ϵ_2 be the nodes among these simple roots, say with ϵ_1 compact and ϵ_2 noncompact. If $n = 2$ and if $\{\alpha\}$ is abbreviated as α, then

$$\pi(\gamma, a \leftrightarrow \alpha) = \overline{\pi}(\gamma, a \leftrightarrow \alpha) + \overline{\pi}(s_{\epsilon_1}\gamma, a \leftrightarrow \alpha) + \overline{\pi}(s_{\epsilon_2}\gamma, a \leftrightarrow \alpha) + \overline{\pi}(s_{\epsilon_2}\gamma, a \leftrightarrow \emptyset), \tag{2.1}$$

while if $n \geq 3$, then

$$\pi(\gamma, a \leftrightarrow \alpha) = \overline{\pi}(\gamma, a \leftrightarrow \alpha) + \overline{\pi}(s_{\epsilon_1}\gamma, a \leftrightarrow \alpha) + \overline{\pi}(s_{\epsilon_2}\gamma, a \leftrightarrow \alpha) + \overline{\pi}(s_{\epsilon_1}s_{\epsilon_2}\gamma, a \leftrightarrow \alpha). \tag{2.2}$$

<u>Proof</u>. We proceed by induction on n, treating $n = 2$ and $n = 3$ separately. First let $n = 2$. The relevant part of the Dynkin diagram is

and we introduce

$\Delta_1^+ = s_{\epsilon_1}\Delta_0^+$ $\qquad$ $\alpha = ②$ $\qquad$ $s_{\epsilon_1}\gamma$ dominant

$\Delta_0^+ \subseteq \Delta^+$ $\qquad$ $\alpha = ① + ②$ $\qquad$ γ dominant

By the Schmid identity (Theorem 1.1) applied to $s_{\epsilon_1}\Delta^+$,

$$\pi(s_{\epsilon_1}\gamma, a \leftrightarrow \alpha) = \overline{\pi}(s_{\epsilon_1}\gamma, a \leftrightarrow \alpha) + \overline{\pi}(s_{\epsilon_1}\gamma, a \leftrightarrow \emptyset) + \overline{\pi}(s_\alpha s_{\epsilon_1}\gamma, a \leftrightarrow \emptyset). \tag{2.3}$$

We shall apply the functor s_1. To compute the τ-invariants of the right side of (2.3), we need one more diagram:

$s_\alpha s_{\epsilon_1} \Delta_0^+$ 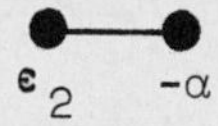 $s_\alpha s_{\epsilon_1}\gamma$ dominant

The τ-invariants within the set $\{1,2\}$ are

$$\tau(\overline{\pi}(s_{\epsilon_1}\gamma\,,\,\mathfrak{a}\leftrightarrow\alpha)) = \{2\}$$

$$\tau(\overline{\pi}(s_{\epsilon_1}\gamma\,,\,\mathfrak{a}\leftrightarrow\emptyset)) = \{1\}$$

$$\tau(\overline{\pi}(s_\alpha s_{\epsilon_1}\,,\,\mathfrak{a}\leftrightarrow\emptyset)) = \emptyset$$

by Theorem 1.4. Then Theorem 1.6 gives

$$s_1\overline{\pi}(s_{\epsilon_1}\gamma\,,\,\mathfrak{a}\leftrightarrow\alpha) = \overline{\pi}(s_{\epsilon_1}\gamma\,,\,\mathfrak{a}\leftrightarrow\alpha) + \overline{\pi}(\gamma\,,\,\mathfrak{a}\leftrightarrow\alpha) + \Theta_1$$

$$s_1\overline{\pi}(s_{\epsilon_1}\gamma\,,\,\mathfrak{a}\leftrightarrow\emptyset) = -\overline{\pi}(s_{\epsilon_1}\gamma\,,\,\mathfrak{a}\leftrightarrow\emptyset)$$

$$s_1\overline{\pi}(s_\alpha s_{\epsilon_1}\gamma\,,\,\mathfrak{a}\leftrightarrow\emptyset) = \overline{\pi}(s_\alpha s_{\epsilon_1}\gamma\,,\,\mathfrak{a}\leftrightarrow\emptyset) + \overline{\pi}(s_\alpha s_{\epsilon_1}\gamma\,,\,\mathfrak{a}\leftrightarrow\epsilon_2) + \Theta_2\,.$$

Here Θ_2 is the sum of constituents of $\pi(s_\alpha s_{\epsilon_1}\gamma\,,\,\mathfrak{a}\leftrightarrow\emptyset) = \overline{\pi}(s_\alpha s_{\epsilon_1}\gamma\,,\,\mathfrak{a}\leftrightarrow\emptyset)$ having 1 in the τ-invariant, and so $\Theta_2 = 0$. Also Θ_1 is the sum of constituents of $\pi(s_{\epsilon_1}\gamma\,,\,\mathfrak{a}\leftrightarrow\alpha)$ having 1 in the τ-invariant, and so (2.1) shows $\Theta_1 = c\overline{\pi}(s_{\epsilon_1}\gamma\,,\,\mathfrak{a}\leftrightarrow\emptyset)$. Now Theorem 1.7 gives

$$c = \text{mult}\ \overline{\pi}(s_{\epsilon_1}\gamma\,,\,\mathfrak{a}\leftrightarrow\emptyset)\ \text{ in }\ U_1(\overline{\pi}(s_{\epsilon_1}\gamma\,,\,\mathfrak{a}\leftrightarrow\alpha))\,, \qquad 1\notin\tau = \{2\}\,,$$

$$= \text{mult}\ \overline{\pi}(s_{\epsilon_1}\gamma\,,\,\mathfrak{a}\leftrightarrow\alpha)\ \text{ in }\ U_2(\overline{\pi}(s_{\epsilon_1}\gamma\,,\,\mathfrak{a}\leftrightarrow\emptyset))\,, \qquad 2\notin\tau = \{1\}\,.$$

For the latter we write

$$U_2(\overline{\pi}(s_{\epsilon_1}\gamma\,,\,\mathfrak{a}\leftrightarrow\emptyset)) = \overline{\pi}(s_{\epsilon_1}\gamma\,,\,\mathfrak{a}\leftrightarrow\alpha) + \Theta_3\,.$$

This shows $c \geq 1$, and Theorem 1.7 says $c = 1$. Applying s_1 to (2.3) and using Theorem 1.5, we obtain

$$\begin{aligned}
\pi(\gamma\,,\alpha\leftrightarrow\alpha) &= s_1\pi(s_{\epsilon_1}\gamma\,,\alpha\leftrightarrow\alpha)\\
&= s_1\overline{\pi}(s_{\epsilon_1}\gamma\,,\alpha\leftrightarrow\alpha) + s_1\overline{\pi}(s_{\epsilon_1}\gamma\,,\alpha\leftrightarrow\emptyset) + s_1\overline{\pi}(s_\alpha s_{\epsilon_1}\gamma\,,\alpha\leftrightarrow\emptyset)\\
&= \overline{\pi}(s_{\epsilon_1}\gamma\,,\alpha\leftrightarrow\alpha) + \overline{\pi}(\gamma\,,\alpha\leftrightarrow\emptyset) + \overline{\pi}(s_{\epsilon_1}\gamma\,,\alpha\leftrightarrow\emptyset)\\
&\quad - \overline{\pi}(s_{\epsilon_1}\gamma\,,\alpha\leftrightarrow\emptyset)\\
&\quad + \overline{\pi}(s_\alpha s_{\epsilon_1}\gamma\,,\alpha\leftrightarrow\emptyset) + \overline{\pi}(s_\alpha s_{\epsilon_1}\gamma\,,\alpha\leftrightarrow\epsilon_2)\,. \qquad (2.4)
\end{aligned}$$

Since ϵ_1 is compact and $s_{\epsilon_1}s_\alpha s_{\epsilon_1} = s_{\epsilon_2}$, we have

$$\overline{\pi}(s_\alpha s_{\epsilon_1}\gamma\,,\alpha\leftrightarrow\emptyset) = \overline{\pi}(s_{\epsilon_1}s_\alpha s_{\epsilon_1}\gamma\,,\alpha\leftrightarrow\emptyset) = \overline{\pi}(s_{\epsilon_2}\gamma\,,\alpha\leftrightarrow\emptyset)$$

and

$$\overline{\pi}(s_\alpha s_{\epsilon_1}\gamma\,,\alpha\leftrightarrow\epsilon_2) = \overline{\pi}(s_{\epsilon_1}s_\alpha s_{\epsilon_1}\gamma\,,\alpha\leftrightarrow s_{\epsilon_1}\epsilon_2) = \overline{\pi}(s_{\epsilon_2}\gamma\,,\alpha\leftrightarrow\alpha)\,.$$

Substitution into (2.4) gives the desired result (2.1).

Next let $n=3$. The corresponding diagrams are

$\Delta_1^+ = s_{\epsilon_1}\Delta_0^+$

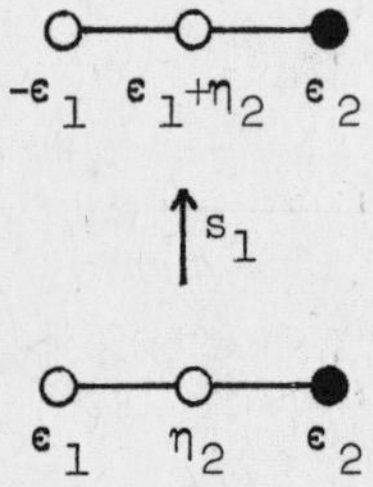

$\alpha =$ ② + ③ $\qquad s_{\epsilon_1}\gamma$ dominant

$\Delta_0^+ \subseteq \Delta^+$ $\qquad \alpha =$ ① + ② + ③ $\qquad \gamma$ dominant

By (2.1) for $s_{\epsilon_1}\gamma$,

$$\begin{aligned}
\pi(s_{\epsilon_1}\gamma\,,\alpha\leftrightarrow\alpha) &= \overline{\pi}(s_{\epsilon_1}\gamma\,,\alpha\leftrightarrow\alpha) + \overline{\pi}(s_{\epsilon_1+\eta_2}s_{\epsilon_1}\gamma\,,\alpha\leftrightarrow\alpha)\\
&\quad + \overline{\pi}(s_{\epsilon_2\epsilon_1}\gamma\,,\alpha\leftrightarrow\alpha) + \overline{\pi}(s_{\epsilon_2\epsilon_1}\gamma\,,\alpha\leftrightarrow\emptyset)\,. \qquad (2.5)
\end{aligned}$$

We shall apply the functor s_1. To compute the τ-invariants of the right side of (2.5), we need the diagrams

$s_{\epsilon_1+\eta_2} s_{\epsilon_1} \Delta_0^+$ 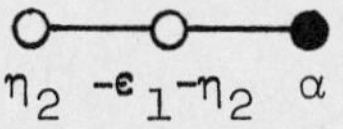

$\alpha =$ ③ $\qquad s_{\epsilon_1+\eta_2} s_{\epsilon_1} \gamma$ dominant

$s_{\epsilon_2} s_{\epsilon_1} \Delta_0^+$ 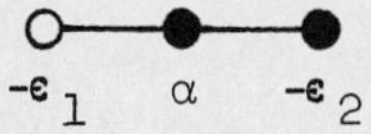

$\alpha =$ ② $\qquad s_{\epsilon_2} s_{\epsilon_1} \gamma$ dominant

The τ-invariants within $\{1,2,3\}$ are

$$\begin{aligned}
\tau(\overline{\pi}(s_{\epsilon_1}\gamma\,,\, a \leftrightarrow \alpha)) &= \{2,3\} \\
\tau(\overline{\pi}(s_{\epsilon_1+\eta_2} s_{\epsilon_1}\gamma\,,\, a \leftrightarrow \alpha)) &= \{1,3\} \\
\tau(\overline{\pi}(s_{\epsilon_2} s_{\epsilon_1}\gamma\,,\, a \leftrightarrow \alpha)) &= \{2\} \\
\tau(\overline{\pi}(s_{\epsilon_2} s_{\epsilon_1}\gamma\,,\, a \leftrightarrow \emptyset)) &= \{1\}\,.
\end{aligned} \tag{2.6}$$

Then

$$\begin{aligned}
s_1\overline{\pi}(s_{\epsilon_1}\gamma\,,\, a \leftrightarrow \alpha) &= \overline{\pi}(s_{\epsilon_1}\gamma\,,\, a \leftrightarrow \alpha) + \overline{\pi}(\gamma\,,\, a \leftrightarrow \alpha) + \Theta_1 \\
s_1\overline{\pi}(s_{\epsilon_1+\eta_2} s_{\epsilon_1}\gamma\,,\, a \leftrightarrow \alpha) &= -\,\overline{\pi}(s_{\epsilon_1+\eta_2} s_{\epsilon_1}\gamma\,,\, a \leftrightarrow \alpha) \\
s_1\overline{\pi}(s_{\epsilon_2} s_{\epsilon_1}\gamma\,,\, a \leftrightarrow \alpha) &= \overline{\pi}(s_{\epsilon_2} s_{\epsilon_1}\gamma\,,\, a \leftrightarrow \alpha) + \overline{\pi}(s_{\epsilon_2}\gamma\,,\, a \leftrightarrow \alpha) + \Theta_2 \\
s_1\overline{\pi}(s_{\epsilon_2} s_{\epsilon_1}\gamma\,,\, a \leftrightarrow \emptyset) &= -\,\overline{\pi}(s_{\epsilon_2} s_{\epsilon_1}\gamma\,,\, a \leftrightarrow \emptyset)\,.
\end{aligned} \tag{2.7}$$

From (2.5) and (2.6)

$$\Theta_1 = c_1\overline{\pi}(s_{\epsilon_1+\eta_2} s_{\epsilon_1}\gamma\,,\, a \leftrightarrow \alpha) + c_2\overline{\pi}(s_{\epsilon_2} s_{\epsilon_1}\gamma\,,\, a \leftrightarrow \emptyset)\,.$$

Now Theorem 1.7 gives

$$\begin{aligned}
c_1 &= \text{mult } \overline{\pi}(s_{\epsilon_1+\eta_2} s_{\epsilon_1}\gamma\,,\, a \leftrightarrow \alpha) \text{ in } U_1(\overline{\pi}(s_{\epsilon_1}\gamma\,,\, a \leftrightarrow \alpha))\,,\ 1 \notin \tau = \{2,3\}, \\
&= \text{mult } \overline{\pi}(s_{\epsilon_1}\gamma\,,\, a \leftrightarrow \alpha) \text{ in } U_2(\overline{\pi}(s_{\epsilon_1+\eta_2} s_{\epsilon_1}\gamma\,,\, a \leftrightarrow \alpha))\,,\ 2 \notin \tau = \{1,3\}.
\end{aligned}$$

For the latter, we write

$$U_2\overline{\pi}(s_{\epsilon_1+\eta_2}s_{\epsilon_1}\gamma\,,\, a \leftrightarrow \alpha) = \overline{\pi}(s_{\epsilon_1}\gamma\,,\, a \leftrightarrow \alpha) + \Theta_3 .$$

This shows $c_1 \geq 1$, and hence $c_1 = 1$. Next,

$$c_2 = \text{mult } \overline{\pi}(s_{\epsilon_2}s_{\epsilon_1}\gamma\,,\, a \leftrightarrow \emptyset) \text{ in } U_1\overline{\pi}(s_{\epsilon_1}\gamma\,,\, a \leftrightarrow \alpha), \qquad 1 \notin \tau = \{2,3\},$$

$$= \text{mult } \overline{\pi}(s_{\epsilon_1}\gamma\,,\, a \leftrightarrow \alpha) \text{ in } U_2\overline{\pi}(s_{\epsilon_2}s_{\epsilon_1}\gamma\,,\, a \leftrightarrow \emptyset), \qquad 2 \notin \tau = \{1\}.$$

Now

$$U_2\overline{\pi}(s_{\epsilon_2}s_{\epsilon_1}\gamma\,,\, a \leftrightarrow \emptyset) = \overline{\pi}(s_{\epsilon_2}s_{\epsilon_1}\gamma\,,\, a \leftrightarrow \alpha) + \Theta_4 .$$

Here Θ_4 is the sum of constituents of $\pi(s_{\epsilon_2}s_{\epsilon_1}\gamma\,,\, a \leftrightarrow \emptyset)$ $= \overline{\pi}(s_{\epsilon_2}s_{\epsilon_1}\gamma\,,\, a \leftrightarrow \emptyset)$ having 2 in the τ-invariant, and so $\Theta_4 = 0$. Since $\overline{\pi}(s_{\epsilon_1}\gamma\,,\, a \leftrightarrow \alpha)$ and $\overline{\pi}(s_{\epsilon_2}s_{\epsilon_1}\gamma\,,\, a \leftrightarrow \alpha)$ have respective τ-invariants $\{2,3\}$ and $\{2\}$, they are unequal characters, and thus $c_2 = 0$. Hence

$$\Theta_1 = \overline{\pi}(s_{\epsilon_1+\eta_2}s_{\epsilon_1}\gamma\,,\, a \leftrightarrow \alpha) . \tag{2.8}$$

To compute Θ_2, we use the Schmid identity

$$\pi(s_{\epsilon_2}s_{\epsilon_1}\gamma\,,\, a \leftrightarrow \alpha) = \overline{\pi}(s_{\epsilon_2}s_{\epsilon_1}\gamma\,,\, a \leftrightarrow \alpha) + \overline{\pi}(s_{\epsilon_2}s_{\epsilon_1}\gamma\,,\, a \leftrightarrow \emptyset) + \overline{\pi}(s_\alpha s_{\epsilon_2}s_{\epsilon_1}\gamma\,,\, a \leftrightarrow \emptyset)$$

and compute that the τ-invariants for the terms on the right are $\{2\}$, $\{1\}$, and $\{3\}$. Then it follows that

$$\Theta_2 = c\overline{\pi}(s_{\epsilon_2}s_{\epsilon_1}\gamma\,,\, a \leftrightarrow \emptyset) .$$

By Theorem 1.7,

$$c = \text{mult } \overline{\pi}(s_{\epsilon_2}s_{\epsilon_1}\gamma\,,\, a \leftrightarrow \emptyset) \text{ in } U_1(\overline{\pi}(s_{\epsilon_2}s_{\epsilon_1}\gamma\,,\, a \leftrightarrow \alpha)), \qquad 1 \notin \tau = \{2\},$$

$$= \text{mult } \overline{\pi}(s_{\epsilon_2}s_{\epsilon_1}\gamma\,,\, a \leftrightarrow \alpha) \text{ in } U_2(\overline{\pi}(s_{\epsilon_2}s_{\epsilon_1}\gamma\,,\, a \leftrightarrow \emptyset)), \qquad 2 \notin \tau = \{1\}.$$

Since

$$U_2(\overline{\pi}(s_{\epsilon_2}s_{\epsilon_1}\gamma\,,\, a \leftrightarrow \emptyset)) = \overline{\pi}(s_{\epsilon_2}s_{\epsilon_1}\gamma\,,\, a \leftrightarrow \alpha) + \Theta_5 ,$$

we have $c \geq 1$. Thus $c = 1$ and

$$\Theta_2 = \overline{\pi}(s_{\epsilon_2} s_{\epsilon_1} \gamma \, , \, a \leftrightarrow \emptyset) \, . \tag{2.9}$$

Finally we apply s_1 to both sides of (2.5), use the identity

$$s_1 \pi(s_{\epsilon_1} \gamma \, , \, a \leftrightarrow \alpha) = \pi(\gamma \, , \, a \leftrightarrow \alpha)$$

given in Theorem 1.5, and substitute from (2.7), (2.8), and (2.9) to obtain (2.2) for $n = 3$.

Now let $n \geq 4$, and assume inductively that (2.1) and (2.2) have been proved for all cases $\leq n-1$. The starting diagrams are

$\Delta_1^+ = s_{\epsilon_1} \Delta_0^+$

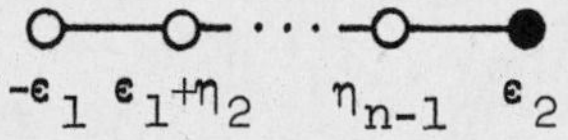

$\alpha =$ ② $+ \dots +$ Ⓝ $\quad s_{\epsilon_1}\gamma$ dominant

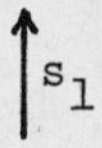

$\Delta_0^+ \subseteq \Delta^+$

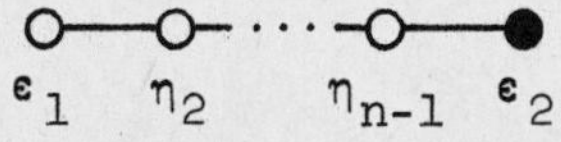

$\alpha =$ ① $+ \dots +$ Ⓝ $\quad \gamma$ dominant

We shall use also the diagrams

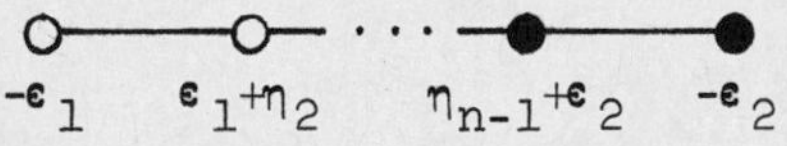

$\alpha =$ ② $+ \dots +$ (n-1) (2.10)

$s_{\epsilon_1} s_{\epsilon_2} \gamma$ dominant

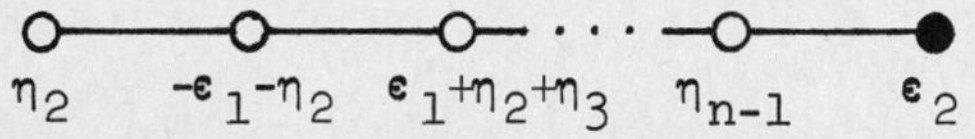

$\alpha =$ ③ $+ \dots +$ Ⓝ (2.11)

$s_{\epsilon_1+\eta_2} s_{\epsilon_1} \gamma$ dominant

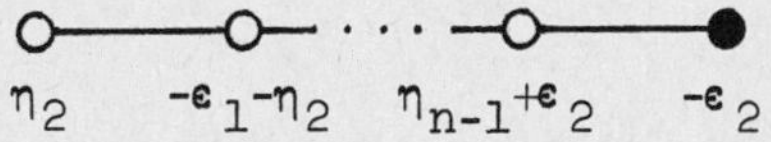

$\alpha =$ ③ $+ \dots +$ (n-1) (2.12)

$s_{\epsilon_1+\eta_2} s_{\epsilon_2} s_{\epsilon_1} \gamma$ dominant

By inductive hypothesis, (2.2) for $n-1$ and $s_{\varepsilon_1}\gamma$ gives us

$$\pi(s_{\varepsilon_1}\gamma, a \leftrightarrow \alpha) = \overline{\pi}(s_{\varepsilon_1}\gamma, a \leftrightarrow \alpha) + \overline{\pi}(s_{\varepsilon_1}s_{\varepsilon_2}, a \leftrightarrow \alpha)$$

$$+ \overline{\pi}(s_{\varepsilon_1+\eta_2}s_{\varepsilon_1}\gamma, a \leftrightarrow \alpha) + \overline{\pi}(s_{\varepsilon_1+\eta_2}s_{\varepsilon_2}s_{\varepsilon_1}\gamma, a \leftrightarrow \alpha), \qquad (2.13)$$

and the respective τ-invariants for the terms on the right are $\{2,\ldots,n\}$, $\{2,\ldots,n-1\}$, $\{1,3,4,\ldots,n\}$, and $\{1,3,4,\ldots,n-1\}$. Then

$$\begin{aligned}
s_1\overline{\pi}(s_{\varepsilon_1}\gamma, a \leftrightarrow \alpha) &= \overline{\pi}(s_{\varepsilon_1}\gamma, a \leftrightarrow \alpha) + \overline{\pi}(\gamma, a \leftrightarrow \alpha) + \Theta_1 \\
s_1\overline{\pi}(s_{\varepsilon_1}s_{\varepsilon_2}\gamma, a \leftrightarrow \alpha) &= \overline{\pi}(s_{\varepsilon_1}s_{\varepsilon_2}\gamma, a \leftrightarrow \alpha) + \overline{\pi}(s_{\varepsilon_2}\gamma, a \leftrightarrow \alpha) + \Theta_2 \\
s_1\overline{\pi}(s_{\varepsilon_1+\eta_2}s_{\varepsilon_1}\gamma, a \leftrightarrow \alpha) &= -\overline{\pi}(s_{\varepsilon_1+\eta_2}s_{\varepsilon_1}\gamma, a \leftrightarrow \alpha) \\
s_1\overline{\pi}(s_{\varepsilon_1+\eta_2}s_{\varepsilon_2}s_{\varepsilon_1}\gamma, a \leftrightarrow \alpha) &= -\overline{\pi}(s_{\varepsilon_1+\eta_2}s_{\varepsilon_2}s_{\varepsilon_1}\gamma, a \leftrightarrow \alpha). \qquad (2.14)
\end{aligned}$$

From (2.13),

$$\Theta_1 = c_1\overline{\pi}(s_{\varepsilon_1+\eta_2}s_{\varepsilon_1}\gamma, a \leftrightarrow \alpha) + c_2\overline{\pi}(s_{\varepsilon_1+\eta_2}s_{\varepsilon_2}s_{\varepsilon_1}\gamma, a \leftrightarrow \alpha).$$

By the same argument as when $n = 3$, we find $c_1 = 1$. Also

$$\begin{aligned}
c_2 &= \text{mult of } \overline{\pi}(s_{\varepsilon_1+\eta_2}s_{\varepsilon_2}s_{\varepsilon_1}\gamma, a \leftrightarrow \alpha) \text{ in } U_1(\overline{\pi}(s_{\varepsilon_1}\gamma, a \leftrightarrow \alpha)), \\
&\qquad\qquad 1 \notin \tau = \{2,\ldots,n-1\}, \\
&= \text{mult of } \overline{\pi}(s_{\varepsilon_1}\gamma, a \leftrightarrow \alpha) \text{ in } U_2(\overline{\pi}(s_{\varepsilon_1+\eta_2}s_{\varepsilon_2}s_{\varepsilon_1}\gamma, a \leftrightarrow \alpha)), \\
&\qquad\qquad 2 \notin \tau = \{1,3,4,\ldots,n-1\}.
\end{aligned}$$

Now

$$U_2(\overline{\pi}(s_{\varepsilon_1+\eta_2}s_{\varepsilon_2}s_{\varepsilon_1}\gamma, a \leftrightarrow \alpha)) = \overline{\pi}(s_{\varepsilon_2}s_{\varepsilon_1}\gamma, a \leftrightarrow \alpha) + \Theta_3,$$

with each constituent of Θ_3 in $\pi(s_{\varepsilon_1+\eta_2}s_{\varepsilon_2}s_{\varepsilon_1}\gamma, a \leftrightarrow \alpha)$. By inductive assumption for $n-3$ (if $n > 4$) or by a Schmid identity if $n=4$, this π has three or four terms in its expansion, all computed within the diagram (2.12), and 1 is in the τ-invariant of each. Since 1 is not in the τ-invariant of $\overline{\pi}(s_{\varepsilon_1}\gamma, a \leftrightarrow \alpha)$,

$\overline{\pi}(s_{\epsilon_1}\gamma, \alpha \leftrightarrow \alpha)$ does not occur in Θ_3. But nor do we have $\overline{\pi}(s_{\epsilon_1}\gamma, \alpha \leftrightarrow \alpha) = \overline{\pi}(s_{\epsilon_2}s_{\epsilon_1}\gamma, \alpha \leftrightarrow \alpha)$, since n is in the τ-invariant of the left side but not the right side. Thus $c_2 = 0$, and

$$\Theta_1 = \overline{\pi}(s_{\epsilon_1+\eta_2}s_{\epsilon_1}\gamma, \alpha \leftrightarrow \alpha) . \tag{2.15}$$

To compute Θ_2, we note that each constituent of Θ_2 must occur in $\pi(s_{\epsilon_1}s_{\epsilon_2}\gamma, \alpha \leftrightarrow \alpha)$ and have 1 in its τ-invariant. By inductive assumption for $n-2$, this π has four terms in its expansion, all computed within the diagram (2.10). A term in which α includes root 2 will not have 1 in its τ-invariant, while a term with α not including root 2 (or in the case $n=4$—in which $\alpha \leftrightarrow \emptyset$) will have 1 in its τ-invariant. There are two terms of the latter kind,

$$\overline{\pi}(s_{\epsilon_1+\eta_2}s_{\epsilon_2}s_{\epsilon_1}\gamma, \alpha \leftrightarrow \alpha)$$

and

$$\begin{cases} \overline{\pi}(s_{\eta_3+\epsilon_2}s_{\epsilon_1}s_{\epsilon_2}\gamma, \alpha \leftrightarrow \emptyset) & \text{if } n = 4 \\ \overline{\pi}(s_{\eta_{n-1}+\epsilon_2}s_{\epsilon_1+\eta_2}s_{\epsilon_1}s_{\epsilon_2}\gamma, \alpha \leftrightarrow \alpha) & \text{if } n > 4 . \end{cases}$$

Let c_3 and c_4 be the respective coefficients of these terms in Θ_2. A familiar argument shows $c_3 = 1$. For the computation of c_4, let us treat $n = 4$ and $n > 4$ separately.

First suppose $n = 4$. Then

$$c_4 = \text{mult } \overline{\pi}(s_{\eta_3+\epsilon_2}s_{\epsilon_1}s_{\epsilon_2}\gamma, \alpha \leftrightarrow \emptyset) \text{ in } U_1(\overline{\pi}(s_{\epsilon_1}s_{\epsilon_2}\gamma, \alpha \leftrightarrow \alpha)),\ 1 \notin \tau = \{2,3\},$$

$$= \text{mult } \overline{\pi}(s_{\epsilon_1}s_{\epsilon_2}\gamma, \alpha \leftrightarrow \alpha) \text{ in } U_3(\overline{\pi}(s_{\eta_3+\epsilon_2}s_{\epsilon_1}s_{\epsilon_2}\gamma, \alpha \leftrightarrow \emptyset)),\ 3 \notin \tau = \{1,2\},$$

and we find that

$$U_3(\overline{\pi}(s_{\eta_3+\epsilon_2}s_{\epsilon_1}s_{\epsilon_2}\gamma, \alpha \leftrightarrow \emptyset)) = \overline{\pi}(s_{\eta_3+\epsilon_2}s_{\epsilon_1}s_{\epsilon_2}\gamma, \alpha \leftrightarrow \eta_3+\epsilon_2) .$$

The τ-invariant of the term on the right turns out to be $\{1,3\}$, while $\tau(\overline{\pi}(s_{\epsilon_1}s_{\epsilon_2}\gamma, \alpha \leftrightarrow \alpha))$ is $\{2,3\}$. Thus $c_4 = 0$ when $n = 4$.

Now suppose $n > 4$. Then

$$c_4 = \text{mult } \overline{\pi}(s_{\eta_{n-1}+\epsilon_2} s_{\epsilon_1+\eta_2} s_{\epsilon_1} s_{\epsilon_2} \gamma, \alpha \leftrightarrow \alpha) \quad \text{in} \quad U_1(\overline{\pi}(s_{\epsilon_1} s_{\epsilon_2} \gamma, \alpha \leftrightarrow \alpha)),$$

$$1 \notin \tau = \{2,3,\ldots,n-1\},$$

$$= \text{mult } \overline{\pi}(s_{\epsilon_1} s_{\epsilon_2} \gamma, \alpha \leftrightarrow \alpha) \quad \text{in} \quad U_2(\overline{\pi}(s_{\eta_{n-1}+\epsilon_2} s_{\epsilon_1+\eta_2} s_{\epsilon_1} s_{\epsilon_2} \gamma, \alpha \leftrightarrow \alpha)),$$

$$2 \notin \tau = \{1,3,\ldots,n-2,n\},$$

the diagram for the latter τ-invariant being

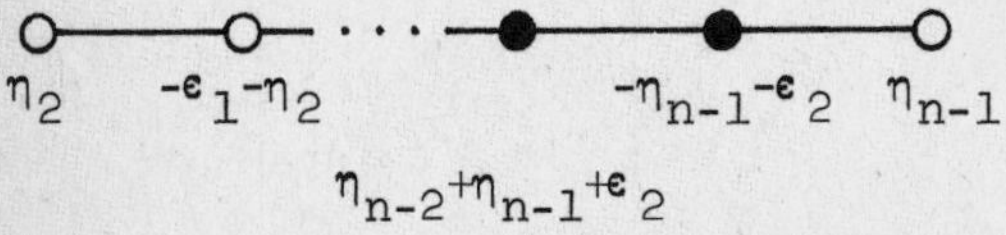

$\alpha = (3) + \ldots + (n-2)$

Here

$$U_2(\overline{\pi}(s_{\eta_{n-1}+\epsilon_2} s_{\epsilon_1+\eta_2} s_{\epsilon_1} s_{\epsilon_2} \gamma, \alpha \leftrightarrow \alpha) = \overline{\pi}(s_{\eta_{n-1}+\epsilon_2} s_{\epsilon_1} s_{\epsilon_2} \gamma, \alpha \leftrightarrow \alpha) + \Theta_3 \tag{2.16}$$

with each term of Θ_3 contained in

$$\pi(s_{\eta_{n-1}+\epsilon_2} s_{\epsilon_1+\eta_2} s_{\epsilon_1} s_{\epsilon_2} \gamma, \alpha \leftrightarrow \alpha). \tag{2.17}$$

Our inductive assumption for $n-4$ (or a Schmid identity if $n=5$) shows that 1 is in the τ-invariant of each term of (2.17), while 1 is not in the τ-invariant of $\overline{\pi}(s_{\epsilon_1} s_{\epsilon_2} \gamma, \alpha \leftrightarrow \alpha)$. So $\overline{\pi}(s_{\epsilon_1} s_{\epsilon_2} \gamma, \alpha \leftrightarrow \alpha)$ does not occur in Θ_3. Finally n is in the τ-invariant of the first term on the right of (2.16) but is not in the τ-invariant of $\overline{\pi}(s_{\epsilon_1} s_{\epsilon_2} \gamma, \alpha \leftrightarrow \alpha)$. Thus $c_4 = 0$ when $n > 4$.

So all cases $n \geq 4$ have

$$\Theta_2 = \overline{\pi}(s_{\epsilon_1+\eta_2} s_{\epsilon_2} s_{\epsilon_1} \gamma, \alpha \leftrightarrow \alpha). \tag{2.18}$$

Finally we apply s_1 to both sides of (2.13), use the identity

$$s_1 \pi(s_{\epsilon_1} \gamma, \alpha \leftrightarrow \alpha) = \pi(\gamma, \alpha \leftrightarrow \alpha)$$

given in Theorem 1.5, and substitute from (2.14), (2.15), and (2.18) to obtain (2.2) for n. This completes the induction and the proof of the theorem.

3. SO(2n-1,2)

For $SO(2n-1,2)$ we shall consider two sets of examples. For each we specify λ by attaching $2\langle\lambda,\beta\rangle/|\beta|^2$ to each simple root β in the Dynkin diagram of type B_n. The black simple roots are the noncompact ones, and P is built from α. We number the simple roots from 1 to n, with n denoting the short simple root. The theorem in each case is that $U(P,\sigma,\nu)$ is irreducible for the indicated value of ν.

a. First set of examples

$n=3$

λ :

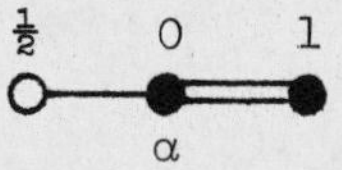

tangent case at $\nu = \frac{1}{2}\alpha$

n odd ≥ 5, $t = \frac{1}{2}(n-1)$

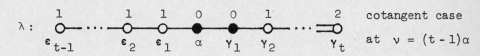

λ :

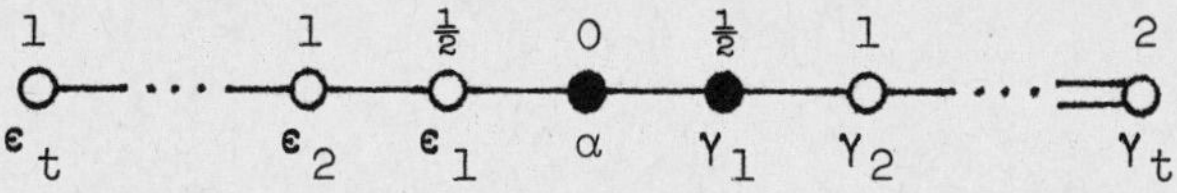

tangent case at $\nu = (t-\frac{1}{2})\alpha$

n even ≥ 4, $t = \frac{1}{2}n$

λ :

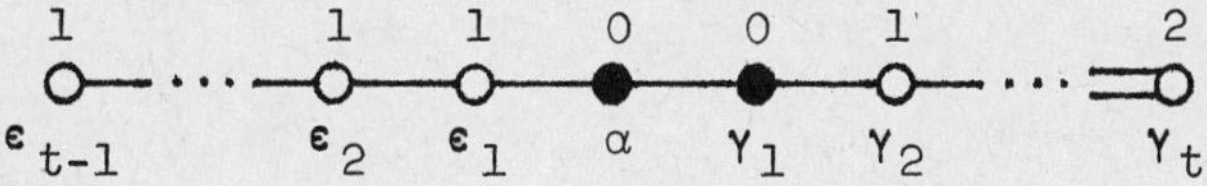

cotangent case at $\nu = (t-1)\alpha$

The diagrams for $\lambda+\nu$ are

$\lambda+\nu$:

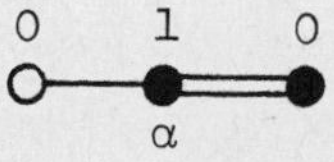

$(n=3)$

$\lambda+\nu$: 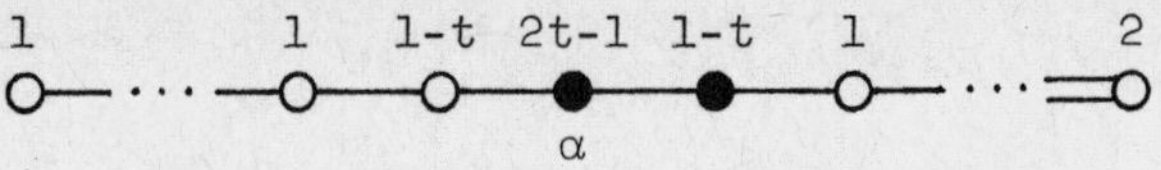

(n odd ≥ 5)

$\lambda+\nu$: 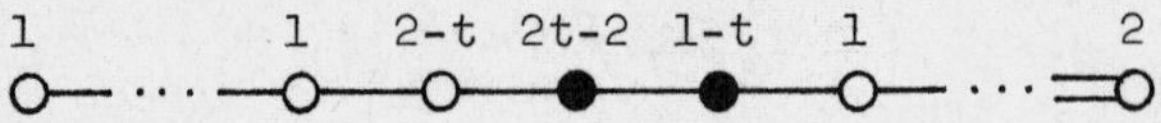

(n even ≥ 4)

For $n=3$, we let $t=1$. In the same way as at the end of Section 1, we apply a sequence of complex root reflections to each of these diagrams. When n is odd, the sequence is $s_{2t}\cdots s_{t+2}s_1\cdots s_{t-1}s_t$. When n is even, the sequence is $s_{2t-1}\cdots s_{t+1}s_1\cdots s_{t-1}$. In all cases the resulting diagram is

Δ_0^+

$\lambda+\nu$: 

$\alpha = (1) + (2) + \ldots + (n-1)$

This Δ_0^+ is a system compatible with $\Delta^+(\mathfrak{m}^{\mathbb{C}},\mathfrak{t}^{\mathbb{C}})$ that makes $\lambda+\nu$ dominant. We take it as the system in which γ_0 is to be dominant. The set of singular roots is $\{1,n\}$.

The nature of α allows us to apply Theorem 2.1 immediately to obtain

$$\begin{aligned}\pi(\gamma_0, \mathfrak{a}\leftrightarrow\alpha) &= \overline{\pi}(\gamma_0, \mathfrak{a}\leftrightarrow\alpha) + \overline{\pi}(s_{\beta_1}\gamma_0, \mathfrak{a}\leftrightarrow\alpha) \\ &\quad + \overline{\pi}(s_{\beta_{n-1}}\gamma_0, \mathfrak{a}\leftrightarrow\alpha) + \overline{\pi}(s_{\beta_1}s_{\beta_{n-1}}\gamma_0, \mathfrak{a}\leftrightarrow\alpha), \qquad (3.1)\end{aligned}$$

except that the last term is replaced by $\overline{\pi}(s_{\beta_{n-1}}\gamma_0, \mathfrak{a}\leftrightarrow\emptyset)$ when $n=3$. The respective τ-invariants for the representations on the right side are $\{1,\ldots,n-1\}$, $\{2,\ldots,n-1\}$, $\{1,\ldots,n-2,n\}$, and $\{2,\ldots,n-2,n\}$. The only one of these that is disjoint from the set $\{1,n\}$ of singular roots is the second one. By Theorem 1.3, the only term on the right side of (3.1) with nonzero image under the ψ functor is the second one, and its image is irreducible. Therefore $U(P,\sigma,\nu)$ is irreducible.

b. Second set of examples

$n=3$

λ :

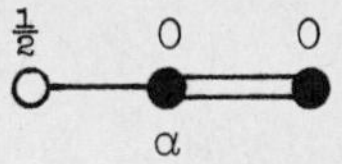

tangent case at $\nu = \frac{1}{2}\alpha$

n odd ≥ 5, $t=\frac{1}{2}(n-1)$

λ :

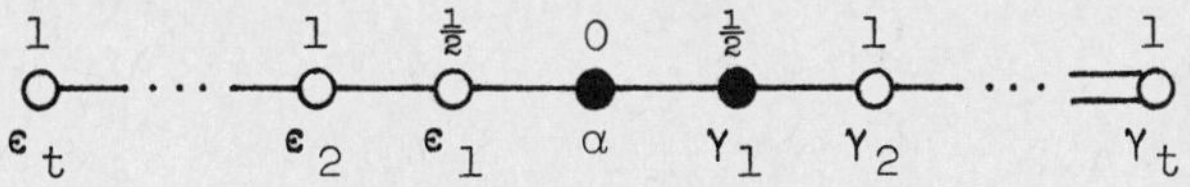

tangent case at $\nu = (t-\frac{1}{2})\alpha$

n even ≥ 4, $t=\frac{1}{2}n$

λ :

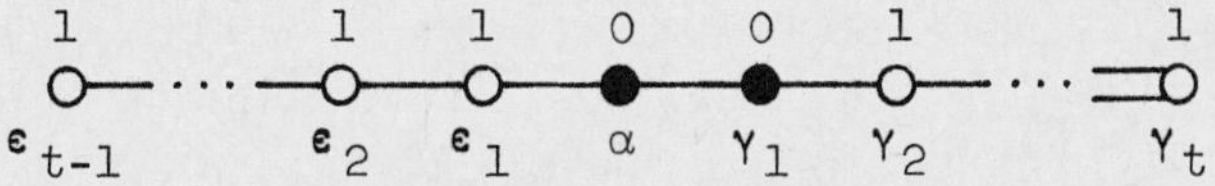

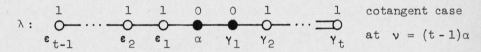

cotangent case at $\nu = (t-1)\alpha$

The diagrams for $\lambda+\nu$ are

$\lambda+\nu$:

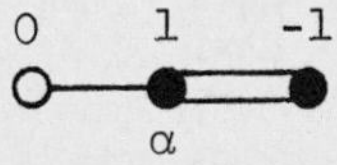

$(n=3)$

$\lambda+\nu$:

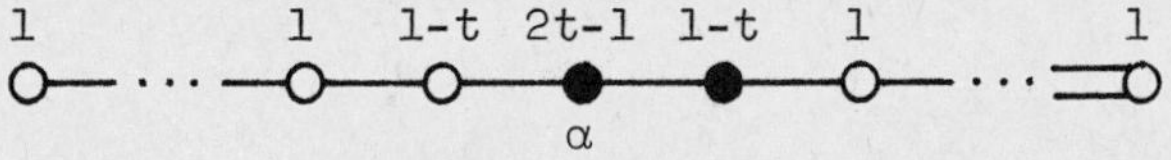

(n odd ≥ 5)

$\lambda+\nu$:

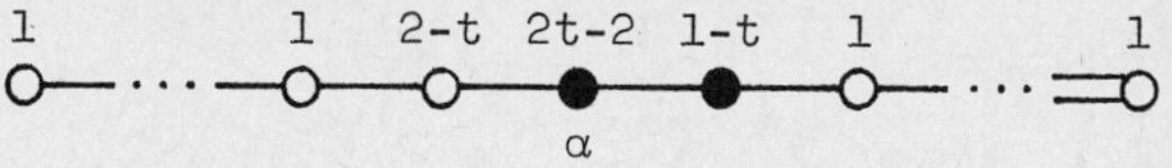

(n even ≥ 4)

For $n=3$, we let $t=1$. We apply the same sequence of complex root reflections as in Section 3a, obtaining a system Δ_1^+, and then we apply the reflection s_n to Δ_1^+ to obtain Δ_0^+. The diagrams in all cases are

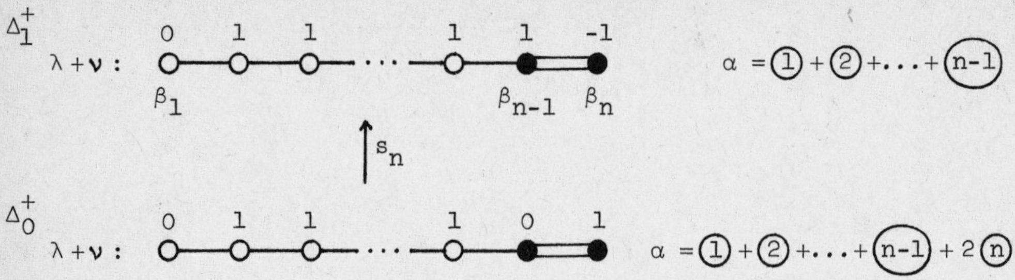

This Δ_0^+ is a system compatible with $\Delta^+(\mathfrak{m}^{\mathbb{C}}, \mathfrak{t}^{\mathbb{C}})$ that makes $\lambda+\nu$ dominant. We take it as the system in which γ_0 is to be dominant. The set of singular roots is $\{1, n-1\}$.

We can apply Theorem 2.1 in the system Δ_1^+ to obtain

$$\begin{aligned}\pi(s_n\gamma_0, \mathfrak{a} \leftrightarrow \alpha) &= \bar{\pi}(s_n\gamma_0, \mathfrak{a} \leftrightarrow \alpha) + \bar{\pi}(s_1 s_n \gamma_0, \mathfrak{a} \leftrightarrow \alpha) \\ &\quad + \bar{\pi}(s_{n-1} s_n \gamma_0, \mathfrak{a} \leftrightarrow \alpha) + \bar{\pi}(s_1 s_{n-1} s_n \gamma_0, \mathfrak{a} \leftrightarrow \alpha),\end{aligned} \tag{3.2}$$

except that the last term is replaced by $\bar{\pi}(s_{n-1}s_n\gamma_0, \mathfrak{a} \leftrightarrow \emptyset)$ when $n=3$. The respective τ-invariants for the representations on the right side are $\{1,\ldots,n-1\}$, $\{2,\ldots,n-1\}$, $\{1,\ldots,n-2,n\}$, and $\{2,\ldots,n-2,n\}$. We apply s_n to both sides of (3.2). We could go through the step-by-step analysis, but it is simpler to observe that everything that happens in the computation is oblivious to the presence of the double line in the diagram. Therefore the answer has to be of the form given in Theorem 2.1:

$$\begin{aligned}\pi(\gamma_0, \mathfrak{a} \leftrightarrow \alpha) &= \bar{\pi}(\gamma_0, \mathfrak{a} \leftrightarrow \alpha) + \bar{\pi}(s_1\gamma_0, \mathfrak{a} \leftrightarrow \alpha) \\ &\quad + \bar{\pi}(s_n\gamma_0, \mathfrak{a} \leftrightarrow \alpha) + \bar{\pi}(s_1 s_n\gamma_0, \mathfrak{a} \leftrightarrow \alpha).\end{aligned}$$

The respective τ-invariants for the representations on the right side are $\{1,\ldots,n-2,n\}$, $\{2,\ldots,n-2,n\}$, $\{1,\ldots,n-1\}$, and $\{2,\ldots,n-1\}$. The only one of these that is disjoint from the set $\{1,n-1\}$ of singular roots is the second one. Thus only the image of

$\overline{\pi}(s_1\gamma_0, \alpha \leftrightarrow \alpha)$ is nonzero when we apply the ψ functor, and its image is irreducible. Therefore $U(P,\sigma,\nu)$ is irreducible.

4. $Sp(3,\mathbb{R})$

For $Sp(3,\mathbb{R})$ we shall specify λ by attaching $2\langle\lambda,\beta\rangle/|\beta|^2$ to each simple root β in the Dynkin diagram of type C_3. The black simple roots are the noncompact ones, and P is built from α.

λ : 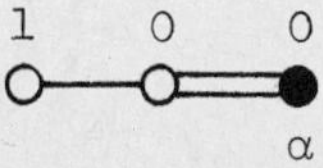

cotangent case
at $\nu = \frac{1}{2}\alpha$

We shall prove that $U(P,\sigma,\nu)$ is irreducible for the indicated value of ν.

We number the roots as $\{1,2,3\}$ from left to right. The diagram of $\lambda+\nu$ is

$\lambda+\nu$: 

$\alpha = ③$ $\qquad \Delta_1^+$

Put $\Delta_0^+ = s_2\Delta_1^+$. The picture is

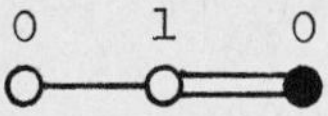

$\alpha = 2②+③ \qquad \Delta_0^+ = s_2\Delta_1^+$

We take Δ_0^+ as the system in which γ_0 is dominant. The set of singular roots is $\{1,3\}$. We do not have an immediately available character identity in Δ_0^+ but have one in Δ_1^+. Here α does not satisfy the parity condition, and Theorem 1.2 says that

$$\pi(s_2\gamma_0, \alpha \leftrightarrow \alpha) = \overline{\pi}(s_2\gamma_0, \alpha \leftrightarrow \alpha). \tag{4.1}$$

The τ-invariant for the right side is $\{1\}$, and we find

$$s_2\overline{\pi}(s_2\gamma_0, \alpha \leftrightarrow \alpha) = \overline{\pi}(s_2\gamma_0, \alpha \leftrightarrow \alpha) + \overline{\pi}(\gamma_0, \alpha \leftrightarrow \alpha) + \Theta .$$

Here the constituents of Θ must occur on the right side of (4.1) and must have 2 in their τ-invariants. So $\Theta = 0$. Therefore

$$\pi(\gamma_0, \alpha \leftrightarrow \alpha) = s_2\pi(s_2\gamma_0, \alpha \leftrightarrow \alpha) = s_2\overline{\pi}(s_2\gamma_0, \alpha \leftrightarrow \alpha)$$
$$= \overline{\pi}(s_2\gamma_0, \alpha \leftrightarrow \alpha) + \overline{\pi}(\gamma_0, \alpha \leftrightarrow \alpha) .$$

The τ-invariants for the two terms on the right side are $\{1\}$ and $\{2\}$. Only the second of these is disjoint from the set $\{1,3\}$ of singular roots, and it follows just as in Section 3 that $U(P,\sigma,\nu)$ is irreducible.

5. $SO^*(10)$

For $SO^*(10)$, we specify λ as at the end of Section 1. This is a cotangent case and we treat $\nu = \alpha$. We prove that $U(P,\sigma,\nu)$ is irreducible for this value of ν.

We number the roots as at the end of Section 1, and the diagram for $\lambda + \nu$ is what is called Δ_2^+ there. It is a little more convenient to define Δ_1^+ as $s_3 s_1 \Delta_2^+$. We continue with $\Delta_0^+ = s_5 \Delta_1^+$. Then our diagrams are

Δ_1^+

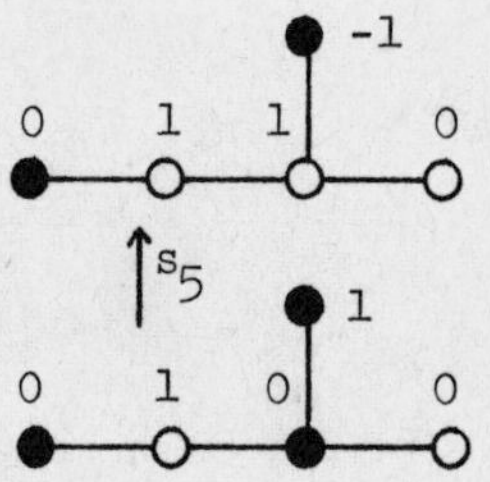

α = ① + ② + ③

Δ_0^+

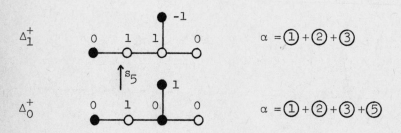

α = ① + ② + ③ + ⑤

We use Δ_0^+ to define γ_0. The set of singular roots is $\{1,3,4\}$.

We can apply Theorem 2.1 in the system Δ_1^+ to obtain

$$\pi(s_5\gamma_0, \mathfrak{a} \leftrightarrow \alpha) = \overline{\pi}(s_5\gamma_0, \mathfrak{a} \leftrightarrow \alpha) + \overline{\pi}(s_1 s_5\gamma_0, \mathfrak{a} \leftrightarrow \alpha)$$
$$+ \overline{\pi}(s_3 s_5\gamma_0, \mathfrak{a} \leftrightarrow \alpha) + \overline{\pi}(s_1 s_3 s_5\gamma_0, \mathfrak{a} \leftrightarrow \alpha) . \quad (5.1)$$

The respective τ-invariants of the terms on the right are $\{1,2,3\}$, $\{2,3\}$, $\{1,2,4\}$, and $\{2,4\}$, and 5 is not in any of these. But even more, we see from Theorem 2.1 that 5 is not in the τ-invariant of any constituent of the corresponding representations $\pi(\ldots, \mathfrak{a} \leftrightarrow \alpha)$ for the members of the right side of (5.1). This says that all the extra Θ terms are 0 when we apply s_5 to the terms on the right side of (5.1). Consequently application of s_5 to (5.1) gives exactly

$$\pi(\gamma_0, \mathfrak{a} \leftrightarrow \alpha) = (\text{same 4 terms as in (5.1)})$$
$$+ \overline{\pi}(\gamma_0, \mathfrak{a} \leftrightarrow \alpha) + \overline{\pi}(s_1\gamma_0, \mathfrak{a} \leftrightarrow \alpha)$$
$$+ \overline{\pi}(s_5 s_3 s_5\gamma_0, \mathfrak{a} \leftrightarrow \alpha, \beta) + \overline{\pi}(s_5 s_1 s_3 s_5\gamma_0, \mathfrak{a} \leftrightarrow \alpha, \beta) .$$

Here β denotes the noncompact root in position 5 for the positive system in question. We readily compute that the respective τ-invariants of our four new terms are $\{1,2,5\}$, $\{2,5\}$, $\{1,2,4,5\}$, and $\{2,4,5\}$. Of our eight τ-invariants, the only one that fails to meet the set $\{1,3,4\}$ of singular roots is $\{2,5\}$. Thus only the image of $\overline{\pi}(s_1\gamma_0, \mathfrak{a} \leftrightarrow \alpha)$ is nonzero when we apply the ψ functor, and it follows as in Sections 3 and 4 that $U(P,\sigma,\nu)$ is irreducible.

6. $SO^*(2n)$

For $SO^*(2n)$ with $n \geq 6$, we shall specify λ by attaching $2\langle \lambda, \beta\rangle / |\beta|^2$ to each simple root β in the Dynkin diagram of type D_n. The black simple roots are the noncompact ones, and P is built from α.

n odd , $t = \frac{1}{2}(n-1)$

λ :

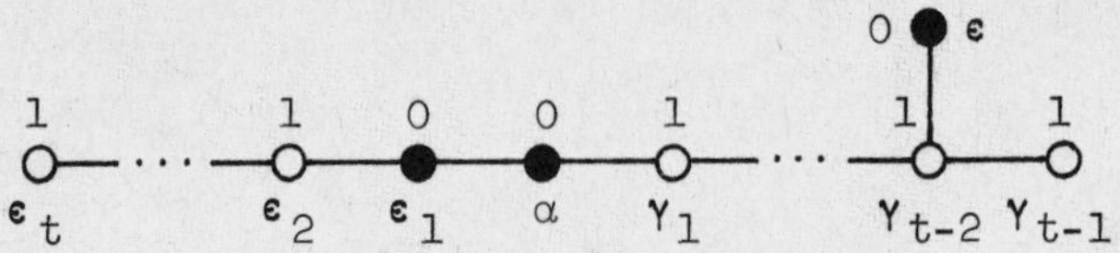

cotangent case at $\nu = (t-1)\alpha$

n even , $t = \frac{1}{2}(n-2)$

λ :

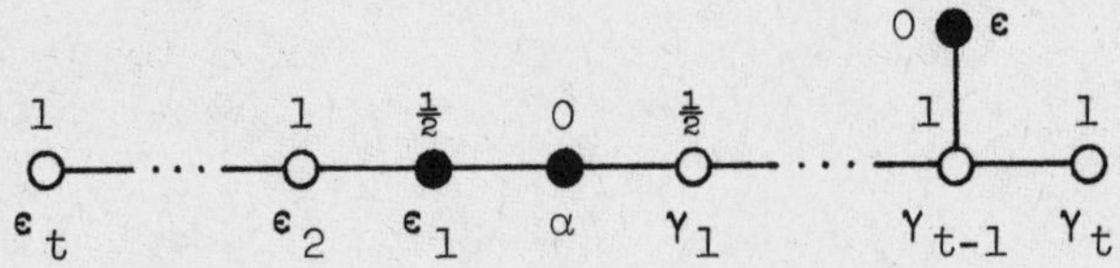

tangent case at $\nu = (t-\frac{1}{2})\alpha$

We shall prove that $U(P,\sigma,\nu)$ is irreducible for the indicated values of ν.

We number the roots on the horizontal as $\{1,\ldots,n-1\}$ and denote by n the root extending upward. The diagrams for $\lambda+\nu$ are

$\lambda+\nu$:

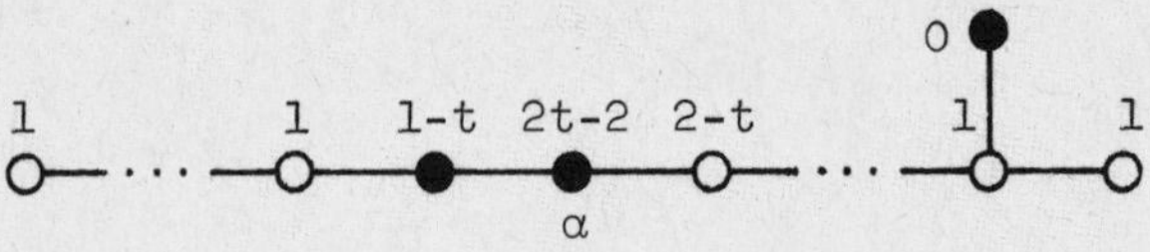

(n odd)

$\lambda+\nu$:

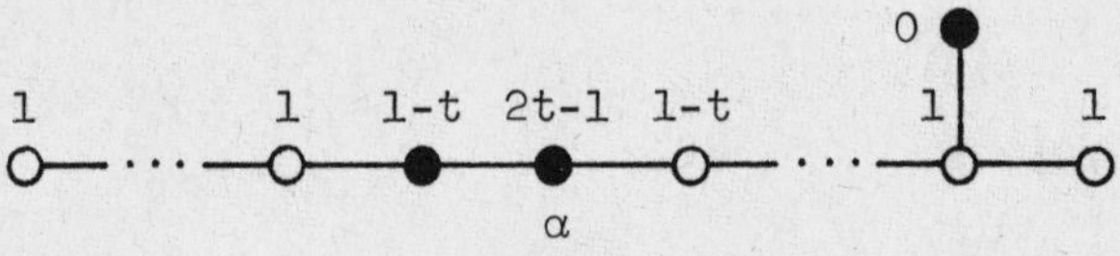

(n even)

We apply a sequence of complex root reflections to one or the other of these diagrams. When n is odd, the sequence is $s_{2t-1}\cdots s_{t+2}s_1\cdots s_t$. When n is even the sequence is $s_{2t}\cdots s_{t+2}s_1\cdots s_t$. In both cases the resulting diagram is

Δ_1^+

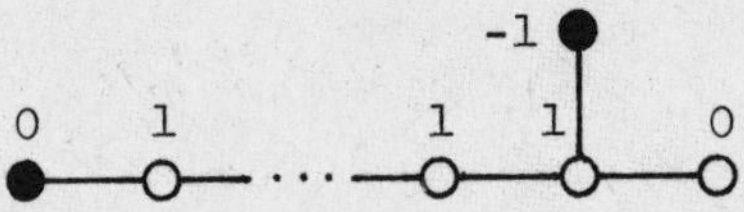

$\alpha = (1) + \ldots + (n-2)$

We let $\Delta_0^+ = s_n \Delta_1^+$ and use Δ_0^+ to define γ_0. The diagram is

Δ_0^+

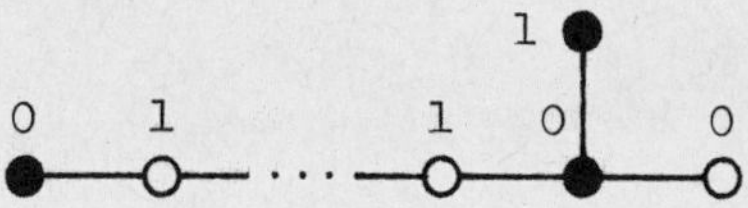

$\alpha =$ (1) $+ \ldots +$ (n-2) $+$ (n)

and the set of singular roots is $\{1, n-2, n-1\}$. The argument now proceeds just as with $SO^*(10)$ in Section 5, and the result is that $U(P,\sigma,\nu)$ is irreducible.

7. Groups of type E

For groups of type E, we shall consider 13 specific examples, of which 2 are in E_6, 5 are in E_7, and 6 are in E_8. The theorem is that $U(P,\sigma,\nu)$ is irreducible in all 13 cases. Our numbering of the roots in E_8 is

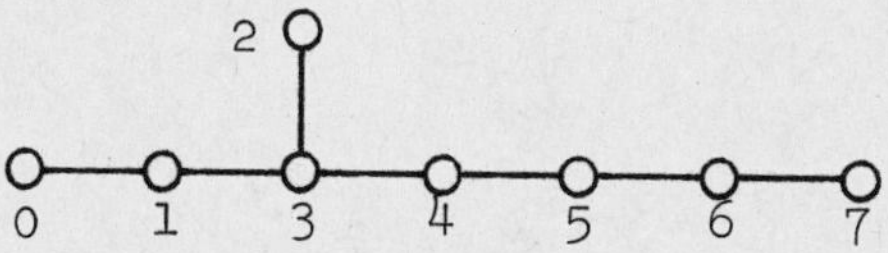

and we drop root 7 or roots 6 and 7 in E_7 or E_6.

For each example, we state what ν is, and we give two diagrams, the left one for the m parameter λ (with $2\langle\lambda,\beta\rangle/|\beta|^2$ attached to the simple root β) and the right one for the infinitesimal character $\lambda + \nu$. The black simple roots are the noncompact ones, α is circled, and P is built from α.

In addition, we give the sequence of reflections used to pass to Δ_0^+. A vertical line in the middle indicates the stage at which we apply Theorem 2.1. (Thus effectively we have only to implement wall crossings for reflections to the left of this line.) With each example we list the set of singular roots.

The 13 examples are listed below. After giving the list, we shall discuss the proof of irreducibility.

(a) E_6 with $\nu = \frac{3}{2}\alpha$.

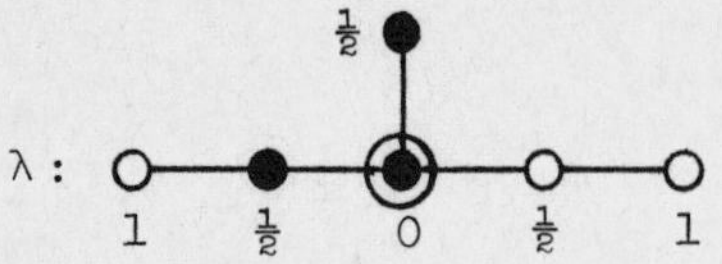

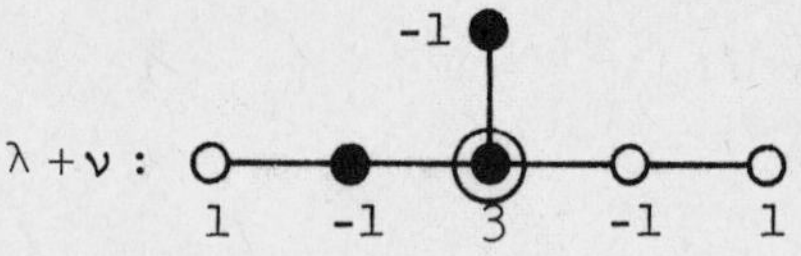

Reflections: $s_2 | s_0 s_1 s_5 s_4$. Singular set: $\{1,3,5\}$.

(b) E_6 with $\nu = 2\alpha$.

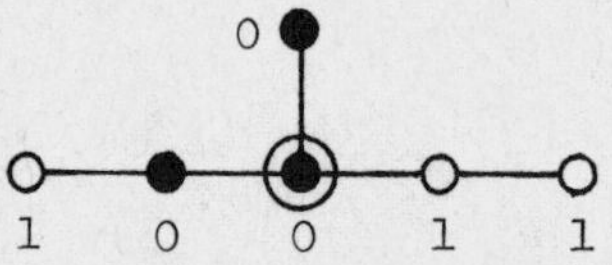

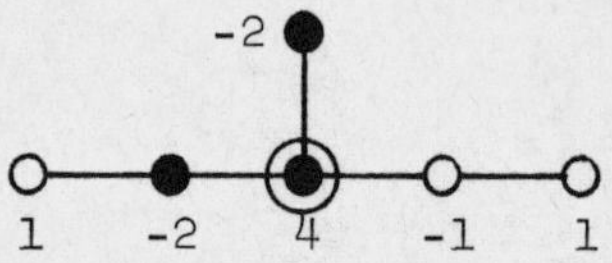

Reflections: $s_3 s_2 | s_0 s_1 s_5 s_4$. Singular set: $\{1,4,5\}$.

(c) E_7 with $\nu = 2\alpha$.

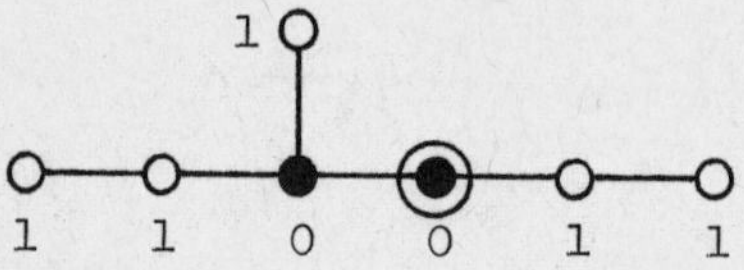

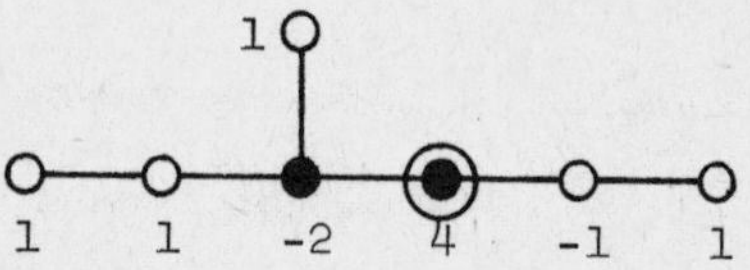

Reflections: $s_2 | s_0 s_1 s_3 s_5$. Singular set: $\{0,3,6\}$.

(d) E_7 with $\nu = 2\alpha$.

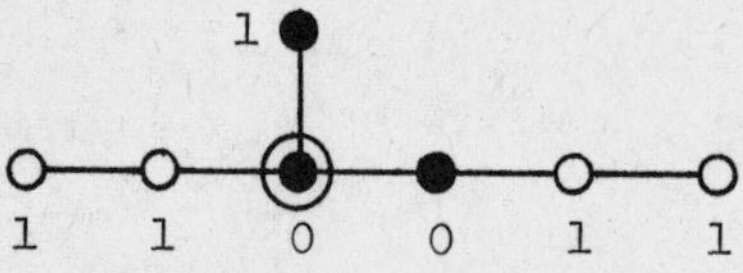

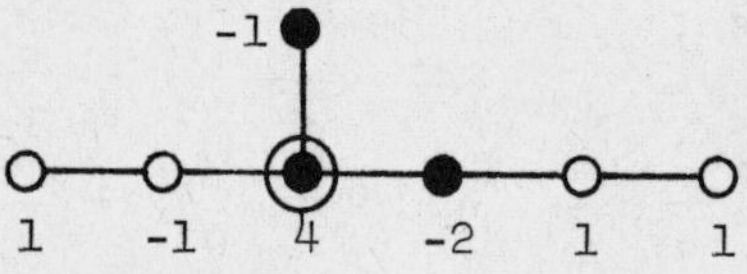

Reflections: $s_2 | s_0 s_1 s_5 s_4$. Singular set: $\{0,3,6\}$.

(e) E_7 with $\nu = \frac{5}{2}\alpha$.

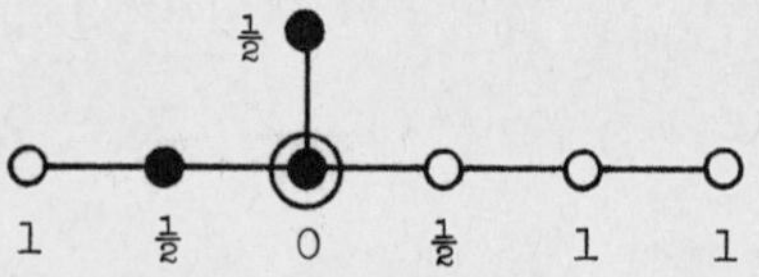

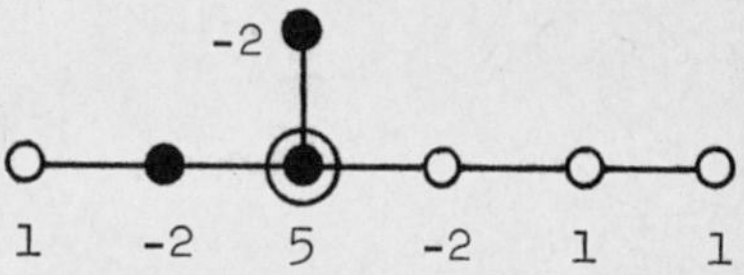

Reflections: $s_3 s_2 | s_0 s_1 s_5 s_4$. Singular set: $\{1,4,6\}$.

(f) E_7 with $\nu = 2\alpha$.

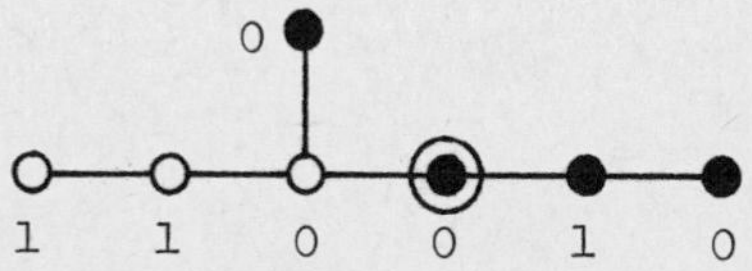

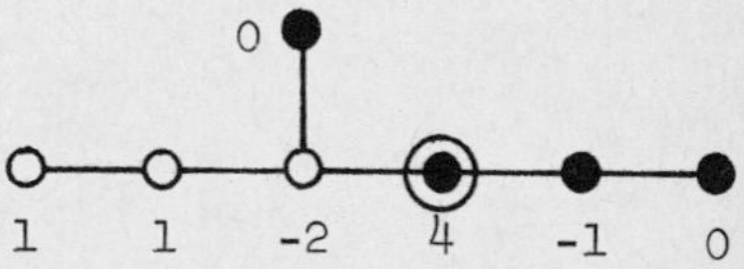

Reflections: $s_6 s_3 s_2 | s_5 s_1 s_3$. Singular set: $\{0,1,4,5\}$.

(g) E_7 with $\nu = 3\alpha$.

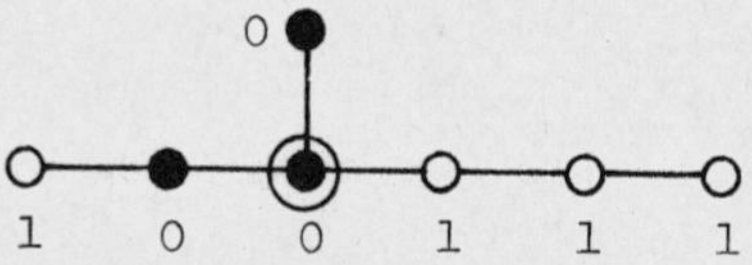

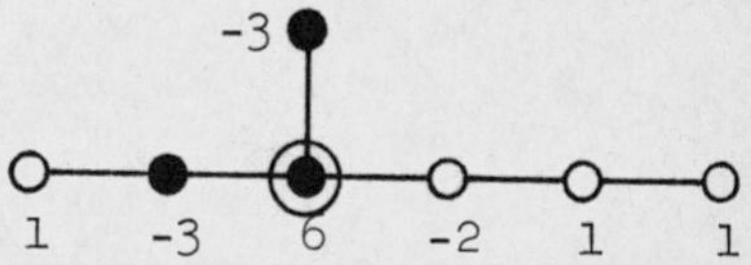

Reflections: $s_4 s_1 s_3 s_2 | s_0 s_1 s_5 s_4$. Singular set: $\{3,5,6\}$.

(h) E_8 with $\nu = \frac{5}{2}\alpha$.

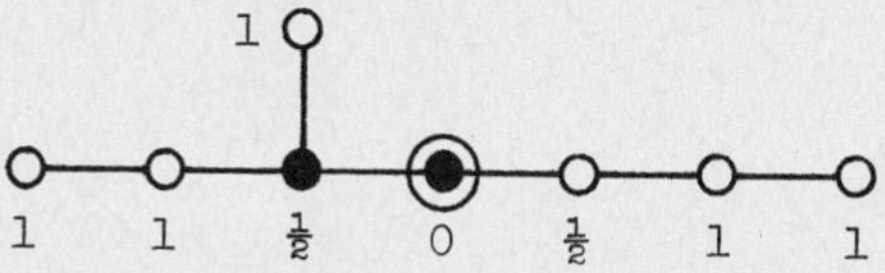

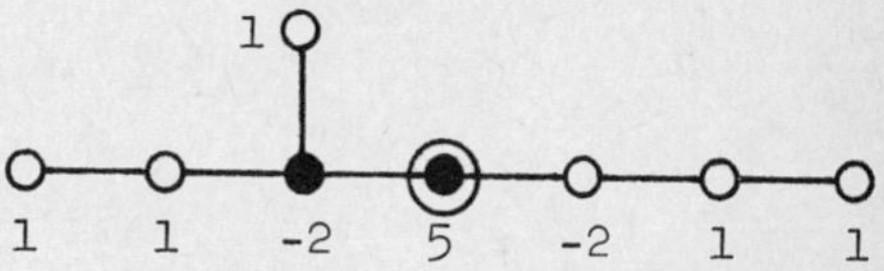

Reflections: $s_2 | s_0 s_1 s_3 s_6 s_5$. Singular set: $\{0,3,7\}$.

(i) E_8 with $\nu = 3\alpha$.

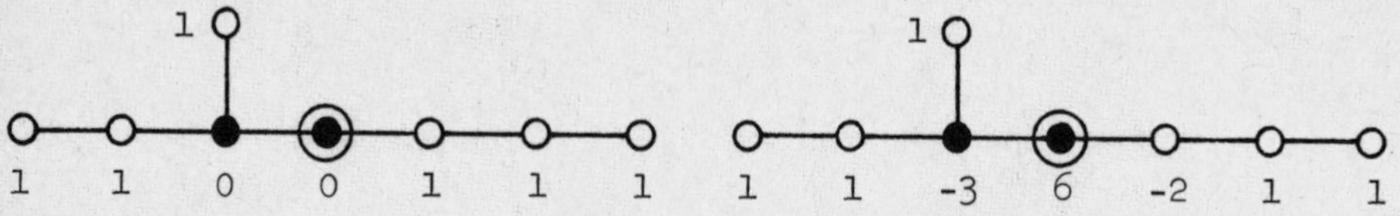

Reflections: $s_3 s_2 | s_0 s_1 s_3 s_6 s_5$. Singular set: $\{1,4,7\}$.

(j) E_8 with $\nu = \frac{5}{2}\alpha$.

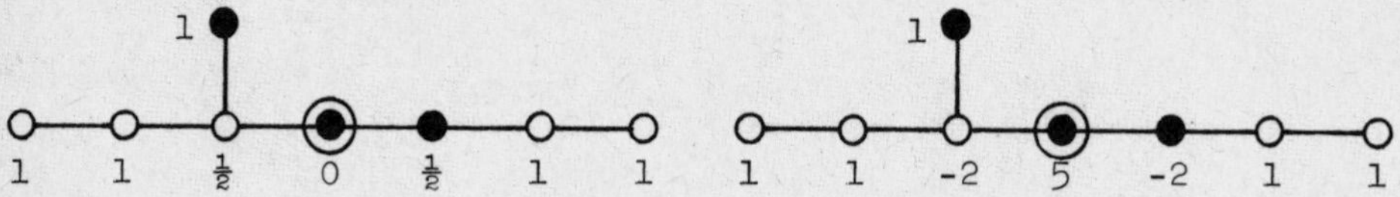

Reflections: $s_2 | s_1 s_3 s_6 s_5$. Singular set: $\{0,3,7\}$.

(k) E_8 with $\nu = \frac{5}{2}\alpha$.

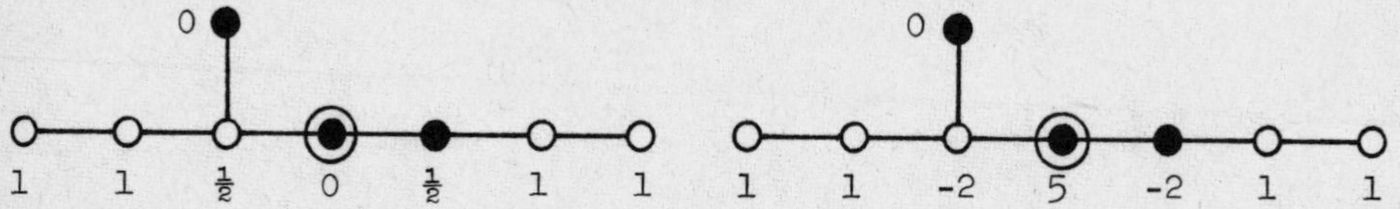

Reflections: $s_3 s_2 | s_0 s_1 s_3 s_6 s_5$. Singular set: $\{0,1,4,7\}$.

(ℓ) E_8 with $\nu = \frac{7}{2}\alpha$.

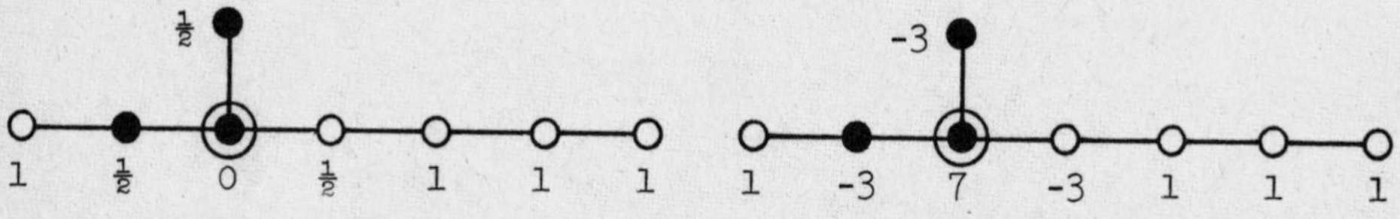

Reflections: $s_4 s_1 s_3 s_2 | s_0 s_1 s_6 s_5 s_4$. Singular set: $\{3,5,7\}$.

(m) E_8 with $\nu = 4\alpha$.

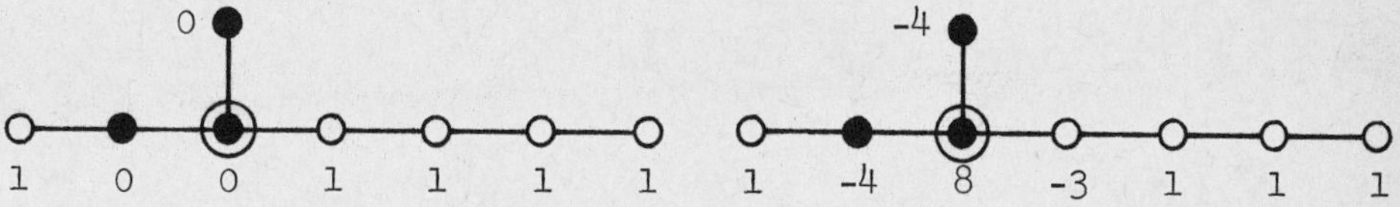

Reflections: $s_3 s_5 s_4 s_1 s_3 s_2 | s_0 s_1 s_6 s_5 s_4$. Singular set: $\{2,4,6,7\}$.

As we said, $U(P,\sigma,\nu)$ is irreducible in all 13 cases. We turn to the proof. Let Δ^+ be the positive system indicated above, let w be the product of the reflections to the right of the vertical line, let n be the number of reflections to the left of the vertical line, and put $\Delta_n^+ = w\Delta^+$. Then define $\Delta_{n-1}^+, \ldots, \Delta_0^+$ by applying one at a time the reflections that are to the left of the vertical line. Then we can define $\gamma_0 = \lambda + \nu + \mu$ as in Section 1, and we can reflect it to obtain γ_j dominant for Δ_j^+, $0 \leq j \leq n$. Theorem 2.1 enables us to decompose $\pi(\gamma_n, \alpha \leftrightarrow \alpha)$ as the sum of four irreducible characters.

In every case, the first reflection to the left of the vertical line is s_2. Moreover, the root 2 is not in the τ-invariant of any of the four irreducible characters, nor is it in the τ-invariant of any constituent of any of the corresponding π's for these four characters. Applying s_2, we then obtain a decomposition of $\pi(\gamma_{n-1}, \alpha \leftrightarrow \alpha)$ as the sum of eight irreducible characters. (This is all very similar to what happened in Section 5.)

When $n = 1$, we have only to check that the singular set meets the τ-invariant of 7 of these 8 characters, and then we have irreducibility for $U(P,\sigma,\nu)$. Thus we are done in cases (a), (c), (d), (h), and (j).

When $n = 2$, the next (and last) reflection is s_3, and we prepare to apply s_3 to our expansion of $\pi(\gamma_1, \alpha \leftrightarrow \alpha)$ into the sum of eight irreducible characters. The four characters that occurred also in $\pi(\gamma_2, \alpha \leftrightarrow \alpha)$ have 3 in their τ-invariants and may be disregarded. In order to handle the Θ terms, it is necessary to decompose the π's that correspond to the four additional characters that first appeared in $\pi(\gamma_1, \alpha \leftrightarrow \alpha)$. Each of these π's is seen to be the sum of eight irreducible characters, all of whose τ-invariants meet the singular set. Therefore all of the Θ terms may be disregarded. We thus need to consider only our four new characters and their main new terms under U_3. Of these 8 characters, 7 have τ-invariants that meet the singular set. Thus $U(P,\sigma,\nu)$ is irreducible in cases (b), (e), (i), and (k).

In cases (g) and (ℓ), we have $n=4$. The reflections to the left of the vertical line are $s_4 s_1 s_3 s_2$, and it is necessary to calculate the decomposition of $\pi(\gamma_2, \alpha \leftrightarrow \alpha) = s_3 s_2 \pi(\gamma_4, \alpha \leftrightarrow \alpha)$ exactly. This calculation is more complicated in case (ℓ) than in case (g) but can be done without tools more advanced than Theorem 1.7. One handles the last two wall crossings in the spirit of the previous paragraph, discarding as early as possible any terms that will not affect the final irreducibility. The details are fairly long and will be omitted, but the result is that $U(P,\sigma,\nu)$ is irreducible in cases (g) and (ℓ).

Case (f) has $n=3$. The reflections to the left of the vertical line are $s_6 s_3 s_2$, and we calculate $\pi(\gamma_1, \alpha \leftrightarrow \alpha) = s_3 s_2 \pi(\gamma_3, \alpha \leftrightarrow \alpha)$ as in case (g) above. The resulting character identity involves ten irreducible characters. The root 6 is not in the τ-invariant of any of these ten characters, nor is it in the τ-invariant of any constituent of the corresponding π's for these ten characters. Applying s_6, we obtain a decomposition of $\pi(\gamma_0, \alpha \leftrightarrow \alpha)$ as the sum of 20 irreducible characters. The singular set meets the τ-invariant of 19 of these 20 characters, and thus $U(P,\sigma,\nu)$ is irreducible in case (f).

Finally we consider case (m), in which $n=6$ and the reflections are $s_3 s_5 s_4 s_1 s_3 s_2$. Here we calculate $\pi(\gamma_3, \alpha \leftrightarrow \alpha) = s_1 s_3 s_2 \pi(\gamma_6, \alpha \leftrightarrow \alpha)$ exactly, using Theorem 1.7, and we calculate $\pi(\gamma_2, \alpha \leftrightarrow \alpha)$

$= s_4 \pi(\gamma_3, \mathfrak{a} \leftrightarrow \alpha)$ except for two $\circledast$ terms. One handles the final two wall crossings in the spirit of the cases $n = 2$ and $n = 4$. The details, however, are much more complicated in this situation. But at any rate the result is that $U(P, \sigma, \nu)$ is irreducible in case (m).

References

[1] Baldoni-Silva, M. W., and A. W. Knapp, Unitary representations induced from maximal parabolic subgroups, preprint, 1985.

[2] Baldoni Silva, M. W., and H. Kraljević, Composition factors of the principal series representations of the group Sp(n,1), Trans. Amer. Math. Soc. 262 (1980), 447-471.

[3] Barbasch, D., and D. A. Vogan, Reducibility of standard representations, Bull. Amer. Math. Soc. 11 (1984), 383-385.

[4] Knapp, A. W., and G. Zuckerman, Classification theorems for representations of semisimple Lie groups, "Non-Commutative Harmonic Analysis," Springer-Verlag Lecture Notes in Math. 587 (1977), 138-159.

[5] Knapp, A. W., and G. J. Zuckerman, Classification of irreducible tempered representations of semisimple groups, Ann. of Math. 116 (1982), 389-501.

[6] Schmid, W., On the characters of the discrete series: the Hermitian symmetric case, Invent. Math. 30 (1975), 47-144.

[7] Schmid, W., Two character identities for semisimple Lie groups, "Non-Commutative Harmonic Analysis," Springer-Verlag Lecture Notes in Math. 587 (1977), 196-225.

[8] Speh, B., and D. A. Vogan, Reducibility of generalized principal series representations, Acta Math. 145 (1980), 227-299.

[9] Vogan, D. A., Irreducible characters of semisimple Lie groups I, Duke Math. J. 46 (1979), 61-108.

[10] Vogan, D. A., Irreducible characters of semisimple Lie groups II. The Kazhdan-Lusztig conjectures, Duke Math. J. 46 (1979), 805-859.

[11] Zuckerman, G., Tensor products of finite and infinite dimensional representations of semisimple Lie groups, Ann. of Math. 106 (1977), 295-308.

Dipartimento di Matematica
Università degli Studi di Trento
38050 Povo (TN), Italy

Department of Mathematics
Cornell University
Ithaca, New York 14853, U.S.A.

UNIPOTENT REPRESENTATIONS AND UNITARITY

by
Dan Barbasch
Department of Mathematics
Rutgers University
New Brunswick, N.J. 08903

1. Introduction

The aim of this paper is to describe the classification of the unitary spectrum for complex reductive Lie group of classical type.

Let G be a complex connected reductive Lie group and $\mathfrak{g}_o$ its Lie algebra. Let $K \subseteq G$ be a maximal compact subgroup of G. Let $\mathfrak{g} = (\mathfrak{g}_o)_c$ be the complexification. In general we will denote by subscript o a real Lie algebra and drop the subscript for the complexification.

A hermitian form $\langle\,,\rangle$ on a $(\mathfrak{g},K)$ module (π,V) is called $\mathfrak{g}$-invariant if for every $X = X_1 + \sqrt{-1}\,X_2$ with $X_1,X_2 \in \mathfrak{g}_o$ the relation

$$(1.1)\qquad \langle\pi(X)v,w\rangle = -\langle v,\pi(X^*)w\rangle \qquad v,w \in V$$

where $X^* = X_1 - \sqrt{-1}\,X_2$, holds.

Then it is a well known result of Harish-Chandra that the aforementioned problem is equivalent to the classification of irreducible $(\mathfrak{g},K)$ modules admitting a $\mathfrak{g}$-invariant positive hermitian form. In particular, suppose (π,V) admits a nondegenerate form. Following [V2], we can define

$$(1.2)\qquad [V] = \sum_{\gamma\in\hat{K}} \dim[\gamma:V]\chi_\gamma$$

$$(1.3)\qquad [V]_\pm = \sum_{\gamma\in\hat{K}} \dim[\gamma:V]_\pm\chi_\gamma$$

where χ_γ is the character of γ and $[\gamma:V]$, $[\gamma:V]_\pm$ are the dimensions of the γ-isotypic component and $\pm$ eigenspace with respect to $\langle\,,\,\rangle$.

In particular, V is unitary if and only if $[V]_- = 0$.

The formal K-characters $[V]_\pm$, which are called signatures, are well behaved under unitary induction and complementary series. Precisely, let $P = MN$ be a real parabolic subgroup and $\chi e^{t\xi}$ a character of the Levi subgroup M depending on the real parameter $t \in \mathbb{R}$ with χ unitary.

Let V_M be an irreducible $(\underline{m}, K \cap M)$ module admitting an invariant nondegenerate form $< , >_M$.

Let

$$(1.4) \qquad V_t = \mathrm{Ind}_P^G(V_M \otimes \chi e^{t\xi} \otimes 1\!\!1) \quad \text{(unitary induction)}$$

Then V_o admits a canonical nondegenerate invariant form $< , >$ and

$$(1.5) \qquad [V_o]_\pm = \mathrm{Ind}_{K \cap M}^K([V_M]_\pm \otimes \chi)$$

(In particular if V_M is unitary, every factor of V_o is unitary.)

Suppose V_t is irreducible for $|t| < \varepsilon$ and admits a nondegenerate form $< , >_t$ depending continuously on t.

Then

$$(1.6) \qquad [V_t]_\pm = \mathrm{Ind}_{K \cap M}^K([V_M]_\pm \otimes \chi)$$

The representations V_t are called complementary series.

It is reasonable to try to find a finite set of representations $\mathcal{U}_o$ such that the unitary spectrum is obtained from $\mathcal{U}_o$ by unitary induction and complementary series.

For integral infinitesimal character such a set is described in [BV] (and the appropriate conjectures are made). (See also [A].) This set of representations which we will denote by $\mathcal{U}_1$ are called special unipotent representations.

A similar set $\mathcal{U}_{1/2}$ can be written down for half-integral infinitesimal character. The main result is

<u>THEOREM 1.1</u>: A $(\mathfrak{g}, K)$ module (π, V) is unitary if and only if V is obtained by unitary induction and/or complementary series from a representation in $\mathcal{U}_o = \mathcal{U}_1 \cup \mathcal{U}_{1/2}$.

Previously, the unitary spectrum for complex groups was known for small rank groups from the work of Duflo. For regular integral infinitesimal character a description is given in the work of Enright. The case of $GL(n,\mathbb{C})$ was treated by Vogan. Various partial results for spherical representations are due to Guillemonat.

One of the difficulties which is not present in the previously known cases is to show that the representations in $\mathcal{U}_0$ are unitary. The majority are not complementary series, endpoints of complementary series or unitarily induced. Nor is the K-spectrum particularly simple.

The proof, sketched in section 4, relies heavily on [BV].

The techniques used in this paper apply to the real case as well. In particular, it seems possible to decide the unitarity of certain spherical representations in the quasisplit case conjectured by Arthur in [A]. In the case of the unitary groups, the reduction techniques described in section 5 can be used to describe the unitary spectrum for spherical representations (see also [B]). I plan to pursue this is a future paper.

2. Special unipotent representations

In this section we review some results from [BV]. Let $\mathfrak{b}_0 \subseteq \mathfrak{g}_0$ be a Borel subalgebra and $\mathfrak{h}_0 \subseteq \mathfrak{b}_0$ a Cartan subgroup. Let $\mathfrak{h}_0 = \mathfrak{t}_0 + \mathfrak{a}_0$ where $\mathfrak{t}_0 = \mathfrak{k}_0 \cap \mathfrak{h}_0$ and $\mathfrak{a}_0$ be the split component. For every $\alpha \in \Delta(\mathfrak{b}_0, \mathfrak{h}_0)$ the coroot $\check{\alpha}$ is defined by the relation

$$\check{\alpha} = \frac{2(\alpha,\ )}{(\alpha,\alpha)} \in \mathfrak{h}_0^* . \tag{2.1}$$

The coroots form a root system ${}^L\Delta(\mathfrak{b}_0, \mathfrak{h}_0)$. Let ${}^L\mathfrak{g}_0$ be the corresponding simple algebra with Cartan subalgebra ${}^L\mathfrak{h}_0 \simeq \mathfrak{h}_0^*$. By the Jacobson-Morozov theorem every nilpotent orbit $\mathcal{O} \subseteq \mathfrak{g}_0$ under the adjoint action is identified with a conjugacy class of a map

$$\phi\colon s\ell(2,\mathbb{C}) \to \mathfrak{g}_0 .$$

Let $\begin{pmatrix}0&1\\0&0\end{pmatrix}$, $\begin{pmatrix}0&0\\1&0\end{pmatrix}$, $\begin{pmatrix}1&0\\0&-1\end{pmatrix}$ be generators for $s\ell(2,\mathbb{C})$. Then $e = \phi\begin{pmatrix}0&1\\0&0\end{pmatrix}$,

$h = \phi\begin{pmatrix}1 & 0\\ 0 & -1\end{pmatrix}$, $f = \phi\begin{pmatrix}0 & 1\\ 0 & 0\end{pmatrix}$ are such that e and f are elements of the nilpotent orbit $\mathcal{O}$ and h is semisimple. $\mathcal{O}$ is called even if ad h has even eigenvalues only.

Let now

$$^L\phi\colon \mathfrak{sl}(2,\mathbb{C}) \to {}^L\mathfrak{g}_0$$

be an even nilpotent. Then $\lambda_{\mathcal{O}} = \frac{1}{2}\,{}^Lh \in {}^L\mathfrak{h}_0 \simeq \mathfrak{h}_0^*$ defines an integral infinitesimal character for G. Let $A(\phi)$ be the connected component group of the centralizer of ^Le. According to [L] there is a canonically defined quotient group $\overline{A}(\phi)$. (In the classical groups case $A(\phi)$ and $\overline{A}(\phi)$ are products of $\mathbb{Z}_2$'s.) In addition [L], attaches to every conjugacy class $[x] \in [\overline{A}(\phi)]$ a Weyl group representation σ_x. In [BV] a correspondence is established

(2.2) $\pi \in \hat{\overline{A}}(\phi) \longleftrightarrow X_\pi$-irreducible $(\mathfrak{g},K)$ module with infinitesimal character $\lambda_{\mathcal{O}}$

(2.3) $[x] \in \overline{A}(\phi) \longleftrightarrow R_x$ virtual character with the following properties

(a) If $\mathcal{O} \longleftrightarrow {}^L\mathcal{O}$ is the inclusion reversing isomorphism on special orbits (in the sense of [L]), then

$$WF(X_\pi) = \overline{\mathcal{O}}.$$

(b) $$X_\pi = \frac{1}{|\overline{A}(\phi)|} \sum_{[x]\in[\overline{A}(\mathcal{O})]} \operatorname{tr} \pi(x)\,|[x]|\,R_x$$

$$R_x = \sum_{\pi\in\hat{\overline{A}}(\phi)} \operatorname{tr} \pi(x) X_\pi.$$

The X_π are called special unipotent representations.

Let $\mathfrak{p} = \underline{m} + \underline{n}$ be a parabolic subalgebra and $\mathcal{O}_M$ a nilpotent orbit in $\underline{m}$. We say that a nilpotent orbit $\mathcal{O}$ is induced and write

$$\mathcal{O} = \operatorname{ind}_{\mathfrak{p}}^{\mathfrak{g}} \mathcal{O}_M$$

if $\mathcal{O}$ is dense in Ad $G\cdot(\mathcal{O}_M + \underline{n})$. Suppose both $^L\mathcal{O}$ and $^L\mathcal{O}_M$ are even. Then we can find a subgroup $A^{(m)}(\phi) \subset \overline{A}(\phi)$ such that $\overline{A}(\phi_m)$ is a quotient of

$A^{(m)}(\phi)(A^{(m)}(\phi) \simeq \overline{A}(\phi_m)$ in the classical groups case) and

(2.4) $$\mathrm{Ind}_P^G(X_\pi^{(M)}) = \sum_{[\eta|_{A^{(m)}}:\ \pi]\neq 0} X_\eta$$

$\mathcal{O}$ is called smoothly induced if $\overline{A}(\phi) \simeq \overline{A}(\phi_m)$.

<u>THEOREM 2.1</u>: Suppose $\mathcal{O}$ is induced from $\mathcal{O}_M$ such that both ${}^L\mathcal{O}$ and ${}^L\mathcal{O}_M$ are even. Then

(i) if $\mathcal{O}$ is smoothly induced, every X_η is unitarily induced irreducible from an $X_\pi^{(m)}$

(ii) if $\mathcal{O}$ is not smoothly induced then every X_η is in a complementary series from an $X_\pi^{(M)}$ possibly using the det representation on some of the simple factors of type A of M.

<u>PROOF</u>: This follows from (2.4) and the description of X_π in [BV].

We now illustrate this theorem with an example. First we parametrize $(\mathfrak{g},K)$ modules.

Identify $\mathfrak{g}$ with $\mathfrak{g}_0 \times \mathfrak{g}_0$ in such a way that

(2.5) $\underline{\mathfrak{k}} = \{(x,\sigma x):\ x \in \mathfrak{g}_0\}$ where σ is the Chevalley automorphism -id on $\mathfrak{h}_0$.

(2.6) $\mathfrak{t} = \{(x,-x):\ x \in \mathfrak{h}_0\}$, $\mathfrak{a}_0 = \{(x,x):\ x \in \mathfrak{h}_0\}$.

For every $(\lambda,\mu) \in \mathfrak{h}^* \times \mathfrak{h}^*$ with $\lambda-\mu$ a weight of a finite dimensional representation, let $\mathbb{C}_{(\lambda,\mu)}$ be the character of H defined by

$$\mathbb{C}_{(\lambda,\mu)}\Big|_T = \mathbb{C}_{\lambda-\mu} \qquad \mathbb{C}_{(\lambda,\mu)}\Big|_A = \mathbb{C}_{\lambda+\mu}.$$

Then write

(2.7) $X(\lambda,\mu)$ = K-finite part of $\mathrm{Ind}_B^G(\mathbb{C}_{(\lambda,\mu)} \otimes \mathbb{1})$

(2.8) $\overline{X}(\lambda,\mu)$ = unique irreducible subquotient of $X(\lambda,\mu)$ containing the K-type with extremal weight $\lambda-\mu$

THEOREM 2.2 (Zhelobenko, Rao-Parthasarathy-Varadarajan): Fix (λ,μ), (λ',μ') as before. The following are equivalent: .

(a) $X(\lambda,\mu)$, $X(\lambda',\mu')$ have the same composition factors

(b) $\overline{X}(\lambda,\mu) \approx \overline{X}(\lambda',\mu')$

(c) there is $w \in W$ such that $(w\lambda,w\mu) = (\lambda',\mu')$

Every irreducible $(\mathfrak{g},K)$ module occurs as an $\overline{X}(\lambda,\mu)$.

PROOF: See [D1].

PROPOSITION 2.3 (Duflo): $\overline{X}(\lambda,\mu)$ admits a nondegenerate hermitian form if and only if there is $w \in W$ such that

$$w(\lambda-\mu) = \lambda-\mu$$
$$w(\lambda+\mu) = -\overline{\lambda}-\overline{\mu}.$$

PROOF: See [D2].

EXAMPLE: Let $\mathfrak{g}_0 = sp(2)$, ${}^L\mathfrak{g}_0 = so(5)$. Let

$${}^L\phi\colon s\ell(2) \to so(5)$$

be the nilpotent corresponding to $V_1+V_1+V_3$ where V_1 is the 1-dimensional, V_3 the 3-dimensional irreducible representation of $s\ell(2)$. Then ${}^Lh = (1,0)$. The nilpotent $\mathcal{O}$ corresponds to V_2+V_2 and $A(\phi) = \overline{A}(\phi) = \mathbb{Z}_2$.

Then $\mathcal{O} = \mathrm{ind}_{g\ell(2)}^{sp(2)}(\mathrm{triv})$ and also $\mathcal{O} = \mathrm{ind}_{sp(1)}^{sp(2)}(\mathrm{triv})$.

$$X_{triv} = \mathrm{Ind}_{g\ell(2)}^{sp(2)}(1 \text{ dim'}\ell) = \overline{X}(10,10)$$
$$X_{sgn} = \mathrm{Ind}_{g\ell(2)}^{sp(2)}(1 \text{ dim'}\ell) = \overline{X}(01,10).$$

We also note the composition series

$$\mathrm{Ind}_{Sp(1)}^{Sp(2)}(\mathrm{triv}) = X_{triv} + X_{sgn}$$

3. The $\frac{1}{2}$ integral case

A possible generalization of $\mathcal{U}_1$ is to consider (as [A] originally did) all maps

$$^L\phi\colon \mathfrak{sl}(2,\mathbb{C}) \to {}^L\mathfrak{g}_0$$

which are not necessarily even. Then $\lambda_{\mathcal{O}} = \frac{1}{2}\,{}^Lh$ is no longer integral. However in this case it is easy to check that the representations obtained this way are unitarily induced irreducible. On the other hand, it is well known that the metaplectic representation has half-integral infinitesimal character but cannot be unitarily induced from any proper parabolic subgroup.

Let $\xi \in \mathfrak{h}_0^*$ denote an infinitesimal character. Then a root α is called integral with respect to ξ if

$$\frac{2(\alpha,\xi)}{(\alpha,\alpha)} = (\check{\alpha},\xi) \in \mathbb{Z}. \tag{3.1}$$

Clearly the set of coroots $\{\check{\alpha}\colon (\check{\alpha},\xi) \in \mathbb{Z}\}$ form a root system ${}^L\Delta(\xi) \subseteq {}^L\Delta({}^L\mathfrak{g},{}^Lb)$, in fact this is the root system of ${}^LS_\xi \subseteq {}^LG$ where ${}^LS_\xi = \mathrm{Cent}_{{}^LG}(e^{2\pi i\xi})$. Let ${}^L\mathfrak{s}_\xi$ be its Lie algebra. S_ξ, the dual group, is called an endoscopic subgroup for G. Let $\mathfrak{s}_\xi$ be the Lie algebra, K_ξ the maximal compact subgroup of S_ξ.

THEOREM 3.1: The character theory for $(\mathfrak{g},K)$ modules with infinitesimal character ξ is equivalent to the character theory of $(\mathfrak{s}_\xi,K_\xi)$ modules with integral infinitesimal character.

PROOF: This can be found in [V1].

${}^L\Delta(\xi)$ is called $\frac{1}{2}$-integral if it is a maximal proper subsystem in ${}^L\Delta({}^L\mathfrak{g},{}^L\mathfrak{b})$.

In view of theorem 3.1, it seems reasonable to expect that a similar statement is valid for the unitary spectrum. More precisely, it would be desirable to have an inclusion

$$\mathcal{U}_1(S_\xi) \to \mathcal{U}_\xi(G) \tag{3.2}$$

where $\mathcal{U}_\xi(G)$ is the set of unitary irreducible representations with infinitesimal character ξ.

This inclusion might possibly involve some twisting of infinitesimal characters to match up with the statements of theorem 3.1.

Some more complicated examples show that in this generality such a map does not exist. One gets either too many or too few representations in $\mathcal{U}_\xi(G)$.

The next section deals with the construction of $\mathcal{U}_0$.

4. Construction and Unitarity of $\mathcal{U}_0$

We define a relation among $(\mathfrak{g},K)$ modules which preserves unitarity. Let $(\mathfrak{g}_0,K_0)$ and $(\mathfrak{g}_1,K_1)$ be reductive algebras of classical type. Let V_i, i=0,1, be irreducible $(\mathfrak{g}_i,K_i)$ modules admitting nondegenerate forms normalized to be positive on the lowest K-type.

DEFINITION 4.1: We say $V_1 \overset{u}{\sim} V_2$ if there are algebras $\mathfrak{a}_i$ and unitary representations X_i, X_i', Y_i such that $\mathfrak{g}_i \times \mathfrak{a}_i$ are Levi components in a Lie algebra $\mathfrak{g}$ and

$$[\mathrm{Ind}_{\mathfrak{g}_i\times\mathfrak{a}_i}^{\mathfrak{g}}(V_i \times Y_i)]_\pm = [\mathrm{Ind}_{\mathfrak{g}_{1-i}\times\mathfrak{a}_{1-i}}^{\mathfrak{g}}(V_{1-i} \times X_{1-i})]_\pm$$
$$+ [\mathrm{Ind}_{\mathfrak{g}_{1-i}\times\mathfrak{a}_{1-i}}^{\mathfrak{g}}(V_{1-i} \times X'_{1-i})]_\pm .$$

REMARKS: 1. If $V \overset{u}{\sim} W$ then V is unitary if and only if W is unitary. Typically $\overset{u}{\sim}$ arises from composition factors of induced representations and complementary series.

2. u has the flavor of reduced K-theory or the classification of quadratic forms over $\mathbb{Z}$ as in Serre's "A course in Arithmetic".

$\mathcal{U}_0$ can then be described in the following way.

THEOREM 4.2: $\mathcal{U}_0$ is the set of representations obtained by u from the trivial and determinant representations of $g\ell(n)$.

Most conspicuously, the trivial representations for the various simple classical groups are missing.

The next example shows how they disappear.

EXAMPLES: 1. Let $\mathfrak{g}_o = Sp(1)$, V_o the trivial representation. Induce unitarily to $Sp(2)$. Then as previously

$$(4.2) \qquad \mathrm{Ind}_{Sp(1)}^{Sp(2)}(\mathrm{triv}) = X_{\mathrm{triv}} + X_{\mathrm{sgn}}.$$

X_{triv} is in a complementary series from

$$(4.3) \qquad \overline{X}(\tfrac{1}{2} - \tfrac{1}{2}, \tfrac{1}{2} - \tfrac{1}{2}) = \mathrm{Ind}_{GL(2)}^{Sp(2)}(\mathrm{triv})$$

$$X_{\mathrm{sgn}} = \mathrm{Ind}_{GL(2)}^{Sp(2)}(\det).$$

To see that $\tilde{u}$ holds it is enough to check that the form on the lowest K-type of X_{sgn} which can be written as $(1,1)$ is induced from $Sp(1)$ and is positive. But the form on the $(1,1)$-isotypic component is induced from (0) on $Sp(1)$. Thus it must be positive.

2. Let $\mathfrak{g} = Sp(n)$ and $V_o = \overline{X}(n - \frac{1}{2} \dots \frac{1}{2}, n - \frac{1}{2} \dots \frac{1}{2})$ which is a factor of the metaplectic representation.

It is enough to see that it can be obtained from representations of the same type in lower groups.

Induce unitarily by $GL(2n-2)$. Then

$$(4.5) \qquad \mathrm{Ind}_{Sp(n)\times GL(2n-2)}^{Sp(3n-2)}(V_o \otimes \mathrm{triv}) = X_{\mathrm{triv}} + X_{\mathrm{sgn}}$$

X_{triv} is a complementary series from $\overline{X}(n - \frac{3}{2} \dots \frac{1}{2}, n - \frac{3}{2} \dots \frac{1}{2})$ tensored with the trivial of $GL(2n-1)$. On the other hand, X_{sgn} is unitarily induced from the nonspherical representation $\overline{X}(n - \frac{3}{2} \dots \frac{3}{2} - \frac{1}{2}, n - \frac{3}{2} \dots \frac{1}{2})$ tensored with the determinant representation on $GL(2n-1)$.

By induction, the form is positive on X_{triv} and definite on X_{sgn}. Remains to see that it is positive on X_{sgn}. It is enough to check it is positive on the lowest K-type which is $(\underbrace{1\ldots1}_{2n}\,0\ldots0)$. Since the form on (4.5) is induced from $V_o \otimes \mathrm{triv}$ it is enough to check positivity on the finite number of K-types occuring in the restriction of $(1\ldots1\ 0\ldots0)$ to $V_o \otimes \mathrm{triv}$. We omit the details.

5. Nonunitarity and the general case

In this section we describe two reduction methods which are sufficient to exclude all non-unitary representations.

We make use of the fact that theorem 1.1 is known for $GL(n,\mathbb{C})$.

Let $\overline{\mu}_o$ be the lowest K-type of $\overline{X}(\lambda,\mu)$. We may assume that $\overline{\mu}_o = (\mu_o,-\mu_o)$ is dominant with respect to a fixed positive root system $\Delta^+(\mathfrak{k},\mathfrak{t}) \subseteq \Delta(\mathfrak{k},\mathfrak{t})$. Let $\mathfrak{q}^o = \underline{\ell}^o + \underline{\mathfrak{u}}^o$ be the parabolic subalgebra defined by $\overline{\mu}_o$. This means

$$\Delta(\underline{\mathfrak{u}}^o,\mathfrak{h}) = \{\alpha \in \Delta(\mathfrak{g},\mathfrak{h}) : (\alpha,\overline{\mu}_o) > 0\} \tag{5.1}$$

$$\Delta(\underline{\ell}^o,\mathfrak{h}) = \{\alpha \in \Delta(\mathfrak{g},\mathfrak{h}) : (\alpha,\overline{\mu}_o) = 0\}. \tag{5.2}$$

Let $\mathfrak{q} \supseteq \mathfrak{q}^o$ be a parabolic subalgebra, $\mathfrak{q} = \underline{\ell} + \underline{\mathfrak{u}}$, $\underline{\ell} \supseteq \underline{\ell}^o$, and $L \subseteq G$ the subgroup with Lie algebra $\underline{\ell}_o$ such that $(\underline{\ell}_o^o)_c = \underline{\ell}$. Let $L^o \subseteq L$ be the similar group corresponding to $\underline{\ell}^o$.

<u>THEOREM 5.1</u>: Assume $\overline{X}(\lambda,\mu)$ and $X^{(L)}(\lambda,\mu)$ admit invariant hermitian forms. Let $\overline{\mu}$ be a K-type in $\overline{X}(\lambda,\mu)$ such that $\overline{\mu} = (\mu,-\mu)$ viewed as a highest weight for L is an $L \cap K$ type of $X^{(L)}(\lambda,\mu)$.
Then $[\overline{X}(\lambda,\mu) : \overline{\mu}]_\pm = [\overline{X}^{(L)}(\lambda,\mu) : \overline{\mu}]_\pm$.

Such a K-type $\overline{\mu}$ is called $\mathfrak{q}$-bottom layer. The proof of 5.1 can be found in [V2].

$\mathfrak{q}$-bottom layer K-types are fairly rare. Nevertheless, using the analog of theorem 1.1 in $GL(n,\mathbb{C})$ we can derive the following.

<u>THEOREM 5.2</u>: Let $\mathfrak{g}$ be simple of classical type, $\overline{X}(\lambda,\mu)$, $\mathfrak{q}^o$ and $X^{(L^o)}(\lambda,\mu)$ as before. Then

$$\underline{\ell}^0 \simeq \underline{\ell}^1\times\ldots\times\underline{\ell}^m$$
$$\overline{X}^{(L_0)} \simeq \overline{X}^1\otimes\ldots\otimes\overline{X}^m$$

where $\underline{\ell}^i$ for $i \geq 2$ are factors of type A.

Then $\overline{X}$ is unitary only if $\overline{X}^2,\ldots,\overline{X}^m$ are unitary.

These two theorems allow us to reduce considerations only to $\overline{X}(\lambda,\mu)$ where the lowest K-type is a fundamental representation. (Note that $\mathcal{U}_0$ is formed of such representations.)

We define a relation on representations admitting $\mathfrak{g}$-invariant hermitian forms.

DEFINITION 5.3: Let V_0, V_1 be $(\mathfrak{g}_0,K_0)$, $(\mathfrak{g}_1,K_1)$ modules. We say $V_0 \underset{h}{\geq} V_1$ if there are $\mathfrak{g}_0'$, $\mathfrak{g}_1'$ of type A such that $\mathfrak{g}_0\times\mathfrak{g}_0'$ and $\mathfrak{g}_1\times\mathfrak{g}_1'$ are Levi factors in a bigger algebra $\mathfrak{g}$ and $\mathrm{Ind}_{\mathfrak{g}_1\times\mathfrak{g}_1'}^{\mathfrak{g}}(V_1\otimes\mathrm{triv})$ contains a factor with signature $[\mathrm{Ind}_{\mathfrak{g}_0\times\mathfrak{g}_0'}^{\mathfrak{g}}(V_0\otimes\mathrm{triv})]_{\pm}$ in its composition series.

Clearly, if we start with V_0 not unitary, then V_1 will not be unitary.

In practice, we choose V_0, $\mathfrak{g}_0'$, $\mathfrak{g}_1'$ such that $\mathrm{Ind}_{\mathfrak{g}_0\times\mathfrak{g}_0'}^{\mathfrak{g}}(V_0\otimes\mathrm{triv})$ is irreducible and $\mathrm{Ind}_{\mathfrak{g}_1\times\mathfrak{g}_1'}^{\mathfrak{g}}(V_1\otimes\mathrm{triv})$ contains it or a complementary series as a factor (most often containing the lowest K-type as well).

THEOREM 5.4: Let $\overline{X}(\lambda,\mu)$ be an irreducible $(\mathfrak{g},K)$ module admitting a hermitian form and having lowest K-type a fundamental weight $(1\ldots1\ 0\ldots0)$. Then $\overline{X}(\lambda,\mu)$ is unitary if and only if it is in $\mathcal{U}_0$.

If $\overline{X}(\lambda,\mu)$ is not unitary then the nondegenerate form is negative on a K-type of the form $(2\ldots2\ 1\ldots1\ 0\ldots0)$. (Here we have identified $\mathfrak{t}_0^*$ with $\mathbb{R}^n$.)

6. Another description of $\mathcal{U}_0$

In this section we describe an alternate set $\mathcal{U}$ generalizing the special unipotent representations which serves the same purpose as $\mathcal{U}_0$. In addition it has the kind of properties with respect to endoscopic groups described in section 3.

Let $\mathcal{O}$ be an arbitrary nilpotent orbit in $\mathfrak{g}$. Let $\sigma(\mathcal{O})$ be the Weyl group representation attached by the Springer correspondence to $\mathcal{O}$ and the trivial character. Let $\tilde{\mathcal{O}}$ be the special nilpotent such that $\sigma(\tilde{\mathcal{O}})$ and $\sigma(\mathcal{O})$ are in the same double cell.

DEFINITION 6.1: An orbit $\mathcal{O}$ is called rigid if it is not induced from any proper parabolic subalgebra.

PROPOSITION 6.2: Fix a rigid orbit $\mathcal{O}$ and let ${}^L\tilde{\mathcal{O}}$ be the dual special orbit. Then ${}^L\tilde{\mathcal{O}}$ is even.

As in section 12 [BV] let $\bar{s} \in \bar{A}({}^L\tilde{\mathcal{O}})$ and $\tilde{s} \in \mathrm{Cent}_{{}^LG}({}^L\phi)$ be a semisimple representative of $\bar{s}$. Let ${}^LS$, ${}^L\underline{\mathfrak{s}}, \underline{\mathfrak{s}}$ and S be as in section 3.

Then ${}^L\tilde{\mathcal{O}}$ meets ${}^L\underline{\mathfrak{s}}$ in the orbit ${}^L\tilde{\mathcal{O}}_s$. Let $\mathcal{O}_{\mathfrak{s}}$ and $\sigma_1^{(\mathfrak{s})}$ be the corresponding dual orbit and special representation.

According to proposition 12.1 in [BV], $\bar{s}$ can be chosen so that

(6.1) $$\mathcal{J}_{W(S)}^{W}(\sigma_1^{(\mathfrak{s})}) = \sigma_{\bar{s}} = \sigma(\mathcal{O})$$

Among the representatives of $\bar{s}$ it is possible to choose one so that ${}^LS$ is maximal. Denote it by s. Let $\lambda_s \in {}^L\mathfrak{h}$ be such that the following are satisfied.

(6.1) $e^{2\pi i\lambda_s}$ is conjugate to s and λ_s is integral for $\Delta(S)$, the roots of S.

(6.2) $\|\lambda_s\|$ is minimal subject to (6.1) and

$$WF(\bar{X}(\lambda_s,\lambda_s)) = \bar{\mathcal{O}}.$$

DEFINITION 6.3: Let $\mathcal{O}$ be rigid, λ_s as in 6.1, 6.2. An irreducible representation X with infinitesimal character λ_s is called unipotent attached to $\mathcal{O}$ if

$$WF(X) = \bar{\mathcal{O}}.$$

The set $\mathcal{U}$ is the union of unipotent representations over all rigid orbits.

THEOREM 6.4: There are bijections

$$[x] \in \bar{A}(\mathcal{O}) \longleftrightarrow \sigma_x \in \hat{W}(S)$$

$$[x] \longleftrightarrow R_x = \frac{1}{|W(\lambda_s)|} \sum_{w \in W(S)} \operatorname{tr} \sigma_x(w) X(\lambda_s, w\lambda_s)$$

$\pi \in \hat{\bar{A}}(\mathcal{O}) \longleftrightarrow X_\pi$ unipotent attached to $\mathcal{O}$ such that

$$X_\pi = \frac{1}{|\bar{A}(\mathcal{O})|} \sum_{[x] \in \bar{A}(\mathcal{O})} \operatorname{tr} \pi(x) |[x]| R_x$$

$$R_x = \sum_{\pi \in \hat{\bar{A}}(\mathcal{O})} \operatorname{tr} \pi(x) X_\pi .$$

REMARK: The definitions in this section make sense for all groups not just the classical ones. In the exceptional groups most of the representations in $\mathcal{U}$ can be recognized to be unitarily induced, complementary series or endpoints of complementary series.

THEOREM 6.5: Theorem 1.1 is true for classical groups with $\mathcal{U}_0$ replaced by $\mathcal{U}$.

CONJECTURE: Theorem 6.5 is true for any complex simple group.

References

[A] J. Arthur, On some problems suggested by the trace formula, Lecture Notes in Math 1041, Springer Verlag 1984.

[B] D. Barbasch, A reduction theorem for the unitary spectrum of U(p,q), preprint Rutgers University 1984.

[BV] D. Barbasch, D. Vogan, Unipotent representations of complex semisimple Lie groups, Ann. of Math., 121, (1985) 41-110.

[D1] M. Duflo, Représentations irréductibles des groupes semisimples complexes, LNM 497 (1975) 26-88.

[D2] Représentations unitaires irréductibles des groupes simples complexes de rang deux, Bull. Soc. Math., France 107 (1979) 55-96.

[L] G. Lusztig, Characters of a reductive group over a finite field, Princeton Univ. Press, Princeton NJ 1984.

[V1] D. Vogan, Representations of real reductive groups, Birkhäuser Boston 1981.

[V2] _______, Unitarizability of certain series of representations, Ann. of Math., 120 (1984) 141-187.

[V3] _______, The unitary dual of GL(n) over an archimedean field, preprint.

BOCHNER-RIESZ MEANS OF H^p FUNCTIONS
$(0 < p < 1)$ ON COMPACT LIE GROUPS

Jean-Louis Clerc (*)
E.R.A. 839 - U.A. 750
Université de Nancy I
B.P. 239
54506 Vandoeuvre lès Nancy Cedex

The H^p-classes, since their introduction by E.M. Stein and G. Weiss have gained a major role in harmonic analysis. In [5], Taibleson and the two named authors studied the convergence of the Bochner-Riesz means for H^p-functions on $\mathbb{R}^n (0 < p < 1)$. Similar results were obtained later on for spheres. Here we study the same problem for compact Lie groups, and obtain weak-type estimates for the maximal Bochner-Riesz operators (cf. theorem 6.1). The proof requires sharp estimates on the corresponding kernels and their derivatives.

Sections 2 to 4 are devoted to obtain estimates for derivatives of a general central function knowing its restriction to a Cartan subgroup, and are of technical nature, but might be of independant interest. In section 5, the relevant estimates for Bochner-Riesz kernels are obtained using our previous work [3]. Once again, the situation involves a term similar to the $\mathbb{R}^n$ case plus a "correction" having mild "peaks" on the singular elements.

In section 6, the weak-type estimates are proved following [5] closely. We believe (but do not prove) that, by analogy with $\mathbb{R}^n$, these results are sharp. The situation for the critical index $\delta = \frac{n-1}{2}$ and the blocks space is studied in [7].

(*) During the time this work was completed, I enjoyed the warm hospitality of the mathematics department at Washington University and it was a pleasure for me to work "in the spirit of St Louis".

1.- Notations

Let G be a compact semi-simple Lie group, which, for convenience, is assumed to be simply connected ; let g be its Lie algebra, with $\dim(g) = n$.

Fix once for all a maximal torus T and let t be its Lie algebra, with $\dim(t) = \text{rank}(G) = \ell$.

Let Δ be the associated root system of the pair $(g^{\mathbb{C}}, t^{\mathbb{C}})$, which we view as a subset of t^* (the real dual of t) , although the roots are pure imaginary on t .

Let $\Gamma = \{H \in t \,, \exp(H) = e\}$ be the unit-lattice, and
$t_r = \{H \in t \,, \alpha(H) \notin 2\pi\,\mathbb{Z} \,, \forall\alpha \in \Delta\}$ be the set of regular elements in t .

Fix a connected component of t_r (an alcove in the terminology of [2]), say A , and assume, as we may, that $0 \in \bar{A}$; A is contained in a unique Weyl chamber, denoted by t^+ ; the corresponding set of positive roots is denoted by Δ^+ , and the set of simple roots by B .

The alcove is a fundamental domain for the exponential map up to conjugacy : any element in G is conjugate to exactly one element in $\exp(\bar{A})$ and any regular element in G is conjugate to one in $\exp(A)$ (see [2]).

A more explicit description of the alcove can be offered (cf. [1], ch. VI, § 2, prop. 5), and will be needed.

Call $\beta \in \Delta^+$ a largest root if β is the largest root for some irreducible component of Δ^+ . Denote by $\tilde{B}$ the set of all largest roots. Now

(1.1) $$A = \{H \in t \,, \alpha(H) > 0 \,, \forall\alpha \in B \,, \beta(H) < 2\pi \,, \forall\beta \in \tilde{B}\} .$$

2.- The domains $\Gamma_H^{(R)}$ and a partition of the alcove

Let R be a large positive number (how large is to be made more precise later on).

For any $H \in A$, introduce the following sets :

(2.1) $$I_H^R = \{\alpha \in B \,, \alpha(H) \le 1/R\} \qquad \tilde{I}_H^R = \{\alpha \in \tilde{B} \,, \alpha(H) \ge 2\pi - 1/R\} .$$

We drop the indices R and H , except when we wish to distinguish different sets for different H or different R.

Elements in I and $\tilde{I}$ are linearly independant, at least for R large enough. In fact the only possible linear relations are $\beta = \sum_{\alpha \in B'} n_\alpha \alpha$, where B' is the set of simple roots corresponding to some irreducible component of Δ, and β is the corresponding largest root. But we know that $n_\alpha \geq 1$, so we have $\beta(H) \leq (\sum_{\alpha \in B'} n_\alpha) 1/R$, which contradicts $\beta(H) \geq 2\pi - 1/R$, for R large enough.

In this context we now introduce the corresponding facet (following again the terminology of [2]) :

$$(2.2) \quad F_{I,\tilde{I}} = \left\{ \begin{array}{l} K \in \bar{A}, \alpha(K) = 0 \quad \text{for} \quad \alpha \in I, \beta(K) = 2\pi \quad \text{for} \quad \beta \in \tilde{I}, \\ 0 < \alpha(K) < 2\pi \quad \text{for} \quad \alpha \in B \setminus I, 0 < \beta(K) < 2\pi \quad \text{for} \quad \beta \in \tilde{B} \setminus \tilde{I} \end{array} \right\}$$

and let $\mathcal{F}_{I,\tilde{I}}$ be the affine subspace generated by $F_{I,\tilde{I}}$, so that

$$(2.3) \quad \mathcal{F}_{I,\tilde{I}} = \{K \in t, \alpha(K) = 0 \quad \text{for} \quad \alpha \in I, \beta(K) = 2\pi \quad \text{for} \quad \beta \in \tilde{I}\}.$$

A positive root γ is <u>R-singular</u> at H, or <u>$(I,\tilde{I})$-singular</u> of type I (resp. of type II), if the following equivalent conditions are satisfied :

(2.4)
(i) $\gamma(F_{I,\tilde{I}}) = \{0\}$ (resp. $= \{2\pi\}$)

(ii) $\gamma(\mathcal{F}_{I,\tilde{I}}) = \{0\}$ (resp. $= \{2\pi\}$)

(iii) γ can be written as $\sum_{\alpha \in I} n_\alpha \alpha$

(resp. $\beta - \sum_{\alpha \in I} n_\alpha \alpha$ for some $\beta \in \tilde{I}$), and with $n_\alpha \in \mathbb{N}$.

The conditions (i) and (ii) are clearly equivalent, and the fact that (iii) ⇒ (ii) is clear. Now if γ vanishes on $\mathcal{F}_{I,\tilde{I}}$, then by linear algebra, γ is a linear combination of elements in I, and this implies (iii). If γ takes the value 2π on $\mathcal{F}_{I,\tilde{I}}$, we consider the largest root β in the irreducible component of Δ to which γ belongs. Then $\beta \in \tilde{I}$ and $\beta - \gamma$ is a linear combination of the α's in I ; so this implies (iii).

If the context is clear enough we just speak of singular roots, droping the prefix. Other roots are called non-singular.

We denote by μ^R the number of singular positive roots.

(2.5) <u>Lemma</u> : <u>Let</u> γ <u>be a positive non-singular root</u> ; <u>then</u>

$$\frac{1}{R} < \gamma(H) < 2\pi - \frac{1}{R} .$$

In fact write $\gamma = \sum_{\alpha\in I} n_\alpha \alpha + \sum_{\alpha'\in B\setminus I} n_{\alpha'} \alpha'$, with $n_\alpha, n_{\alpha'}, \in \mathbb{N}$.

Clearly at least one of the $n_{\alpha'}$ is non zero, say n_{α_o} , with $\alpha_o \in B\setminus I$. Then $\gamma(H) > n_{\alpha_o} \alpha_o(H) > 1/R$.

So far we used only the fact that γ is not singular of type I ; the other inequality is obtained similarly by using the fact that γ is not singular of type II.

For α singular of type I (resp. type II) let s_α (resp. $\tilde{s}_\alpha$) be the orthogonal symmetry with respect to the hyperplane $\alpha = 0$ (resp. $\alpha = 2\pi$) . Let $W_{I,\tilde{I}}$ be the group generated by $\{s_\alpha\}_{\alpha\in I} \cup \{\tilde{s}_\beta\}_{\beta\in\tilde{I}}$.

It is a finite subgroup of the affine Weyl group (it is exactly the subgroup which fixes $\mathcal{F}_{I,\tilde{I}}$ pointwise, but this fact will not be needed).

Now define

$$\Gamma_H^{(R)} = \text{convex hull of } \{wH\}_{w\in W_{I,\tilde{I}}} . \tag{2.6}$$

(2.7) <u>Proposition</u> : <u>Let</u> γ <u>be a non-singular positive root</u> ; <u>then</u>

$$\frac{1}{R} < \gamma(K) < 2\pi - \frac{1}{R} , \text{ for } K \in \Gamma_H^{(R)} .$$

<u>Proof</u> : Clearly it suffices to show $\frac{1}{R} < \gamma(wH) < 2\pi - \frac{1}{R}$, for $w \in W_{I,\tilde{I}}$.

For $\gamma \in \Delta$, and $p \in \mathbb{Z}$, let $f_{\gamma,p}$ be the affine function defined on t by

$$f_{\gamma,p}(L) = \gamma(L) + 2p\pi .$$

These affine functions are permuted among themselves under the action of the affine Weyl group, and in particular under the action of $W_{I,\tilde{I}}$.

Such a function is said to be $(I,\tilde{I})$-convenient if

$$f_{\gamma,p}(F_{I,\tilde{I}}) \subset]0\ 2\pi[.$$

(2.8) Lemma : $f_{\gamma,p}$ is $(I,\tilde{I})$ convenient if and only if

either $p = 0$ and γ is a positive non-singular root

or $p = 1$ and $-\gamma$ is a positive non-singular root.

The "if" part being an immediate consequence of the definitions, we prove the other part. Now as $2\pi - f_{\gamma,p} = f_{-\gamma,1-p}$, we may always assume that γ is a positive root ; but as $\gamma(F_{I,\tilde{I}}) \subset [0\ 2\pi]$ for any $\gamma \in \Delta^+$ this forces $p = 0$; but now clearly γ has to be non-singular.

Now the proof of (2.7) follows easily : in fact let $w \in W_{I,\tilde{I}}$, and let γ be a non-singular positive root ; then $\gamma \circ w$ is one of the affine functions considered, and clearly, because $W_{I,\tilde{I}}$ fixes $F_{I,\tilde{I}}$, $\gamma \circ w$ is $(I,\tilde{I})$-convenient ; so thanks to lemma (2.8), we see that $\gamma \circ w$ is either a positive non-singular root, or 2π minus a positive non-singular root. But now we use lemma (2.5) to prove the desired inequalities.

Let D be the "Weyl denominator", i. e.

$$D(.) = \prod_{\alpha \in R^+} \sin \frac{\alpha(.)}{2} ,$$

and set for R as before and $H \in A$

$$D^R(.) = D_{I,\tilde{I}}(.) = \prod_{\alpha} \sin \frac{\alpha(.)}{2} ,$$ where α runs though all positive non-singular roots.

(2.9) Lemma : There exists a constant $c > 0$, such that

$$D^R(K) \geq c\ D^R(H) \text{ , for every } K \text{ in } \Gamma_H^{(R)} .$$

Set, for a while $f_\alpha(.) = \sin \frac{\alpha(.)}{2}$, and denote by ν the number of positive non singular roots.

As $t \mapsto \sin \frac{t}{2}$ is a positive concave function on $[0\ 2\pi]$, f_α is positive concave function on $\Gamma_H^{(R)}$. So, if $K \in \Gamma_H^{(R)}$, say

$$K = \sum_{w \in W_{I,\tilde{I}}} \lambda_w (w.H) \text{ , with } \lambda_w \geq 0 \text{ and } \sum_{w \in W_{I,\tilde{I}}} \lambda_w = 1 ,$$

then for each non singular positive root ,

$$f_\alpha(K) \geq \sum_{w \in W_{I,\tilde{I}}} \lambda_w f_\alpha(w.H) .$$

Consequently

$$\prod_\alpha f_\alpha(K) \geq \prod_\alpha(\Sigma \lambda_w f_\alpha(w.H)) \geq \sum_{w \in W_{I,\tilde{I}}} \lambda_w^\nu(\prod_\alpha f_\alpha(w.H)) ,$$

where α runs over all positive non singular roots. But D^R is clearly invariant by $W_{I,\tilde{I}}$; so we proved

$$D^R(K) \geq (\sum_{w \in W_{I,\tilde{I}}} \lambda_w^\nu) D^R(H)$$

and we conclude by using the classical inequality

$$\left(\frac{1}{|W_{I,\tilde{I}}|} \sum_{w \in W_{I,\tilde{I}}} \lambda_w^\nu\right)^{1/\nu} \geq \frac{1}{|W_{I,\tilde{I}}|} \sum_{w \in W_{I,\tilde{I}}} \lambda_w = \frac{1}{|W_{I,\tilde{I}}|} .$$

(2.10) Lemma : Suppose $I \neq B$; then there exists a constant $c > 0$, such that

$$|K| \geq c\ |H| \qquad \forall K \in \Gamma_H^{(R)} .$$

We may (and do) assume that $\tilde{I} = \emptyset$; otherwise $\Gamma_H^{(R)}$ stays far away from the origin. Now

$$|K|^2 = 2 \sum_{\alpha \in \Delta^+} \alpha(K)^2 \geq 2 \sum_{\substack{\alpha \text{ non-singular} \\ \alpha \in \Delta^+}} \alpha(K)^2 \geq c(\sum_{\substack{\alpha \text{ non-singular} \\ \alpha \in \Delta^+}} \alpha(K))^2 .$$

But $\sum_{\substack{\alpha \in \Delta^+ \\ \alpha \text{ non-singular}}} \alpha$ is invariant by W_I , and so

$$|K|^2 \geq c(\sum_{\alpha \text{ non-singular}} \alpha(H))^2 .$$

On the other hand $|H|^2 \leq c(\sum_{\alpha \text{ non-singular}} \alpha(H)^2)$,

because any singular root is dominated by N times a non-singular root, where N is the maximal height of singular roots. And so the lemma follows.

The domains $\Gamma_H^{(R)}$ have the following heriditarity property : if $K \in \Gamma_H^{(R)} \cap A$, then $\Gamma_K^{(R)} \subset \Gamma_H^{(R)}$; in fact, as a consequence of (2.7), if $\alpha \notin I_H$, then certainly $\alpha \notin I_K$, and similarly for $\tilde{I}_H$ and $\tilde{I}_K$; so $I_K \subset I_H$ and $\tilde{I}_K \subset \tilde{I}_H$; so $W_{I_K,\tilde{I}_K} \subset W_{I_H,\tilde{I}_H}$, which implies the statement.

Related to the definition of the sets $\Gamma_H^{(R)}$ is the following partition of A . For $\tilde{I}$ any subset of $\tilde{B}$ and I any subset of B such that $I \cup \tilde{I}$ is linearly independant, define

$$(2.11)\quad A^R_{I,\tilde{I}} = \{H \in A\ ,\ \alpha(H) \leq \frac{1}{R}\ \text{ for }\ \alpha \in I\ ,\ \alpha(H) > \frac{1}{R}\ \text{ for }\ \alpha \in B \setminus I$$

$$\beta(H) \geq 2\pi - \frac{1}{R}\ \text{ for }\ \beta \in \tilde{I}\ ,\ \beta(H) < 2\pi - \frac{1}{R}\ \text{ for }\ \beta \in \tilde{B} \setminus \tilde{I}\}\ .$$

For R large enough (same condition as before), these sets are disjoint and form a partition of A . Moreover, if $H \in A^{(R)}_{I,\tilde{I}}$, then I and $\tilde{I}$ are precisely what they should be following definition (2.1), so that notation is coherent.

All these notions can also be defined on the torus T itself ; in fact, if $h \in T$, then $h = \exp H$ for <u>one</u> element H in $\bar{A}$. So we define R-singular roots at h and similar notions by reference to H . In particular set $\Gamma_h^{(R)} = \exp \Gamma_H^{(R)}$, and define d^R by

$$d^R(\exp(.)) = D^R(.)\ .$$

Notice that d (corresponding to D) is (up to a constant) the Weyl denominator of the Weyl formula for characters of the compact Lie group G .

If α is a root (more generally an element in the dual lattice of Γ) , we denote by e_α the corresponding character of T .

3.- Reduction to differential operators on t

Let $U(g)$ be the complex universal envelopping algebra of g , which can be viewed as the algebra of all (say) left-invariant differential operators on G ; if m is a positive integer, we denote by $U_m(g)$ the operators with degree less than or equal to m . Fixing a vector basis of $g^{\mathbb{C}}$, say $Y_1, \ldots, Y_n$, we denote by Y^J , where J is a n-uple of positive integers, the element $Y_1^{j_1} \ldots Y^{j_n}$. As J varies over all possible n-uples, the $\{Y^J\}$ form a basis of $U(g)$ by Poincaré-Birkhoff-Witt theorem. Moreover if $|J| = j_1 + \ldots + j_n$, then the $\{Y^J\}_{|J| \leq m}$ are a basis of $U_m(g)$.

We use similar notations for the universal envelopping algebra of t , and we fix a basis $H_1, \ldots, H_\ell$ of $t_{\mathbb{C}}$, and use the notation

H^I , where I is a ℓ-uple of positive integers, for $H_1^{i_1} \dots H_\ell^{i_\ell}$. We also identify U(t) with a subalgebra of U(g) by extending the injection from t into g .

By J^∞ we denote the vector space of central C^∞ functions on G ; if $f \in J^\infty$, then f is determined by its restriction to T , or even to $\exp(\bar{A})$.

(3.1) Proposition : Let $m \in \mathbb{N}$; there exists a constant C such that for each I , with $|I| \leq m$

$$|Y^I f(x)| \leq C \Big| \sum_{|J| \leq m} (Y^J f)(h) \Big|$$

where h is the unique point in $\exp \bar{A}$ conjugate to x , x arbitrary in G and f in J^∞ .

This proposition is an easy consequence of the compactness of G , and is well-known.

The next result is a powerful tool for finding estimates of the derivatives of a function f in J^∞ .

(3.2) Theorem : Let $p \in \mathbb{N}$. There exists a constant C such that, for each J , with $|J| \leq p$,

$$|Y^J f(h)| \leq C \Big(\sum_{j=0}^{p} R^j \sum_{|I|=p-j} \sup_{k \in \Gamma_h^{(R)}} |H^I f(k)| \Big)$$

(R is a large enough number, $h \in \exp(A)$, and $f \in J^\infty$) .

In order to prove this theorem, we will prove by induction on q the more complicated result :

(3.3) Theorem : Let $p,q \in \mathbb{N}$. There exists a constant C such that for each I with $|I| \leq p$, and J , $|J| \leq q$

$$|(H^I Y^J f)(h)| \leq C \Big(\sum_{j=0}^{p+q} R^j \sum_{|K|=p+q-j} \sup_{k \in \Gamma_h^{(R)}} |H^K f(k)| \Big)$$

(R is a large enough number, $h \in \exp(A)$, and $f \in J^\infty$) .

Clearly theorem (3.3) with p = 0 is nothing else but (3.2). Now we prove theorem (3.3) by using induction on q . Notice that if q = 0 , the statement is obvious.

We need here a digression into the notion of radial part of a differential operator.

Let φ be any smooth function on G, and set $\phi(x,g) = \varphi(x g x^{-1})$, for $x, g \in G$; ϕ belongs to $C^\infty(G \times G)$.

If $X_1, \ldots, X_r$ are elements of g, and $D \in U(g)$, we have the following formula (see [6], p. 105)

(3.4) $$\phi(x, X_1 \ldots X_r ; g, D) = (W^x \varphi)(x g x^{-1}) ,$$

where $$W = \left[\left(L_{Ad(g^{-1})X_1} - R_{X_1}\right) \ldots \left(L_{Ad(g^{-1})X_r} - R_{X_r}\right)\right](D) .$$

Here R_X and L_X for $X \in g$ denote the right and left multiplication in $U(g)$ by X. And E^x, where $x \in G$ and $E \in U(g)$ is the differential operator defined by

$$(E^x \varphi)(g) = (E(\varphi(x^{-1}.x))(x g x^{-1})) .$$

Let Γ_g be the linear map from $U(g) \otimes U(g)$ into $U(g)$ given by

$$\Gamma_g(X_1 \ldots X_r \otimes D) = \left[\left(L_{Ad(g^{-1})X_1} - R_{X_1}\right) \ldots \left(L_{Ad(g^{-1})X_r} - R_{X_r}\right)\right](D) .$$

Let $U'(g)$ be the set of all differential operators D in $U(g)$ such that $D1 = 0$.

Now if φ is in J^∞, and if $D \in \Gamma_g(U'(g) \times U(g))$, then

$D\varphi(g) = 0$ as a consequence of (3.4).

(3.5) <u>Lemma</u> : <u>Let</u> $m \in \mathbb{N}$, $D \in U_m(g)$; <u>then there exists</u> $D_0 \in U_m(t)$ <u>and</u> $(E_\alpha)_{\alpha \in \Delta}$, $E_\alpha \in U_{m-1}(g)$ <u>for each</u> $\alpha \in \Delta$, <u>such that for every</u> C^∞ <u>central function</u> f <u>on</u> G,

$$Df(h) = (D_0 f)(h) + \sum_{\alpha \in \Delta} (e_\alpha(h^{-1}) - 1)^{-1} (E_\alpha f)(h) \quad (h \in T') .$$

Using the fact that $g^{\mathbb{C}} = t^{\mathbb{C}} \oplus \sum_{\alpha \in \Delta} f^{\mathbb{C}}_\alpha$, where $g^{\mathbb{C}}_\alpha$ is the root space for α, we may write

$$D = D_0 + \sum_{\alpha \in \Delta} Y_\alpha D_\alpha , \text{ with } D_0 \in U_m(t^{\mathbb{C}}) , Y_\alpha \in g^{\mathbb{C}}_\alpha \text{ and } D_\alpha \in U_{m-1}(g^{\mathbb{C}}) .$$

Now for $h \in T'$,

$$\Gamma_h(Y_\alpha \otimes D_\alpha) = e_\alpha(h^{-1}) Y_\alpha D_\alpha - D_\alpha Y_\alpha = (e_\alpha(h^{-1}) - 1) Y_\alpha D_\alpha + [Y_\alpha D_\alpha]$$

so $Y_\alpha D_\alpha = (e_\alpha(h^{-1})-1)^{-1}\ \Gamma_h(Y_\alpha \otimes D_\alpha) - (e_\alpha(h^{-1})-1)^{-1}\ [Y_\alpha D_\alpha]$

which, together with preceding remark, proves the lemma, as $[Y_\alpha D_\alpha] \in U_{m-1}(g)$.

Now apply (3.5) to Y^J , which we rewrite as

$$(3.6) \qquad (Y^J f)(h) = (D_0 f)(h) + \sum_{\alpha\in\Delta^+} g_\alpha(h)(e_\alpha(h^{-1})-1)^{-1} ,$$

with $g_\alpha(h) = E_\alpha f(h) - e_\alpha(h^{-1})\ E_{-\alpha} f(h)$, where we used the fact that

$$(e_{-\alpha}(h^{-1})-1)^{-1} = -e_\alpha(h^{-1})(e_\alpha(h^{-1})-1)^{-1} .$$

Now $H^I D_0 \in U_{p+q}(t)$, so that, with an appropriate constant

$$(3.7) \qquad |H^I D_0 f(h)| \le C\Big(\sum_{|K|\le p+q} |H^K f(h)|\Big) .$$

Further, let $H \in A$ such that $h = \exp H$.

If $\alpha \in \Delta^+$ is non-singular (in the sense of 2.4), then it is easily shown by induction on $|K|$ that

$$|H^K(e_\alpha(.^{-1})-1)^{-1}(h)| \le C\ R^{|K|+1} ,$$

where C is a constant independant of R .

Using Leibnitz rule, it is then clear that

$$|H^I((e_\alpha(.^{-1})-1)^{-1} g_\alpha)(h)| \le C\Big(\sum_{j=o}^{p} R^{p-j+1}\Big(\sum_{|K|=j} |H^K g_\alpha(h)|\Big)\Big) .$$

Now use the induction assumption for $H^K E_\alpha$ and $H^K E_{-\alpha}$, to obtain

$$H^K g(h) \le C\Big(\sum_{j=o}^{|K|+q-1} R^j \sum_{|L|=|K|+q-1-j} \sup_{k\in\Gamma_h(R)} |H^L f(k)|\Big) .$$

So finally, for α non-singular

$$(3.8) \quad |H^I((e_\alpha(.^{-1})-1)^{-1} g_\alpha)(h)| \le C\Big(\sum_{j=o}^{p+q} R^j \sum_{|K|=p+q-j} \sup_{k\in\Gamma_h(R)} |H^K f(k)|\Big) .$$

It remains only to estimate terms like

$|H^I((e_\alpha(.^{-1})-1)^{-1} g_\alpha)(h)|$, where α is singular.

Rewrite (3.6) as

$$(e_\alpha(h^{-1})-1)^{-1}\ g_\alpha(h) = (H^J f)(h) - \sum_{\substack{\beta\in\Delta^+ \\ \beta\neq\alpha}} g_\beta(h)(e_\beta(h^{-1})-1)^{-1} ,$$

and let h_0 be a point where $e_{\pm\alpha}(h_0) = 1$, but where $e_\beta(h_0) \neq 1$ for all other roots. Letting h tend to h_0 , the right hand-side has a limit ; so must the left hand-side ; hence $g_\alpha(h_0) = 0$; by continuity, this is true on all the "wall" (subgroup) where $e_\alpha(h) = 1$.

For convenience call G_α the function on t defined by

$$G_\alpha(H) = g_\alpha(\exp H) .$$

G_α vanishes on the hyperplane $\alpha = 0$ (resp. $\alpha = 2\pi$) if α is singular of type I (resp. of type II) . Let assume α is singular of type I , the type II case being handled in a similar way. Let p_α be the projection on the hyperplane $\alpha = 0$, and let H be the unique vector orthogonal to $\{\alpha = 0\}$ and such that $\alpha(H_\alpha) = 1$. Then

$$\begin{aligned} G_\alpha(H) = G_\alpha(H) - G_\alpha(p_\alpha(H)) &= \int_0^1 \frac{d}{dt}(G_\alpha(p_\alpha(H) + t(H-p_\alpha(H))dt \\ &= \alpha(H) \int_0^1 (H_\alpha G_\alpha)(p_\alpha(H) + t\alpha(H)H_\alpha)dt . \end{aligned}$$

By differentiating under the integral sign, holds

$$|H^I(G_\alpha \alpha^{-1})(H)| \le C \sum_{|K|\le p+1} \sup_{0\le t\le 1} |H^K G_\alpha(p_\alpha(H) + t\alpha(H)H_\alpha)| .$$

Now $p_\alpha(H)$ is the middle of H and $s_\alpha(H)$, so belongs to $\Gamma_H^{(R)}$. So, going back to the torus, and using the fact that $(e_\alpha(h)-1)$ equals $\alpha(H)$ times an analytic function, one obtains that $(e_\alpha(h^{-1})-1)^{-1}g_\alpha$ is a C^∞ function and

$$|H^I((e_\alpha(.^{-1})-1)^{-1}g_\alpha)(h)| \le C \sum_{|K|\le p+1} \sup_{k\in\Gamma_h^{(R)}} |H^K g_\alpha(k)| .$$

Now use the induction hypothesis, and the fact that $k \in \Gamma_h^{(R)}$ implies $\Gamma_k^{(R)} \subset \Gamma_h^{(R)}$, to obtain

$$(3.9)\quad |H^I((e_\alpha(.)^{-1}-1)^{-1}g_\alpha)(h)| \le C\Big(\sum_{j=0}^{p+q} R^j \sum_{|K|=p+q-j} \sup_{k\in\Gamma_h^{(R)}} |H^K f(k)|\Big) .$$

By adding together inequalities (3.7), (3.8) and (3.9) we eventually end up with the proof of (3.3).

4.- Invariant versus skew-invariant functions on T

Assume that g is a C^∞ function on T which is skew-invariant by the Weyl group, id est

$$g(w.h) = \varepsilon(w)\, g(h) \quad , \text{ for every } \quad w \in W$$

where ε denotes the signature of the element in the Weyl group.

Then it is known (see for instance [2] that $f(h) = d(h)^{-1} g(h)$ (a priori defined only on T') extends as a C^∞ function on T and of course is invariant by W.

We wish to estimate f and its derivatives at some point h in terms of derivatives of g in a neighbourhood of h (estimates involving only the derivatives of g at h do not hold, as it is easily seen, e. g. near singular points).

(4.1) Theorem : Let p be a positive integer. There exists a constant C, such that for each I, with $|I| \le p$,

$$|H^I f(h)| \le C\, |d^R(h)^{-1}| \sum_{j=o}^{p} R^j \sum_{|J| \le \mu^R + p - j} \sup_{k \in \Gamma_h^{(R)}} |H^I g(k)| \ ,$$

for each skew-invariant $g \in C^\infty(T)$, and $f = d^{-1} g$.

Conventions made at the end of section 2 are in force ; so write $h = \exp H$, with $H \in A$, and let F, G be defined by $F(.) = f(\exp(.))$ and $G(.) = g(\exp(.))$. Then

$$(4.2) \qquad F(.) = D^R(.)^{-1} \left(\prod_\alpha \sin \frac{\alpha(.)}{2} \right)^{-1} G(.) \ ,$$

where the product is over all positive singular roots.

Now G vanishes on the walls $\{\alpha = 0\}$ (resp. $\{\alpha = 2\pi\}$ if α is singular of type I (resp. of type II), because G satisfies $G o s_\alpha = -G$ (resp. $G o \tilde{s}_\alpha = -G$). As a consequence, using several times the technique we used earlier (see proof of (3.9) and also [4]), we set

$$\tilde{G}(.) = \left(\prod_{\text{sing}} \sin \frac{\alpha(.)}{2} \right)^{-1} G(.) \ , \text{ and deduce}$$

the following inequalities (recall that μ^R is the number of positive singular roots) :

$$(4.3) \qquad |H^I\tilde{G}(H)| \leq C\Big(\sum_{|J|\leq p+\mu^R} \sup_{K\in\Gamma_H^{(R)}} |H^J G(K)|\Big) ,$$

with a constant C independant of g and of R .

As we now have $F(.) = (D^R(.))^{-1}\,\tilde{G}(.)$, the theorem (4.1) follows, using the Leibnitz formula, from the following lemma.

(4.4) Lemma : Let q be a positive integer ; there exists a constant C , such that for each I , with $|I| \leq q$,

$$|(H^I(D^R)^{-1})(H)| \leq C\, R^q\, |(D^R)^{-1}(H)| ,$$

with a constant C independant of R .

By induction, it is easily seen that $H^I(D^R)^{-1}$ can be written as $(D^R)^{-1}$ times a sum of terms, each of which has the following aspect :

$$\left(\sin\frac{\alpha_1(H)}{2}\ldots\sin\frac{\alpha_q(H)}{2}\right)^{-1} P(H) ,$$

where P is some polynomial in $\sin\frac{\alpha_i}{2}$ and $\cos\frac{\alpha_i}{2}$, with α_i ranging over the non-singular roots. As $\frac{1}{R} < \alpha(H) < 2\pi - \frac{1}{R}$ for α non-singular (see (2.5)), conclusion follows.

5.- Estimates for the Bochner-Riesz kernels

Let Λ be the set of dominant integrals weights (viewed as a subset of the dual lattice of Γ) ; Λ parametrizes (up to equivalence) the irreducible unitary representations of G . For $\lambda \in \Lambda$, denote by χ_λ (resp. d_λ) the character (resp. dimension) of the corresponding representation.

In [3] were introduced the Bochner-Riesz kernels : if $\delta \geq 0$, and $R > 0$

$$(5.1) \qquad s_R^\delta = \sum_{\lambda\in\Lambda}\left(1 - \frac{<\lambda+\rho,\lambda+\rho>}{R^2}\right)_+^\delta d_\lambda\, \chi_\lambda .$$

Using the Weyl formula for characters and the Poisson summation formula on T , an expression was obtained for these kernels on T .

$$(5.2)\qquad s_R^\delta(\exp H) = C.R^n\, D(H)^{-1} \sum_{\xi\in\Gamma} \Big(\prod_{\alpha\in\Delta^+} \alpha(H+\xi)\Big) \mathcal{J}_{n/2+\delta}\,(R\,|H+\xi|)\ ,$$

where $\mathcal{J}_\nu(t) = t^{-\nu}\, J_\nu(t)$ and J_ν is the ordinary Bessel function.

Using the machinery developped in the previous chapters, we are able to prove estimates for the derivatives of s_R^δ .

Let $\delta(.)$ denotes the geodesic distance to the origin in G , so that for H in $\bar{A}$ we have $\delta(\exp H) = |H|$, and for R a (large enough) positive number define $\Delta^{(R)}$ as being the invariant function on G such that

$$\Delta^{(R)}(\exp H) = \prod_{\alpha\in\Delta^+} \sup\Big(\frac{1}{R}\,, \sin\frac{\alpha(H)}{2}\Big) \quad \text{for} \quad H\in\bar{A}\ .$$

<u>Remark</u> : Using previous notation observe that

$$\Delta^{(R)}(\exp H) \approx R^{-\mu^R}\, d^R(\exp H) = R^{-\mu^R}\, D^R(H)\ .$$

(5.3) <u>Theorem</u> : <u>Let</u> p <u>be a positive integer</u> ; <u>there exists a constant</u> C , <u>such that, for each</u> J , $|J| \le p$,

$$|Y^J\, s_R^\delta(x)| \le C.R^p \begin{cases} R^n & \underline{\text{if}} \quad \delta(x) < \frac{1}{R} \\ R^{(n/2-1/2)-\delta}\,(\delta(x)^{-n/2-1/2-\delta} + \Delta^{(R)}(x)^{-1}) & \underline{\text{if not}}. \end{cases}$$

We give an other (more explicit) version, which thanks to the remark preceding the theorem is clearly equivalent (see also (3.1)).

(5.4) <u>Theorem</u> : <u>Let</u> p <u>be a positive integer</u> ; <u>there exists a constant</u> C , <u>such that, for each</u> J , $|J| \le p$,

$$|Y^J s_R^\delta(\exp H)| \le C.R^p \begin{cases} R^n & \underline{\text{if}} \quad |H| \le \frac{1}{R} \\ R^{n/2-1/2-\delta}\,(|H|^{-n/2-\delta-1/2} + R^{\mu^R}\, D^R(H)^{-1}) & \underline{\text{if not}}. \end{cases}$$

We now embark on a long list of inequalities.

Recall first that, as a consequence of a classical property for Bessel functions :

$$(5.5)\qquad \Big(\frac{1}{r}\frac{d}{dr}\Big)^k \mathcal{J}_\nu = \mathcal{J}_{\nu+k}\ ,\quad k\in\mathbb{N}\ .$$

Let, for a while, denote by j_ξ the function defined by

$$j_\xi(.) = \mathcal{J}_{n/2+\delta}(R\,|.+\xi|)\ .$$

Now

$$(5.6)\qquad H^I j_\xi(.) = (H^I j_0)(.+\xi) = \sum_{k=o}^{|I|} p_k(.+\xi)\left(\left(\frac{1}{r}\frac{d}{dr}\right)^k j_0\right)(.+\xi)\ ,$$

where p_k is a homogeneous polynomial of degree k.

Hence combining (5.5) and (5.6)

$$(5.7)\qquad (H^I j_\xi)(.) = \sum_{k=o}^{|I|} R^{2k} p_k(.+\xi)\,\mathcal{J}_{n/2+\delta+k}(R\,|.+\xi|)\ .$$

Recall now the inequality

$$(5.8)\qquad |\mathcal{J}_\nu(t)| \le C\ \inf(1,t^{-\nu-1/2})\ .$$

Now we distinguish, between $\xi = 0$ and $\xi \ne 0$.

If $\xi = 0$, then

$$(5.9)\qquad \text{if}\ \ |H| \le 1/R\ ,\ |H^I j_0(.)| \le C.R^{|I|}\ ,$$

$$(5.10)\qquad \text{if}\ \ |H| \ge 1/R\ ,\ |H^I j_0(.)| \le C.R^{|I|}(R\,|H|)^{-n/2-\delta-1/2}\ .$$

If $\xi \ne 0$, then $|\xi+H| \approx |\xi|$ so

$$(5.11)\qquad \text{if}\ \ \xi \ne 0\ ,\ |H^I j_\xi(.)| \le C.R^{|I|}\ |\xi|^{-n/2-\delta-1/2}\ .$$

Let now
$$g(h) = \sum_{\xi\in\Gamma}\left(\prod_{\alpha\in\Delta^+}\alpha(H+\xi)\right)\mathcal{J}_{n/2+\delta}(R(H+\xi))$$
$$= \sum_{\xi\in\Gamma}\left(\prod_{\alpha\in\Delta^+}\alpha(H+\xi)\right) j_\xi(H) = \sum_{\xi\in\Gamma} g_\xi(H)$$
(as usual $h = \exp H$).

Fix a number $a > 0$, such that $|H| \le a$ implies $\alpha(H) \le \pi$ for all $\alpha \in \Delta^+$. In such a region $D(H)$ and $\prod_{\alpha\in\Delta^+}\alpha(H)$ are equivalent (their quotient is analytic) ; we take advantage of this to estimate $D^{-1}g$ directly

$$(5.12)\qquad \text{if}\ \ |H| \le \frac{1}{R}\ ,\ |H^I(D^{-1}g_0)(h)| \le C.R^{|I|}$$

$$(5.13)\qquad \text{if}\ \ \frac{1}{R} \le |H| < a\ ,\ |H^I(D^{-1}g_0)(h)| \le C.R^{|I|}(R\,|H|)^{-n/2-\delta-1/2}\ .$$

Now observe that $\sum_{\xi \neq o} g_\xi$ is still invariant by W, and that the condition $|H| \leq a$ implies that there is no singular root of type II at H, so $W_{I,\tilde{I}} \subset W$ $(\tilde{I} = \emptyset)$. The proof of (4.3) requires only invariance by $W_{I,\tilde{I}}$ so the corresponding estimate is valid for $\sum_{\xi \neq o} g_\xi$.

(5.14) If $|H| \leq a$

$$|H^I(d^{-1} \sum_{\xi \neq o} g_\xi)(h)| \leq C.R^{|I|} d^{(R)}(h)^{-1} R^{-n/2-\delta-1/2} R^{\mu^R} \left(\sum_{\xi \neq o} |\xi|^{m - \frac{n}{2} - \delta - \frac{1}{2}}\right).$$

In proving (5.14) one uses (4.3), Leibnitz rule and (5.11). Of course the sum over the lattice is convergent.

Now if $|H| \geq a$, we use (4.3) for the full sum and so

(5.15) if $|H| \geq a$

$$|H^I(d^{-1} \Sigma g_\xi)(h)| \leq C.R^{|I|} d^{(R)}(h)^{-1} R^{-n/2-\delta-1/2} R^{\mu^R} \left(\sum_\xi (1+|\xi|)^{m - \frac{n}{2} - \delta - \frac{1}{2}}\right).$$

To end up with the proof of (5.3), we make the crucial remark that, thanks to lemmas (2.9) and (2.10), the inequalities (5.12), (5.13), (5.14) and (5.15) are still valid, keeping right-hand side unchanged, but replacing h by any element k in $\Gamma_h^{(R)}$. This remark, together with (3.2) imply eventually (5.3).

6.- Weak-type estimates for the maximal Bochner-Riesz means on H^p-classes

The estimates for the Bochner-Riesz kernels allow us to study convergence problems for distributions in the H^p-class $(0 < p < 1)$. Here we follow [5].

We first need a short description of atomic $H^p(G)$. The description we are going to give is fitted for our purposes, but lacks of intrinsicality and should of course be related to other possible definitions. But we won't pursue this discussion here.

Let $0 < p \leq 1$; the H^p space is generated by two kinds of p-atoms, regular and exceptional. Exceptional atoms are just smooth functions bounded by 1. For regular atoms, we first consider a faithful representation of G, say $\pi : G \to \mathbb{U}(L, \mathbb{C})$ and we look at G as a submanifold in the real vector space E underlying $\mathrm{End}(\mathbb{C}^L)$. Now a regular atom is a function a, which is supported in some ball

$B(G_0,\rho)$, satisfies the size condition

$$|a(g)| \leq C\,\rho^{-n/p} \quad \text{for} \quad g \in B(g_0,\rho)$$

and the following cancellation property

$$\int_G a(g)\, P(\pi(g))dg = 0 ,$$

where P is any polynomial on E , of degree less than $[n(1/p-1)]$.

Two observations will be needed later. First the translate of an atom is an atom ; it is clear for an exceptional atom, and the support and size conditions are also trivial. We only have to check that the cancellation property is invariant ; but if $g_0 \in G$, then

$$\int_G a(gg_0^{-1})\, P(\pi(g))dg = \int_G a(\gamma)\, P(\pi(\gamma g_0))dg = \int_G a(\gamma)\, P(\pi(\gamma)\pi(g_0))dg .$$

But as right multiplication by $\pi(g_0)$ is a linear map on $E, P(.\pi(g_0))$ is a polynomial on E , of the same degree. The cancellation property follows immediatly.

The second observation is that for a smooth function f on G , it is possible to find Taylor approximations using polynomials on E . More specifically, let g_0 be a given point on G , and n_0 a positive integer. Then there exists a polynomial P on E (not unique in general however) such that, for g in some neighbourhood O_0 of g_0 ,

$$|f(g) - P(\pi(g))| \leq C_0 d(g,g_0)^{n_0+1} \sup |Y^J f(\gamma)| ,$$

where the sup is taken over all n-uples J , with $|J| \leq n_0+1$ and all γ in $\overline{O}_0$. In fact, as $\pi(G)$ is a submanifold, we can find a ball O_0 centered at g_0 , a neighbourhood Ω_0 of $\pi(g_0)$ in E , and a positive number T_0 , such that

$$\Omega_0 \approx O_0 \times\,]-T_0,T_0[^d , \text{ where } d = \operatorname{codim}_E \pi(G) .$$

Now we extend f trivially on the fibers of the projection map $\Omega \to O_0$, and use the classical Taylor expansion in E . Notice that, because of the compactness of G , it is possible to choose the size of the ball O_0 , T_0 and C_0 in a uniform way as g_0 varies in G .

The space $H^p(G)$, $0 < p \leq 1$, is the space of distributions having the form $f = \sum_{k=1}^{\infty} c_k a_k$, where each a_k is an atom (regular or exceptional), and the coefficients satisfy $\Sigma |c_k|^p < \infty$. The "norm"

$\|f\|_{H^p}$ is the infimum of all expressions $(\Sigma\ |c_k|^p)^{1/p}$ for which we have such a representation of f .

The Bochner-Riesz means of a function (or distribution) f are defined as

$$S_R^\delta f = s_R^\delta * f \ ,$$

and we further define the associated maximal operator as

$$S_*^\delta f(g) = \sup_R |S_R^\delta f(g)| \ .$$

Our main theorem is (compare [5])

(6.1) Theorem : Let $0 < p < 1$ and $\delta = n/p - (n+1)/2$. Then there exists a constant $C_p > 0$ such that

$$\operatorname{mes}\left(\left\{g \in G \ |S_*^\delta f(g)| > \lambda\right\}\right) \le C_p\left(\frac{\|f\|_{H^p}}{\lambda}\right)^p ,$$

for all $\lambda > 0$ and $f \in H^p(G)$.

As usual it implies the almost every where convergence of $S_R^\delta f(x)$ as R tend to infinity.

Arguing as in [5], it is sufficient to prove the following result.

(6.2) Proposition : Let $0 < p < 1$ and $\delta = n/p - (n+1)/2$. Then there exists a constant $C_p' > 0$, such that

$$\operatorname{mes}(\{g \in G \ S_*^\delta a(g) > \lambda\}) \le C_p' \lambda^{-p} ,$$

for any atom a .

If a is an exceptional atom, we just use the fact that $\|s_R^\delta\|_1 \le C < \infty$, so $S_R^\delta a$ is bounded by C and an estimate like (6.2) is then trivial.

We need two lemmas before beginning the proof of (6.2) for regular atom.

(6.3) Lemma : Let $\mathcal{H} : G \to \bar{A}$ be the map which sends an element $g \in G$ to the unique H in $\bar{A}$ such that g is conjugate to $\exp H$.

Then

$$|\mathcal{H}(g') - \mathcal{H}(g)| \leq d(g,g') .$$

Proof : The statement is clearly equivalent to the following : let $H, H' \in \bar{A}$; then

$$|H-H'| \leq \inf_{\gamma,\gamma' \in G} d(\gamma \exp H \gamma^{-1}, \gamma' \exp H' \gamma'^{-1}) .$$

We shall prove in fact the equality ; by continuity, it is enough to prove this inequality when H and H' are regular (so $H, H' \in A$) . Suppose this ; now the inf in the right hand-side is reached for some $\gamma_0 \in G$ and $\gamma_0' \in G'$; of course by using conjugacy by γ_0^{-1} , we may assume that $\gamma_0 = e$; now there is a geodesic with endpoints H and $\gamma_0' \exp H' \gamma_0'^{-1}$, which length is exactly $d(\exp H, \gamma_0' \exp H' \gamma_0'^{-1})$; but this geodesic must be orthogonal to the tangent plane of the orbit at the point $\exp H$. As H is regular this implies that the tangent vector to the geodesic at $\exp H$ is in t , and so the geodesic lies entirely in T . So its endpoint is a point in T conjugate to $\exp H'$. So we only need to show that for $H, H' \in A$

$$|H-H'| = \inf_{w \in W_a} |H-wH'|$$

(W_a the affine Weyl group).

But it is known (and in fact easy to show) that this is true.

(6.4) Lemma : Let $N = \sup_{\alpha \in \Delta} |\alpha|$ (in the dual norm for linear forms), for H and K in $\bar{A}$, with $|K-H| \leq \frac{1}{2N} \cdot \frac{1}{R}$

$$\Delta^{(R)}(K)^{-1} \leq C\, \Delta^{(R)}(H)^{-1}$$

with C a constant independant of R, H and K .

Let $I, \tilde{I}$ be the usual sets associated to R and H ; by definition

$$\Delta^{(R)}(H) \approx R^{-\mu^R} \left(\prod_\alpha \sin \frac{\alpha(H)}{2} \right)^{-1} ,$$

where α runs over the non-singular roots only.

If α is singular of type I,

$$\alpha(K) = \alpha(H) + \alpha(K-H) \le \frac{C}{R} + |\alpha| \frac{1}{2NR} \le \frac{C'}{R} ,$$

and similarly if α is singular of type II, we get

$$2\pi - \alpha(K) \le \frac{C''}{R} .$$

If α is non-singular, then

$$\alpha(K) = \alpha(H) + \alpha(K-H) \ge \alpha(H) - \frac{1}{2N}\frac{1}{R} |\alpha| \ge \alpha(H) - \frac{1}{2} R \ge \frac{\alpha(H)}{2} ,$$

and similarly $2\pi - \alpha(K) \ge \frac{1}{2}(2\pi - \alpha(H))$, so that in any case

$$\sin \alpha(K) \ge c \sin \alpha(H) \quad (c > 0) .$$

The lemma follows from these inequalities.

Now we come back to (6.2) ; the translate of any regular atom being an atom, we may always assume that the support of the atom is in a ball centered at the origin and with radius ρ .

Observe first that, as $\delta > \frac{n-1}{2}$, we have $\| s_R^\delta \|_1 \le C < +\infty$ (see [3]).

So

$$|S_R(g)| \le C \rho^{-n/p} , \text{ and so}$$

$$S_*^\delta(g) \le C \rho^{-n/p} . \tag{6.5}$$

Now assume $d(e,g) \ge 2\rho$.

Assume first that $R \ge \frac{1}{2N\rho}$ (the N is the constant of lemma (6.4)). Then

$$|(S_R^\delta a)(g)| \le \left| \int_{B(e,\rho)} a(\gamma) \; s_R^\delta(g\gamma^{-1}) d\gamma \right|$$

$$\le C\left(\sup_{\gamma \in B(e,\rho)} |a(\gamma)| \right) \int_{B(g,\rho)} |s_R^\delta(\gamma)| \, d\gamma \le C \rho^{-n/p} \int_{B(g,\rho)} |s_R^\delta(\gamma)| \, d\gamma .$$

As $\gamma \in B(g,\rho)$ $d(e,\gamma) \ge d(e,g) - \rho \ge \frac{1}{2} d(e,g) \ge \rho \ge \frac{1}{2NR}$.

So if $H = \mathcal{H}(\gamma)$, $|H| = d(e,\gamma) \ge \frac{1}{R}$; hence

$$|s_R^\delta(\gamma)| < C R^{(n/2-1/2)-\delta} \left(d(e,\gamma)^{-n/2-\delta-1/2} + \Delta(\gamma)^{-1} \right) .$$

Using the fact that $\delta = n/p - (n+1)/2$, we get $(n/2-1/2) - \delta = -n(1/p-1)$.

So

$$|(S_R^\delta a)(g)| \le C\,\rho^{-n/p}\,\rho^{n}R^{-n(1/p-1)}\left(d(e,g)^{-n/p} + \frac{1}{B(g,\rho)}\int_{B(g,\rho)} \Delta(\gamma)^{-1}d\gamma\right) .$$

Let $(\Delta^{-1})^*$ be the Hardy-Littlewood maximal function of Δ^{-1} ; then

$$(6.6) \qquad |(S_R^\delta a)(g)| \le C(d(e,g)^{-n/p} + (\Delta^{-1})^*(g)) ,$$

where we used the fact that $R\rho \ge \frac{1}{2N}$.

So far we have not used the vanishing moments property for the atom. Let $n_0 = [n(1/p)-1]$, and suppose now that $R \le \frac{1}{2N\rho}$;

$$(S_R^\delta a)(g) = \int_{B(e,\rho)} a(\gamma)\left[s_R^\delta(g\gamma^{-1}) - T_{n_0}^g(s_R^\delta)(\gamma)\right] d\gamma ,$$

where $T_{n_0}^g(s_R^\delta)$ is a Taylor polynomial (*) of s_R^δ at g . Hence

$$|(S_R^\delta a)(g)| \le C\,\rho^n\,\|a\|_\infty\,\rho^{n_0+1} \sup_{\substack{\gamma\in B(g,\rho)\\ |J|\le n_0+1}} |Y^J s_R^\delta(\gamma)| .$$

As $\gamma \in B(g,\rho)$, observe that $|\mathcal{H}(\gamma) - \mathcal{H}(g)| \le d(g,\gamma)$ by (6.3), and so $|\mathcal{H}(\gamma) - \mathcal{H}(g)| \le \rho \le \frac{1}{2NR}$; so use the inequalities (5.3) together with lemma (6.4) to obtain

$$|S_R^\delta a(g)| \le C\,\rho^{n+n_0+1-n/p}\,R^{n_0+1}\,\Big|\,R^{-n(1/p-1)}\,(d(e,g)^{-n/p} + {}^{(R)}(g)^{-1}) .$$

As $-n(1/p-1) + n_0+1 \ge 0$, and $R\rho \le \frac{1}{2N}$, this implies

$$(6.7) \qquad |S_R^\delta a(g)| \le C(d(e,g)^{-n/p} + \Delta(g)^{-1}) .$$

Now we put (6.5), (6.6) and (6.7) together to obtain

$$(S_*^\delta a)(g) \le \begin{cases} C\,\rho^{-n/p} & \text{if } d(e,g) \le 2\rho \\ C(d(e,g)^{-n/p} + (\Delta^{-1})^*(g) + \Delta^{-1}(g)) & \text{if } d(e,g) \ge 2\rho . \end{cases}$$

As $\Delta^{-1} \in L^1$ (even to L^2) , it is now easy to find the desired weak-type estimate. This finishes the proof of (6.2).

Jean-Louis CLERC

(*) Here we need to assume that ρ is smaller than some fixed ρ_0 . This is no serious restriction however.

Références

[1] BOURBAKI : Groupes et algèbres de Lie. Ch. 4, 5, 6.

[2] BOURBAKI : Groupes et algèbres de Lie. Ch. 9.

[3] CLERC J.L. : Sommes de Riesz et multiplicateurs sur un groupe de Lie compact. Ann. Inst. Fourier 24 (1974), 149-172.

[4] CLERC J.L. : Localisation des sommes de Riesz sur un groupe de Lie compact. Studia Math. 55 (1976), 21-26.

[5] STEIN E.M. : TAIBLESON M.H., WEISS G. : Weak-type estimates for maximal operators on certain H^p-classes. Suppl. Rendiconti Circ. Math. Palermo 1 (1981), 81-97.

[6] WARNER G. : Harmonic Analysis on semi-simple Lie groups II. Springer-Verlag, Berlin (1972).

[7] ZALOZNIK A. : Preprint.

INJECTION DE MODULES SPHERIQUES POUR LES ESPACES SYMETRIQUES REDUCTIFS DANS CERTAINES REPRESENTATIONS INDUITES

par Patrick DELORME (*)

Avec un appendice d'Erik van den BAN (* *) et Patrick DELORME.

0. INTRODUCTION.

Soit G un groupe réel réductif dans la classe de Harish Chandra (cf. [H.C.]), σ une involution de G et H un sous-groupe ouvert du groupe G^{σ} des points fixes de σ. Soit θ une involution de Cartan de G commutant à σ (cf. [v.d.B], prop. 1.1) et soit K le groupe des points fixes de θ. C'est un sous-groupe compact maximal de G, σ - stable. Soit $\underline{g}$ l'algèbre de Lie de G. Dans cet article on appellera $\underline{g}$ - module H - sphérique un sous $(\underline{g}, K)$ - module admissible de longueur finie (en abrégé G - module de Harish Chandra) de $C^{\infty}(G/H)$, ou, ce qui revient au même, d'après [v.d.B, D], la donnée d'un G - module de Harish Chandra et d'un vecteur dans son dual algébrique, fixé par $K \cap H$ et l'algèbre de Lie $\underline{h}$ de H.

Nous établissons pour les modules H - sphériques irréductibles des résultats analogues aux premières étapes de la classification de Langlands des G - modules simples. Pour cela, nous utilisons les propriétés asymptotiques des fonctions K - finies et $Z(\underline{g})$ - finies de $C^{\infty}(G/H)$, établies dans [v.d.B]. Ici $Z(\underline{g})$ est le centre de l'algèbre enveloppante de $\underline{g}$. Nous établissons (th. 1) un analogue du théorème du sous-module de Casselman, résultat annoncé il y a plusieurs années par Oshima (cf. [O$_1$], prop. 12). Puis nous montrons (Corollaire du théorème 3) que tout module H - sphérique peut être réalisé comme sous-module d'une induite $\operatorname{Ind}_{MAN\uparrow G} \delta \otimes e^{\lambda} \otimes 1_N$ d'un sous-groupe parabolique $P = MAN$ de G vérifiant les propriétés suivantes :

$L = MA$ est un sous-groupe de Levi de P, σ et θ stable, A est le sous-groupe vectoriel du centre de L constitué des éléments g de celui-ci vérifiant $\theta(g) = g^{-1}$, $\sigma(g) = g^{-1}$ et N le radical unipotent de P ; en outre, δ est une sous-représentation irréductible de $L^2(M/M \cap H)$ (série discrète de l'espace symétrique $M/M \cap H$) et λ une forme linéaire sur $\underline{a} = \text{Lie } A$ tel que $\text{Re } \lambda$ soit dans la fermeture de la chambre de Weyl négative déterminée par les racines de $\underline{a}$ dans $\underline{n} = \text{Lie } N$. Au passage nous montrons que les représentations H-sphériques irréductibles tempérées peuvent être réalisées comme sous-représentations d'une induite du type précédent avec λ unitaire (résultat déjà annoncé par Oshima). Ceci fait l'objet du théorème 2.

Nos démonstrations doivent beaucoup à l'article de Hecht et Schmid ([H.S.]) ainsi qu'à l'article de Carmona ([Car]).

En outre nous utilisons un résultat d'extension de fonctions analytiques de A à G tout entier. Ce résultat est une conséquence de l'appendice qui est un travail commun avec Erik van den Ban.(*)

1. NOTATIONS.

1.1

Si V est un espace vectoriel, on notera V^* son dual. Si V est réel, on note $V_{\mathbb{C}}$ son complexifié, $S(V)$ l'algèbre symétrique de $V_{\mathbb{C}}$ que l'on identifie aux fonctions polynomiales sur V^* lorsque V est de dimension finie.

Si S est un groupe de Lie réel, S^0 désignera sa composante neutre, $\underline{s}$ son algèbre de Lie, $U(\underline{s})$ l'algèbre enveloppante de la complexifiée $\underline{s}_{\mathbb{C}}$ de $\underline{s}$ et $Z(\underline{s})$ le centre de $U(\underline{s})$. On notera $s \to L_s$ (resp. $s \to R_s$) la représentation régulière gauche (resp. droite) de S et $X \to L_X$ (resp. $X \to R_X$) la représentation de $U(\underline{s})$ obtenue par différentiation.

(*) voir note après la bibliographie de l'article.

Si S est un groupe compact, on notera $\hat{S}$ son dual unitaire. Si $\delta \in \hat{S}$ et (μ, E) est une représentation de S, on notera E^{δ} la composante isotypique de type δ et E^{S} les invariants de E sous S. Enfin, si X est un espace muni d'une action de S on appellera fonction μ - sphérique sur X toute fonction f sur X à valeurs dans E telle que :

$$\forall x \in X \ , \ \forall s \in S \ , \ f(s \, . \, x) = \mu(s) f(x) \ .$$

1.2

On retient les notations de l'introduction. Soit $\underline{p}$ (resp. $\underline{q}$) le sous-espace propre pour la valeur propre -1 de l'endomorphisme θ (resp. σ) de $\underline{g}$. On note $\underline{c}$ le centre de $\underline{g}$ et $\underline{g}_1 = [\underline{g}, \underline{g}]$. On fixe une forme bilinéaire B sur $\underline{g}$, négative définie sur $\underline{k}$ et positive définie sur $\underline{p}$, qui coïncide avec la forme de Killing sur $\underline{g}_1$ et telle que $\underline{g}_1$ et $\underline{c}$ soient orthogonaux.

Soit $\underline{a}_{\emptyset}$ un sous-espace abélien maximal de $\underline{p} \cap \underline{q}$. On note $\underline{g}_{\sigma\theta} = \underline{k} \cap \underline{h} \oplus \underline{p} \cap \underline{q}$ i.e. $\underline{g}_{\sigma\theta}$ est l'ensemble des points fixes de l'involution $\sigma\theta$ de $\underline{g}$. On note $\Delta_{\sigma\theta}$ (resp. $\Delta = \Delta(\underline{g}, \underline{a}_{\emptyset})$) l'ensemble des racines de $\underline{a}_{\emptyset}$ dans $\underline{g}_{\sigma\theta}$ (resp. $\underline{g}$). Pour $\alpha \in \Delta$, le sous-espace radiciel correspondant est noté $\underline{g}^{\alpha}$. Il est stable par σ et θ et l'on note $\underline{g}^{\alpha}_{+}$ (resp. $\underline{g}^{\alpha}_{-}$) le sous-espace propre de l'endomorphisme $\sigma\theta$ de $\underline{g}^{\alpha}$ pour la valeur propre $+1$ (resp. -1). On a $\alpha \in \Delta_{\sigma\theta}$ si et seulement si $\underline{g}^{\alpha}_{+} \neq 0$ et en général $\underline{g}^{\alpha}_{+} = \underline{g}_{\sigma\theta} \cap \underline{g}^{\alpha}$. Notez que $\Delta_{\sigma\theta}$ est le système de racines de $(\underline{g}_{\sigma\theta}, \underline{a}_{\emptyset})$ puisque $\underline{a}_{\emptyset}$ est un sous-espace de Cartan de $\underline{g}_{\sigma\theta}$. On se fixe une fois pour toutes un ensemble de racines positives $\Delta^{+}_{\sigma\theta}$ de $\Delta_{\sigma\theta}$. On définit $\underline{a}^{-}_{\emptyset}$ la chambre de Weyl négative correspondante i.e. $\underline{a}_\emptyset^- = \{H \in \underline{a}_\emptyset | \alpha(H) < 0, \forall\, \alpha \in \Delta_{\sigma\theta}^+\}$.

On rappelle que l'application de $K \times (\underline{p} \cap \underline{q}) \times (\underline{p} \cap \underline{h})$ dans G définie par $(h, X, Y) \rightarrow k \exp X \exp Y$ est un difféomorphisme sur G . En outre, pour tout $X \in \underline{p} \cap \underline{q}$, il existe Y dans $\mathrm{cl}\, \underline{a}_\emptyset^-$ (fermeture de $\underline{a}_\emptyset^-$) tel que $X = \mathrm{Ad}\, k\, Y$ pour un $k \in K \cap H^0$. Enfin, pour tout $g \in G$, il existe un unique $a \in \exp \underline{a}_\emptyset^-$ tel que $g \in K a H^0$ (cf. [v.d.B.], prop. 1.3 et cor. 1.4).

On note $\underline{\ell}_\emptyset$ le centralisateur de $\underline{a}_\emptyset$ dans $\underline{g}$. On a
$\underline{\ell}_\emptyset = \underline{\ell}_{\emptyset,kq} \oplus \underline{\ell}_{\emptyset,kh} \oplus \underline{a}_\emptyset \oplus \underline{\ell}_{\emptyset,ph}$, où $\underline{\ell}_{\emptyset,kq} = \underline{\ell}_\emptyset \cap \underline{k} \cap \underline{q}$, etc...
C'est une algèbre de Lie réductive qui contient $\underline{a}_\emptyset$ dans son centre.
Montrons que $\underline{m}_\emptyset = \underline{\ell}_{\emptyset,kq} \oplus \underline{\ell}_{\emptyset,kh} \oplus \underline{\ell}_{\emptyset,ph}$ est une sous-algèbre de $\underline{\ell}_\emptyset$. Le seul point non trivial est que :
$[\underline{\ell}_{\emptyset,kq}, \underline{\ell}_{\emptyset,ph}] \subset \underline{m}_\emptyset$. Mais on a : $[\underline{\ell}_{\emptyset,kq}, \underline{\ell}_{\emptyset,ph}] \subset \underline{\ell}_{\emptyset,pq} = \underline{a}_\emptyset$.
Or $\underline{\ell}_\emptyset$ est réductive et $\underline{a}_\emptyset$ dans le centre de $\underline{\ell}_\emptyset$. Donc $[\underline{\ell}_\emptyset, \underline{\ell}_\emptyset] \cap \underline{a}_\emptyset = \{0\}$. Donc $[\underline{\ell}_{\emptyset,kq}, \underline{\ell}_{\emptyset,ph}] = 0$. Ce qui démontre notre assertion.

On note $\mathfrak{F}$ la famille des ensembles de racines positives de Δ contenant $\Delta_{\sigma\theta}^+$. Un tel ensemble sera dit compatible avec $\Delta_{\sigma\theta}^+$.
Soit $\mathcal{P}$ un élément de $\mathfrak{F}$. On notera alors $\underline{n}_\emptyset(\mathcal{P}) = \bigoplus_{\alpha \in \mathcal{P}} \underline{g}^\alpha$ et $\underline{p}_\emptyset(\mathcal{P}) = \underline{m}_\emptyset \oplus \underline{a}_\emptyset \oplus \underline{n}_\emptyset(\mathcal{P})$. Soit $P_\emptyset(\mathcal{P})$ le sous-groupe parabolique de G d'algèbre de Lie $\underline{p}_\emptyset(\mathcal{P})$. On note $M_\emptyset$ le sous-groupe de G engendré par le sous-groupe analytique de G d'algèbre de Lie $\underline{m}_\emptyset$ et le centralisateur dans K de $\underline{a}_\emptyset$. Alors on a $P_\emptyset(\mathcal{P}) = M_\emptyset A_\emptyset N_\emptyset(\mathcal{P})$ où $A_\emptyset$ et $N_\emptyset(\mathcal{P})$ sont les sous-groupes analytiques de G d'algèbres de Lie $\underline{a}_\emptyset$ et $\underline{n}_\emptyset(\mathcal{P})$. On notera $\underline{a}_\emptyset^-(\mathcal{P}) = \{H \in \underline{a}_\emptyset \,|\, \alpha(H) < 0 , \forall\, \alpha \in \mathcal{P}\}$.

On a $\mathrm{Cl}\,\underline{a}_{\emptyset}^{-} = \bigcup_{\mathcal{P} \in \mathcal{F}} \mathrm{Cl}\,\underline{a}_{\emptyset}^{-}(\mathcal{P})$. Enfin $M_{\emptyset}$ est laissé stable par σ et θ et l'espace symétrique $M_{\emptyset} / M_{\emptyset} \cap H$ est dans la classe d'Harish Chandra.

On notera, pour $\mathcal{P} \in \mathcal{F}$, $\rho_{\mathcal{P}} = \frac{1}{2} \sum_{\alpha \in \mathcal{P}} (\dim \underline{g}_{\alpha})\alpha$,

$\mathcal{L}^{+}(\mathcal{P}) = \{\lambda \in \underline{a}_{\emptyset}^{*} \mid \lambda = \sum_{i} n_i \alpha_i$, où $n_i \in \mathbb{N}$ et $\alpha_i \in \mathcal{P}\}$, $\Sigma_{\mathcal{P}}$ l'ensemble des racines simples de $\mathcal{P}$. Pour $\Theta \subset \Sigma_{\mathcal{P}}$ on notera $\underline{a}_{\Theta} = \bigcap_{\alpha \in \Theta} \mathrm{Ker}\,\alpha$, $\underline{\ell}_{\Theta}$ le centralisateur dans $\underline{g}$ de $\underline{a}_{\Theta}$. On note $\underline{m}_{\Theta}$ l'orthogonal dans $\underline{\ell}_{\Theta}$ de $\underline{a}_{\Theta}$ pour la forme bilinéaire B restreinte à $\underline{\ell}_{\Theta}$. Comme $\underline{a}_{\Theta}$ et donc $\underline{\ell}_{\Theta}$ sont σ et θ stables, il est aisé de voir que B restreinte à $\underline{\ell}_{\Theta}$ est non dégénérée, que $\underline{m}_{\Theta}$ est une sous-algèbre de Lie de $\underline{\ell}_{\Theta}$, σ et θ stable qui vérifie $\underline{\ell}_{\Theta} = \underline{m}_{\Theta} \oplus \underline{a}_{\Theta}$. On note $\underline{a}^{\Theta} = \underline{m}_{\Theta} \cap \underline{a}_{\emptyset}$,

$\underline{n}_{\Theta}(\mathcal{P}) = \bigoplus_{\alpha \in \mathcal{P},\, \alpha_{|\underline{a}_{\Theta}} \neq 0} \underline{g}^{\alpha}$, $\rho_{\mathcal{P},\Theta} = \rho_{\mathcal{P}\,|\underline{a}_{\Theta}}$,

$\mathcal{L}^{+}(\mathcal{P}, \Theta) = \{\lambda_{|\underline{a}_{\Theta}} \mid \lambda \in \mathcal{L}^{+}(\mathcal{P})\}$. On notera $P_{\Theta}(\mathcal{P})$ le sous-groupe parabolique d'algèbre de Lie $\underline{p}_{\Theta}(\mathcal{P}) = \underline{m}_{\Theta} \oplus \underline{a}_{\Theta} \oplus \underline{n}_{\Theta}(\mathcal{P})$. On a alors $P_{\Theta}(\mathcal{P}) = M_{\Theta} A_{\Theta} N_{\Theta}(\mathcal{P})$, où A_{Θ} (resp. $N_{\Theta}(\mathcal{P})$) est le sous-groupe analytique de G d'algèbre de Lie $\underline{a}_{\Theta}$ (resp. $\underline{n}_{\Theta}(\mathcal{P})$) et M_{Θ} est le sous-groupe de G engendré par $M_{\emptyset}$ et le sous-groupe analytique de G d'algèbre de Lie $\underline{m}_{\Theta}$. Alors M_{Θ} est stable sous σ et θ et l'espace symétrique $M_{\Theta} / M_{\Theta} \cap H$ est dans la classe d'Harish Chandra.

2. PROPRIETES ASYMPTOTIQUES DES FONCTIONS K - FINIES ET Z ($\underline{g}$) - FINIES SUR G/H.

2.1 Nous allons maintenant rappeler, sous une forme appropriée, des résultats de van den BAN ([v.d.B]).

Soit I un idéal de Z ($\underline{g}$) de codimension finie. Alors il existe un ensemble fini X (I) dans $(\underline{a}_\emptyset^*)_{\mathbb{C}}$ possédant la propriété suivante : pour toute fonction F, K finie et C^∞ sur G/H, annulée par I, il existe des fonctions sur $K \times \underline{a}_\emptyset$ notées $(k, X) \to P_{\lambda,\rho}(k, X, F)$, où λ décrit $X(I) + \mathcal{L}^+(\rho)$, vérifiant :

(i) Pour k fixé, les fonctions $X \to P_{\lambda,\rho}(k, X, F)$ sont polynomiales sur $\underline{a}_\emptyset$. De plus,

$$\forall X \in \underline{a}_\emptyset^-(\rho),\ F(k \exp X) = \sum_{\lambda \in X(I) + \mathcal{L}^+(\rho)} P_{\lambda,\rho}(k, X, F)\, e^{(\lambda + \rho_\rho)(X)} .$$

En outre, la convergence est valable aussi lorsqu'on développe les polynomes $P_{\lambda,\rho}(k, X, F)$ en monomes, en utilisant des coordonnées sur $\underline{a}_\emptyset$.

(ii) Pour k fixé, les fonctions polynomiales $X \to P_{\lambda,\rho}(k, X, F)$ sont entièrement déterminées par (i). Il suffit même que la convergence soit valable sur un ouvert de $\underline{a}_\emptyset^-$ stable par dilatation.

(iii) La convergence de la série dans (i) est absolue et uniforme, y compris après développement des P_λ en monomes, sur tout translaté de $\underline{a}_\emptyset^-(\rho)$ dont la fermeture est entièrement contenue dans $\underline{a}_\emptyset^-(\rho)$.

(iv) Les fonctions $X \to P_{\lambda,\mathcal{P}}(.,X,F)$ sont des fonctions polynomiales sur $\underline{a}_{\emptyset}$ à image dans un sous-espace de dimension finie et K - fini de $C^{\infty}(K)$ (si F est de type $\mu \in \hat{K}$, $P_{\lambda,\mathcal{P}}(.,X,F)$ est également de type μ). En outre, pour F fixé, leur degré est borné indépendamment de $\lambda \in X(I) + \mathcal{L}^{+}(\mathcal{P})$.

(v) On peut dériver terme à terme par tout élément de $U(\underline{a}_{\emptyset})$ et de $U(\underline{k})$ la série de (i), y compris après développement en monome, les modes de convergence de (i) et (iii) étant préservés.

Références :

Pour exhiber $X(I)$ et les $P_{\lambda,\mathcal{P}}$ satisfaisant (i), il suffit d'utiliser les théorèmes 3.4 et 3.5 de [v.d.B.] et de développer en série de Taylor à l'origine les fonctions $F^{\Sigma}_{s,m}$ du théorème 3.5 de cet article. Alors (ii) résulte du lemme A. 1.7 de [Ca.M.]. Enfin (iii), (iv) et (v) résultent des points précédents, des théorèmes 3.4 et 3.5 de [v.d.B.], ainsi que des propriétés élémentaires des séries entières.

2.2

Nous aurons besoin aussi des développements asymptotiques le long des murs. On conserve les notations de 2.1. On se fixe $\mathcal{P} \in \mathcal{F}$ et Θ un sous-ensemble de l'ensemble des racines simples de $\mathcal{P}$. On notera $X(I,\Theta)$ (resp. $\mathcal{L}^{+}(\mathcal{P},\Theta)$) l'ensemble des restrictions à $\underline{a}_{\Theta}$ des éléments de $X(I)$ (resp. $\mathcal{L}^{+}(\mathcal{P})$) . Alors, pour tout F comme en 2.1, il existe des fonctions $(k,a,X) \to Q_{\mu,\mathcal{P},\Theta}(k,a,X,F)$ où μ décrit $X(I,\Theta) + \mathcal{L}^{+}(\mathcal{P},\Theta)$, définies sur $K \times A^{\Theta} \times \underline{a}_{\Theta}$ (ici $A^{\Theta} = \exp \underline{a}^{\Theta}$)

vérifiant :

(i) Les fonctions $Q_{\mu,\mathcal{P},\mathfrak{F}}$ sont polynomiales sur $\underline{a}_\Theta$ pour k et a fixés et analytiques en $a \in A_\Theta$ pour k et X fixés. En outre, pour k et a fixés, on a :

$$\forall\, X \in \underline{a}_\Theta^-(\mathcal{P}) = \{H \in \underline{a}_\Theta \mid \forall\, \alpha \in \mathcal{P}\,,\ \alpha_{|\underline{a}_\Theta} \neq 0\,,\ \alpha(H) < 0\}\ ,$$

$$F(k\, a \exp X) = \sum_\mu Q_{\mu,\Theta,\mathcal{P}}(k,a,X)\, e^{(\mu + \rho_{\mathcal{P},\Theta})(X)}\ ,$$

la convergence étant également valable après développement des polynomes sur $\underline{a}_\Theta$, $Q_{\mu,\Theta}(k,a,.,F)$ en monomes (en utilisant des coordonnées sur $\underline{a}_\Theta$).

(ii) Les fonctions $Q_{\mu,\Theta,\mathcal{P}}$ sont entièrement déterminées par (i). Il suffit même que la convergence (après développement en monomes) soit valable sur un ouvert de $\underline{a}_\Theta^-(\mathcal{P})$ stable par dilatation. En outre, la convergence est uniforme et absolue sur tout translaté de $\underline{a}_\Theta^-(\mathcal{P})$ dont la fermeture est contenue dans $\underline{a}_\Theta^-(\mathcal{P})$.

(iii) Soit $\mathcal{P}_1$ l'ensemble des racines de $\mathcal{P}$ s'annulant sur $\underline{a}_\Theta$. C'est un ensemble de racines positives du système des racines de $\underline{a}^\Theta$ dans $\underline{m}_\Theta$. Soit $\mathcal{P}'_1$ un autre ensemble de racines positives de ce système tel que si $\alpha \in \Delta^+_{\sigma\theta}$ est nul sur $\underline{a}_\Theta$ on ait $\alpha \in \mathcal{P}'_1$.
Soit $\mathcal{P}' = \mathcal{P}'_1 \cup \{\alpha \in \mathcal{P} \mid \alpha_{|\underline{a}_\Theta} \neq 0\}$. Alors $\mathcal{P}'$ est ensemble de racines positives compatible avec $\Delta^+_{\sigma\theta}$, i.e. $\mathcal{P}' \in \mathfrak{F}$. On notera Θ' les racines simples de $\mathcal{P}'_1$ (qui sont simples aussi dans $\mathcal{P}'$) . Alors $\underline{a}_\Theta = \underline{a}_{\Theta'}$,

$\underline{a}^{\Theta} = \underline{a}^{\Theta'}$, $\underline{m}_{\Theta} = \underline{m}_{\Theta'}$, etc... En outre, $\underline{a}^{-}_{\Theta'}(\mathcal{P}') = \underline{a}^{-}_{\Theta}(\mathcal{P})$. Enfin, si $X \in (\underline{a}^{\Theta})^{-}(\mathcal{P}'_1) = \{H \in \underline{a}^{\Theta} \mid \forall \alpha \in \mathcal{P}'_1 , \alpha(H) < 0\}$ et $Y \in \underline{a}^{-}_{\Theta}(\mathcal{P}')$, on a les relations :

$$Q_{\mu,\mathcal{P},\Theta}(k, \exp X, Y, F) = \sum_{\lambda \in X(I) + \mathcal{L}^{+}(\mathcal{P}),\ \lambda_{|\underline{a}_{\Theta}} = \mu} P_{\lambda,\mathcal{P}'}(k, X+Y, F) e^{(\lambda + \rho_{\mathcal{P}'_1})(X)}$$

où $\rho_{\mathcal{P}'_1}$ est la demi somme des racines de $\mathcal{P}'_1$ comptées avec multiplicités, valable y compris après développement des $P_{\lambda,\mathcal{P}'}$ en monomes.

(iv) On peut dériver terme à terme la série donnant $F(k\, a \exp X)$ par tout élément de $U(\underline{k})$ et $S(\underline{a}_{\Theta})$ tout en préservant le mode de convergence décrit en (i). De même on peut dériver terme à terme le développement de $Q_{\mu,\mathcal{P},\Theta}(k, \exp X, Y, F)$ de (iii) par tout élément de $S(\underline{a}^{\Theta})$ tout en préservant le mode de convergence décrit en (iii).

Références :

Pour (i), cf. [v.d.B], § 7 et 8. Le point (ii) résulte de [Ca.M.], lemme A. 1.7. Pour (iii), on utilise le développement relatif à $\mathcal{P}_1$ de $F(k \exp (X + Y))$ donné en 2.1, puis on procède à un regroupement des termes correspondant à des λ ayant même restriction à $\underline{a}_{\Theta}$. On obtient alors un développement du type de celui décrit en (i) pour $F(k\, a \exp X)$ valable pour $\log a$ dans $(\underline{a}^{\Theta})^{-}(\mathcal{P}'_1)$ et X dans $\underline{a}^{-}_{\Theta'}(\mathcal{P}') = \underline{a}^{-}_{\Theta}(\mathcal{P})$. D'après l'unicité, décrite en (ii), d' un tel développement, on obtient l'identité voulue. Pour les dérivations terme à terme de (iv), les propriétés élémentaires des séries entières permettent de conclure.

2.3 Soit F une fonction comme ci-dessus. Alors on notera

$e(\mathcal{P}, F) = \{\lambda \mid P_{\lambda,\mathcal{P}}(.,.,F) \not\equiv 0\}$ et $e(\mathcal{P},\Theta,F) = \{\mu \mid Q_{\mu,\mathcal{P},\Theta}(.,.,.,F) \not\equiv 0\}$.

On remarquera que pour tout $\mathcal{P}'$ comme en 2.2 (iii), on a aussi l'égalité :

$e(\mathcal{P},\Theta,F) = \{\mu \mid \mu = \lambda_{|\underline{a}_\Theta}, \lambda \in e(\mathcal{P}',F)\}$. En effet, utilisant le dévelop-

pement des $Q_{\mu,\mathcal{P},\Theta}$ donné en 2.2 (iii) dont l'unicité est assurée grâce au lemme A.1.7 de [Ca.M.], on voit que si $Q_{\mu,\mathcal{P},\Theta} \equiv 0$ on a $P_{\lambda,\mathcal{P}'} \equiv 0$ dès que $\lambda_{|\underline{a}_\Theta} = \mu$.

Le lemme suivant est une conséquence immédiate de l'unicité des fonctions $Q_{\mu,\mathcal{P},\Theta}$ (cf. 2.3 (iii)).

<u>Lemme 1</u>.

Avec les notations ci-dessus on a

(i) $\forall\, k_0, k \in K$, $\forall\, a \in A^\Theta$, $\forall\, X \in \underline{a}_\Theta$,

$$Q_{\mu,\mathcal{P},\Theta}(k_0 k, a, X, F) = Q_{\mu,\mathcal{P},\Theta}(k, a, X, L_{k_0^{-1}} F) .$$

(ii) D'autre part, si $m \in M_\Theta \cap K \cap H$ vérifie $m^{-1} a m \in A^\Theta$ pour un $a \in A^\Theta$ on a :

$$\forall\, k \in K, \ \forall\, X \in \underline{a}_\Theta, \ Q_{\mu,\mathcal{P},\Theta}(k m, a, X, F) = Q_{\mu,\mathcal{P},\Theta}(k, \mathrm{Ad}m^{-1} a, X, F) .$$

<u>Remarque</u> :

Les résultats concernant les polynomes Q s'appliquent aux polynomes P en faisant $\Theta = \emptyset$, car $P_{\lambda,\mathcal{P}} = Q_{\lambda,\mathcal{P},\emptyset}$.

3. CONSTRUCTION DE MORPHISMES ENTRE REPRESENTATIONS H - SPHERIQUES ET REPRESENTATIONS INDUITES.

3.1 On se donne un sous module de Harish Chandra V de $C^{\infty}(G/H)$, annulé par l'idéal de codimension finie $Z(\underline{g})$, I. On notera $e(\mathbb{P},V) = \bigcup_{F \in V} e(\mathbb{P},F)$ et $e(\mathbb{P},\Theta,V)$ l'ensemble des restrictions à $\underline{a}_{\Theta}$ des éléments de $e(\mathbb{P},V)$. On introduit un ordre $\underset{\mathbb{P}}{\leq}$ sur $\underline{a}^{*}_{\emptyset,\mathbb{C}}$ (resp. $\underset{\mathbb{P},\Theta}{\leq}$ sur $\underline{a}^{*}_{\Theta,\mathbb{C}}$) de la façon suivante :

$$\lambda \underset{\mathbb{P}}{\leq} \lambda' \quad \Leftrightarrow \quad \lambda' - \lambda \in \mathcal{L}^{+}(\mathbb{P})$$

$$\mu \underset{\mathbb{P},\Theta}{\leq} \mu' \quad \Leftrightarrow \quad \mu' - \mu \in \mathcal{L}^{+}(\mathbb{P},\Theta)$$

On notera $e_{\ell}(\mathbb{P},V)$ (resp. $e_{\ell}(\mathbb{P},\Theta,V)$) les éléments minimaux pour $\underset{\mathbb{P}}{\leq}$ (resp. $\underset{\mathbb{P},\Theta}{\leq}$) de $e(\mathbb{P},V)$ (resp. $e(\mathbb{P},\Theta,V)$. D'après 2.1 et 2.2 ces ensembles sont non vides, discrets et, d'après [v.d.B.], th. 3.4, $e_{\ell}(\mathbb{P},V)$ est fini.

Lemme 2.

Soit $\lambda_0 \in e_{\ell}(\mathbb{P},\Theta,V)$. Pour $F \in V$ on note $j_{\lambda_0}(F)$ la fonction sur A^{Θ} à valeurs dans $S(\underline{a}_{\Theta})$ qui à $a \in A^{\Theta}$ associe la fonction polynomiale sur $\underline{a}_{\Theta}$, $X \to Q_{\lambda_0,\mathbb{P},\Theta}(e,a,X,F)$.

Alors :

(i) Si $D \in \underline{n}_{\Theta}(\mathbb{P})$, $j_{\lambda_0}(DF) = 0$, pour tout $F \in V$.

(ii) Soit ad la représentation adjointe de $\underline{a}_{\Theta}$ dans $S(\underline{a}_{\Theta})$ et $\lambda_0 + \rho_{P,\Theta}$ le caractère de $\underline{a}_{\Theta}$ correspondant à cet élément de $\underline{a}^*_{\Theta,\mathbb{C}}$.
Alors, si $D \in S(\underline{a}_{\Theta})$, $j_{\lambda_0}(DF) = [\mathrm{ad} \otimes (\lambda_0 + \rho_{P,\Theta})] (D) (j_{\lambda_0}(F))$.

(iii) Il existe un entier n tel que pour tout $F \in V$ et $a \in A^{\Theta}$, $j_{\lambda_0}(F)(a)$ est de degré inférieur à n dans $S(\underline{a}_{\Theta})$.

(iv) Pour tout $F \in V$ et tout $D \in S(\underline{a}^{\Theta})$, on a :

$$L_D (j_{\lambda_0}(F)) = j_{\lambda_0}(L_D F) \quad .$$

(v) Pour tout $F \in V$ et $D \in \underline{m}_{\Theta} \cap \underline{h}$,

$$j_{\lambda_0}(L_D F)(0) = 0 \quad .$$

Démonstration :

(i) Pour prouver (i), on peut supposer, par linéarité que $D \in g^{\alpha}_{+}$ ou g^{α}_{-} avec $\alpha \in P$ et $\alpha_{|\underline{a}_{\Theta}} \neq 0$. Alors, pour tout $a \in A_{\emptyset}$ avec $a^{\alpha} \neq 1$, on a :

$$D = f_1(a)(D + \theta D) + f_2(a) \,\mathrm{Ad}\, a^{-1} (D + \sigma D)$$

avec $f_1(a) = \dfrac{\mp a^{2\alpha}}{1 \mp a^{2\alpha}}$, $f_2(a) = \dfrac{1}{a^{-\alpha} \mp a^{\alpha}}$.

Alors on peut calculer $(L_D F)(\exp X)$ pour $X \in \underline{a}^{-}_{\emptyset}(P)$ en différentiant terme à terme la série donnant $F(\exp X)$ (cf. 2.1), y compris après développement des polynomes $P_{\lambda,P}$ en monomes. On obtient ainsi pour $X \in \underline{a}^-_{\emptyset}(\mathcal{P})$:

$$(L_D F)(\exp X) = \sum_{\lambda \in e(\mathcal{P},F)} f_1(\exp X)(L_{D+\theta D} P_{\lambda,\mathcal{P}})(e, \exp X, F) e^{\lambda + \rho_{\mathcal{P}}(X)}$$

avec convergence absolue, y compris après développement des $P_{\lambda,\mathcal{P}}$ en monomes et donc aussi des $L_{D+\theta D} P_{\lambda,\mathcal{P}}$.

Par ailleurs on peut développer $f_1(\exp X)$:

$$\forall X \in \underline{a}^-_{\emptyset}(\mathcal{P}) \ , \ f_1(\exp X) = \mp \sum_{n=0}^{\infty} (\pm 1)^n e^{(2n+1)\alpha(X)} \ ,$$

avec convergence absolue. On peut alors réécrire le produit des deux séries donnant $(L_D F)(\exp X)$ en groupant les termes différemment :

$$\forall X \in \underline{a}^-_{\emptyset}(\mathcal{P}) \ , \ (L_D F)(\exp X) = \sum_{\nu} \ \sum_{\lambda \in e(\mathcal{P},F)\,,\ n \in \mathbb{N}\,,\ \lambda + (2n+1)\alpha = \nu}$$

$$(L_{D+\theta D} P_{\lambda,\mathcal{P}})(e, \exp X, F)\, e^{(\nu + \rho_{\mathcal{P}})(X)}$$

avec convergence y compris après développement des $L_{D+\theta D} P_{\lambda,\mathcal{P}}$ en monomes. Ce développement doit coïncider avec le développement de $L_D F$ rappelé en 2.1 (i) d'après son unicité (cf. 2.1.(ii)). Il en résulte que si $P_{\nu,\mathcal{P}}(e, \exp X, L_D F)$ est non identiquement nul, ν est de la forme $\lambda + (2n+1)\alpha$ avec $\lambda \in e(\mathcal{P},F)$. Soit alors $\lambda_0 \in e_{\ell}(\mathcal{P},\Theta,V)$ et $\lambda \in (\underline{a}^*_{\emptyset})_{\mathbb{C}}$ avec $\lambda_{|\underline{a}_{\Theta}} = \lambda_0$. Si on avait $P_{\lambda,\mathcal{P}}(e, \exp X, L_D F) \not\equiv 0$, on aurait, d'après ce qui précède, $\lambda - (2n+1)\alpha \in e(\mathcal{P},F)$ pour au moins un $n \in \mathbb{N}$. En utilisant le lien entre $e(\mathcal{P},\Theta,F)$ et $e(\mathcal{P},F)$ (cf. 2.3), cela contredirait la minimalité de λ_0.

D'où P_λ (e, exp X, D F) $\equiv 0$ pour tout $\lambda \in (\underline{a}^*_\emptyset)_{\mathbb{C}}$ tel que $\lambda_{|\underline{a}_\Theta} = \lambda_0$.

Alors (i) résulte des relations entre les polynomes $Q_{\lambda_0, P, \Theta}$ et $P_{\lambda, P}$ (cf. 2.2 (iii)).

Démontrons (ii). Pour cela on dérive terme à terme par $D \in S(\underline{a}_\Theta)$ le développement de F (a exp X), $a \in A^\Theta$, $X \in \underline{a}^-_\Theta$ (P) de 2.3 (i), grâce à 2.3 (iv). L'unicité du développement de $(L_D\ F)$ (a exp X) implique immédiatement (ii).

Démontrons (iii). D'après (i) l'application j_{λ_0} passe au quotient par $\underline{n}_\Theta$ (P) V . Or $V_1 = V/\underline{n}_\Theta$ (P) V est un module de Harish Chandra pour $M_\Theta\ A_\Theta$ (cf. [H.S.], prop. 2.24). En particulier, V_1 est annulé par un idéal de codimension finie de $S(\underline{a}_\Theta)$. Alors, grâce à (ii), on en déduit qu'il existe $n \in \mathbb{N}$ tel que pour tout $D \in \underline{a}_\Theta$, $F \in V$ et $a \in A^\Theta$

$$([\mathrm{ad} \otimes (\lambda_0 + \rho_P)]\ (D) - (\lambda_0 + \rho_P)\ (D))^n\ (j_{\lambda_0}\ (F)\ (a)) \equiv 0 \quad .$$

Cela signifie que j_{λ_0} (F) (a) est annulé par $(\mathrm{ad}\ D)^n$ pour tout a et $D \in \underline{a}_\Theta$, ce qui implique que j_{λ_0} (F) (a) est de degré inférieur ou égal à n et ceci achève de prouver (iii).

Prouvons (iv). En dérivant terme à terme le développement de 2.1 (i) par $D \in S(\underline{a}^\Theta)$, grâce à 2.1 (v), on obtient $P_{\lambda, P}(e, X, L_D\ F)$ en fonction de $P_{\lambda, P}(e, X, F)$ pour $X \in \underline{a}_\emptyset$ et $\lambda \in (\underline{a}^*_\emptyset)_{\mathbb{C}}$ avec $\lambda_{|\underline{a}_\Theta} = \mu$. En reportant cette formule dans l'expression de $Q_{\mu, P, \Theta}(e, \exp X, Y, L_D\ F)$ donnée par 2.2 (iii), on peut exprimer cette quantité à l'aide des $P_{\lambda, P}(e, X + Y, F)$. Par ailleurs, on obtient le même résultat en dérivant terme à terme par $D \in S(\underline{a}^\Theta)$ l'expression de $Q_{\mu, P, \Theta}$ (e, exp X, Y, F) donnée par 2.2 (iii), en fonction des $P_{\lambda,\rho}(e, X+Y, F)$ (grâce à 2.2 (iv)). D'où l'identité pour $X \in (\underline{a}^{\Theta})^{-}(\rho_1)$ et $Y \in a_{\Theta}^{-}$ de $L_D(Q_{\mu,\rho,\Theta}(e, \exp X, Y, F)$ et $Q_{\mu,\rho,\Theta}(e, \exp X, Y, L_D F)$. D'où le résultat voulu par prolongement analytique des identités.

Prouvons (v). Il nous faut étudier $(L_D F)(\exp X)$ pour $D \in \underline{m}_{\Theta} \cap \underline{h}$ et $X \in \underline{a}_{\Theta}$. Mais $\underline{m}_{\Theta}$ et $\underline{a}_{\Theta}$ commutent et F est invariante à droite par H. D'où $(L_D F)(\exp X) = 0$ pour tout X dans $\underline{a}_{\Theta}$ et l'assertion en résulte.

3.2 On note $V_1 = V/\underline{n}_{\Theta}(\rho)V$. Notons V_2 le sous-espace de $S(\underline{a}_{\Theta})$ engendré par les $j_{\lambda_0}(F)(a)$, où F décrit V et a décrit A_{Θ}. D'après le lemme 2 (iii), V_2 est de dimension finie. On choisit une forme linéaire non nulle sur V_2, ℓ, qui se transforme sous la contragrediente de $\mathrm{ad} \otimes (\lambda_0 + \rho_{\rho,\Theta})$ par le caractère $(-\lambda_0 - \rho_{\rho,\Theta})$ de $\underline{a}_{\Theta}$. Alors à F dans V on associe la fonction $i_{\lambda_0}(F)$ sur $M_{\Theta}/M_{\Theta} \cap H$ définie de la façon suivante :

Si $m \in M_{\Theta}$ s'écrit $m = m_1 a h$ avec $m_1 \in M_{\Theta} \cap K$, $a \in A^{\Theta}$ et $h \in H$, on pose $i_{\lambda_0}(F)(m) = \ell(j_{\lambda_0}(L_{m_1^{-1}} F)(a))$.

On va montrer que $i_{\lambda_0}(F)$ est analytique sur M_{Θ}(*). Il résulte de la définition et du fait que F est K - finie que $i_{\lambda_0}(F)$ est $M_{\Theta} \cap K$ - finie Comme, d'autre part, pour $k \in M_{\Theta} \cap K$, $X \in \underline{p} \cap \underline{q} \cap \underline{m}_{\Theta}$, $i_{\lambda_0}(F)(k \exp X) = i_{\lambda_0}(L_{k^{-1}} F)(\exp X)$, pour démontrer l'analyticité de $i_{\lambda_0}(F)$ pour tout F dans V, il suffit de prouver l'analyticité sur $\underline{p} \cap \underline{q} \cap \underline{m}_{\Theta}$ de $X \to i_{\lambda_0}(F)(\exp X)$ pour tout F dans V.

(*) Voir note après la bibliographie de l'article.

On introduit la fonction Ψ_F de $\underline{p} \cap \underline{q}$ dans $C^\infty(M_\Theta \cap H \cap K)$ définie par

$$(\Phi_F(X))(m) = (i_{\lambda_0}(F))(\exp(\operatorname{Ad} m \,.\, X))$$

$$= (i_{\lambda_0}(F))(m \exp X) = (i_{\lambda_0}(L_{m^{-1}} F))(\exp X)$$

pour $m \in M_\Theta \cap H \cap K$ et $X \in \underline{p} \cap \underline{q} \cap \underline{m}_\Theta$.

Le fait que $i_{\lambda_0}(F)$ soit $M_\Theta \cap K$ - finie implique que Φ_F est à valeurs dans un sous-espace de dimension finie E de $C^\infty(M_\Theta \cap K \cap H)$ stable par les représentations régulières de $M_\Theta \cap K \cap H$. On note μ la représentation régulière droite de $M_\Theta \cap K \cap H$ dans E . Munissant $\underline{p} \cap \underline{q} \cap \underline{m}_\Theta$ de l'action adjointe de $M_\Theta \cap K \cap H$, on voit que Ψ_F est μ - sphérique. En outre Φ_F restreinte à $\underline{a}^\Theta$ coïncide avec $X \to \ell\,(j_{\lambda_0}(L_{m^{-1}} F)(\exp X))$ qui est clairement analytique. On veut calculer $(L_D \Phi)(0)$ pour $D \in S(\underline{a}^\Theta)$. Il résulte du lemme 2 (iv) que $L_D(j_{\lambda_0}(L_{m^{-1}} F)) = j_{\lambda_0}(L_D L_{m^{-1}} F)$.

D'où $((L_D \Psi)(0))(m) = i_{\lambda_0}(L_D L_{m^{-1}} F)(0)$. Or $m \in M_\Theta \cap K \cap H$ et $i_{\lambda_0}(L_D L_{m^{-1}} F)$ est invariante à droite par $M_\Theta \cap H$.

Donc $(L_D \Phi_F(0))(m) = i_{\lambda_0}(L_D L_{m^{-1}} F)(m^{-1})$, soit encore :

$$(L_D \Phi_F(0))(m) = i_{\lambda_0}(L_m L_D L_{m^{-1}} F)(0) \ .$$

On pose alors $\varphi(D) = (L_D \Phi_F)(0)$ pour $D \in S(\underline{a}^\Theta)$. D'autre part on note β la symétrisation de $S(\underline{g})$ dans $U(\underline{g})$. Alors on définit une application linéaire ψ de $S(\underline{p} \cap \underline{q} \cap \underline{m}_\Theta)$ dans E par :

$$\forall\, D \in S(\underline{p} \cap \underline{q} \cap \underline{m}_\Theta),\ \forall\, m \in M_\Theta \cap H \cap K,\ (\psi(D))(m) = i_{\lambda_0}(L_m L_{\beta(D)} L_{m^{-1}} F)(0)$$

Il est clair que ψ est μ - sphérique et prolonge φ . On peut alors appliquer la proposition A.1 de l'appendice pour conclure que Φ_F est analytique et ceci achève de prouver que $i_{\lambda_0}(F)$ est analytique. On note au passage que, toujours d'après la proposition A.1, on a

$(L_{\beta(D)} F)(0) = (\psi(D))(0) = i_{\lambda_0}(L_{\beta(D)} F)(0)$

pour tout $D \in S(\underline{p} \cap \underline{q} \cap \underline{m}_\Theta)$.

On va montrer $i_{\lambda_0}(L_D F) = L_D\, i_{\lambda_0}(F)$ pour tout D dans $U(\underline{m}_\Theta)$.

Pour cela il suffit de comparer les séries de Taylor à l'origine, puisque ces fonctions sont analytiques. Il faut donc voir que :

$\forall D \in U(\underline{m}_\Theta)\ ,\quad i_{\lambda_0}(L_D F)(0) = (L_D\, i_{\lambda_0}(F))(0)$.

La décomposition $\underline{m}_\Theta = (\underline{m}_\Theta \cap \underline{k}) \oplus (\underline{p} \cap \underline{q} \cap \underline{m}_\Theta) \oplus (\underline{p} \cap \underline{h} \cap \underline{m}_\Theta)$ montre que l'on a :

$$U(\underline{m}_\Theta) = U(\underline{h} \cap \underline{m}_\Theta)\ \beta\ (S(\underline{p} \cap \underline{q} \cap \underline{m}_\Theta))\ U(\underline{k} \cap \underline{m}_\Theta)\ .$$

Soit $D = D_1 D_2 \in U(\underline{m}_\Theta)$ avec $D_1 \in \underline{h} \cap \underline{m}_\Theta$. Il est alors clair que $(L_D\, i_{\lambda_0}(F))(0) = 0$, puisque $L_{D_2}\, i_{\lambda_0}(F)$ est invariante à droite par $M_\Theta \cap H$. D'autre part, d'après le lemme 2 (v), on a

$j_{\lambda_0}(L_{D_1}(L_{D_2} F))(0) = 0$ d'où $i_{\lambda_0}(L_D F)(0) = 0$. On a donc l'égalité voulue pour $D \in (\underline{h} \cap \underline{m}_\Theta)\, U(\underline{m}_\Theta)$. D'autre part on a facilement :

$$\forall D \in U(\underline{m}_\Theta \cap \underline{k})\ ,\quad i_{\lambda_0}(L_D F) = L_D\, i_{\lambda_0}(F)\ .$$

Il suffit donc de prouver l'égalité pour tout F et tout $D \in \beta(S(\underline{p} \cap \underline{q} \cap \underline{m}_\Theta))$, mais cela a déjà été vu plus haut. On a donc bien :

$$\forall D \in U(\underline{m}_\Theta)\ ,\quad i_{\lambda_0}(L_D F) = L_D\, i_{\lambda_0}(F)\ .$$

D'autre part, d'après le lemme 2 (ii), on a :

$$\forall\, D \in S(\underline{a}_{\Theta}) \quad , \quad j_{\lambda_0}(D\,F) = (\mathrm{ad} \otimes \lambda_0 + \rho_{\mathcal{P},\Theta})\,(D)\; j_{\lambda_0}(F) \quad .$$

D'où : $i_{\lambda_0}(L_D\,F) = (\lambda + \rho_{\mathcal{P},\Theta})\,(D)\; i_{\lambda_0}(F)$.

En résumé i_{λ_0} est un morphisme de $M_{\Theta}\,A_{\Theta}$ - modules de Harish Chandra de $V/\underline{n}_{\Theta}(\mathcal{P})\,V$ dans $C^{\infty}(M_{\Theta}\,/\,M_{\Theta} \cap H) \otimes \mathbb{C}_{\lambda_0 + \rho_{\mathcal{P},\Theta}}$.

On note $V'_1 \subset C^{\infty}(M_{\Theta}\,/\,M_{\Theta} \cap H)$ l'image de i_{λ_0} . Alors, d'après la réciprocité de Frobenius (cf. [H.S.], 4.11, par exemple) on vient de construire un morphisme (non nul) de V_1 dans

$$\mathrm{Ind}_{M_{\Theta} A_{\Theta} N_{\Theta}(\mathcal{P}) \uparrow G}\; V'_1 \otimes \mathbb{C}_{\lambda_0 + \rho_{\mathcal{P},\Theta}} \otimes 1_{N_{\Theta}(\mathcal{P})} \quad .$$

3.3 Lemme 3 :

Avec les notations ci-dessus, soit $\mathcal{P}'_1$ un ensemble de racines positives du système de racines de $\underline{a}^{\Theta}$ dans $\underline{m}_{\Theta}$ choisi comme en 2.2 (iii). Alors, si $\lambda \in e(\mathcal{P}'_1, V'_1)$, (où $V'_1 = i_{\lambda_0}(V_1)$) , on a $\lambda + \lambda_0 \in e(\mathcal{P}', V)$.

Démonstration :

Pour $F \in V$, on déduit du développement de 2.2 (iii) de $Q_{\lambda_0,\mathcal{P},\Theta}$, un développement de $i_{\lambda_0}(F)$ du type 2.1 (i). L'unicité de celui-ci permet de conclure.

4. RESULTATS PRINCIPAUX.

4.1 Le résultat suivant a été annoncé il y a plusieurs années par Oshima (cf. [O]).

Théorème 1 :

Soit V un sous-module de Harish Chandra de $C^\infty(G/H)$. Alors pour tout $\mathfrak{p} \in \mathfrak{F}$ et tout $\lambda_0 \in e_\ell(\mathfrak{p}, V)$, il existe un sous-module irréductible de dimension finie du $M_\emptyset$ - module $C^\infty(M_\emptyset / M_\emptyset \cap H)$, σ, et un morphisme non nul de $(\underline{g}, K)$ - modules de V dans $\operatorname{Ind}_{P_\emptyset(\mathfrak{p}) \uparrow G} \sigma \otimes e^{\lambda_0} \otimes 1_{N_\emptyset(\mathfrak{p})}$. En particulier, si V est irréductible, ce morphisme est injectif.

Démonstration :

Il suffit d'appliquer la construction de 3.2 avec $\Theta = \emptyset$ pour obtenir un morphisme non nul de modules de Harish Chandra pour $M_\emptyset A_\emptyset$, i_{λ_0}, de $V/\underline{n}_\emptyset(\mathfrak{p}) V = V_1$ dans $C^\infty(M_\emptyset/M_\emptyset \cap H) \otimes \mathbb{C}_{\lambda_0 + \rho_{\mathfrak{p}}}$. Or $M_\emptyset / M_\emptyset \cap H$ est compact. Donc tout sous-module de Harish Chandra pour $M_\emptyset$ de $C^\infty(M_\emptyset / M_\emptyset \cap H)$ est unitarisable, car $C^\infty(M_\emptyset/M_\emptyset \cap H)$ est contenu dans $L^2(M_\emptyset/ M_\emptyset \cap H)$. On en déduit que $i_{\lambda_0}(V_1)$ est semi-simple et l'on peut choisir pour σ un facteur direct non nul de $i_{\lambda_0}(V_1)$. Montrons que σ est de dimension finie. Pour cela rappelons (cf. [O.S.], § 8) que $\underline{m}_\emptyset = \underline{m}_1 \oplus \underline{m}_2$ (produit d'algèbres de Lie) avec $\underline{m}_1 \subset \underline{k}$ et $\underline{m}_2 \subset h$. Il en résulte facilement que tout sous-module de Harish Chandra pour la composante neutre $M^0_\emptyset$ de $M_\emptyset$ de $C^\infty(M^0_\emptyset / M^0_\emptyset \cap H)$ est de dimension finie. Comme, en outre $M_\emptyset = F M^0_\emptyset$

avec F sous groupe fini de K, cela implique la même assertion pour tout sous-module de Harish Chandra de $C^\infty(M_\emptyset / M_\emptyset \cap H)$ pour $M_\emptyset$ et donc σ est de dimension finie. On achève la démonstration du théorème 1 en utilisant la réciprocité de Frobenius (cf. [H.S.] 4.11).

4.2 On dit qu'un sous-module de Harish Chandra, V, de $C^\infty(G/H)$ est H - tempéré si et seulement si pour tout $\mathcal{P} \in \mathcal{F}$ et $\lambda \in e(\mathcal{P}, V)$ on a $\mathrm{Re}\,\lambda(X) \leq 0$ pour tout $X \in a^-_\emptyset(\mathcal{P})$. Rappelons que V est contenu dans $L^2(G/H)$ dès que les inégalités ci-dessus sont strictes pour tout X non central dans la cloture de $\underline{a}^-_\emptyset(\mathcal{P})$ (cf. [v.d.B.], th. 9.4). Le théorème suivant a été également annoncé par Oshima.

Théorème 2 :

Soit V un sous-module de Harish Chandra H - tempéré de $C^\infty(G/H)$. Alors il existe un sous-groupe parabolique de G contenant $P_\emptyset(\mathcal{P})$ pour un $\mathcal{P} \in \mathcal{F}$, de la forme $P_\Theta(\mathcal{P}) = M_\Theta A_\Theta N_\Theta(\mathcal{P})$, une série discrète, δ, pour $M_\Theta / M_\Theta \cap H$ (i.e. un sous-module de Harish Chandra irréductible de $C^\infty(M_\Theta / M_\Theta \cap H)$ contenu dans $L^2(M_\Theta / M_\Theta \cap H)$), $\nu_0 \in i\,\underline{a}^*_\Theta$, et un morphisme non nul de $(\underline{g}, K)$ - modules de V dans $\mathrm{Ind}_{P_\Theta(\mathcal{P}) \uparrow G}\ \delta \otimes e^{\nu_0} \otimes 1_{N_\Theta(\mathcal{P})}$.

Démonstration :

Soit $\mathcal{P} \in \mathcal{F}$ et $\Sigma_\mathcal{P} = \{\alpha_1, \ldots, \alpha_\ell\}$ l'ensemble des racines simples de $\mathcal{P}$. Soient $\beta_1, \ldots, \beta_\ell \in \underline{a}^*_\emptyset$ définies par $(\alpha_i, \beta_j) = \delta_{ij}$ et $\beta_j(\underline{c}) = 0$ où $(.,.)$ est la forme duale de $B_{|\underline{a}_\emptyset \times \underline{a}_\emptyset}$ et $\underline{c}$ est l'intersection du centre de $\underline{g}$ avec $\underline{a}_{\emptyset}$. On se fixe alors $\mathcal{P} \in \mathfrak{F}$ et $\lambda_0 \in e(\mathcal{P}, V)$ tel que l'ensemble des β_i orthogonaux à $\operatorname{Re} \lambda_0$ soit de cardinal maximal. On note alors Θ l'ensemble des $\alpha_i \in \Sigma_{\mathcal{P}}$ tels que $(\operatorname{Re} \lambda_0, \beta_i) \neq 0$. On note $\nu_0 = \lambda_{0|\underline{a}_\Theta}$. On va voir que ν_0 est un élément de $e_\ell(\mathcal{P},\Theta,V)$. Soit $\nu \in e(\mathcal{P},\Theta,V)$ avec $\nu \underset{\mathcal{P},\Theta}{\leq} \nu_0$.

On va voir qu'alors $\nu = \nu_0$. Pour cela on choisit $\lambda \in e(\mathcal{P}, V)$ tel que $\lambda_{|\underline{a}_\Theta} = \nu$ (cf. 2.4). L'hypothèse $\nu \underset{\mathcal{P},\Theta}{\leq} \nu_0$ se traduit par

$\lambda - \lambda_0 = \sum_{i=1}^{\ell} x_i \alpha_i$ avec $x_i \in -\mathbb{N}$ si $\alpha_i \notin \Theta$. L'hypothèse que V est tempérée implique que pout tout i, $(\operatorname{Re} \lambda, \beta_i) \geq 0$.

Calculons alors $(\operatorname{Re} \lambda, \beta_i)$ pour $\beta_i \notin \Theta$. On a $(\operatorname{Re} \lambda_0, \beta_i) = 0$.

D'où

$$(\operatorname{Re} \lambda, \beta_i) = (\operatorname{Re} (\lambda - \lambda_0), \beta_i) = x_i \in -\mathbb{N}.$$

On en déduit $(\operatorname{Re} \lambda, \beta_i) = 0$ si $\beta_i \notin \Theta$.

Or $\nu - \nu_0 = \operatorname{Re} (\nu - \nu_0) = \operatorname{Re} (\lambda - \lambda_0)_{|\underline{a}_\Theta}$. D'où

$$\nu - \nu_0 = \sum_{\alpha_i \notin \Theta} (\operatorname{Re} (\lambda - \lambda_0), \beta_i)\, \alpha_i ,$$

et $\nu = \nu_0$ d'après ce qui précède. Ce qui achève de prouver que $\nu_0 \in e_\ell(\mathcal{P},\Theta,V)$. Alors on peut construire un morphisme non nul i_{ν_0} de $V/\underline{n}_\Theta(\mathcal{P})V$ dans $C^\infty(M_\Theta / M_\Theta \cap H) \otimes \mathbb{C}_{\nu_0 + \rho_{\mathcal{P},\Theta}}$ (cf. 3.2).

Il reste à voir que l'image de i_{ν_0} est dans L^2 et ν_0 imaginaire pur. D'abord, comme V est tempérée, on a $\operatorname{Re} \lambda_0$ nulle sur $\underline{c}$. Alors $\operatorname{Re} \lambda_0 = \sum_{i=1}^{\ell} y_i \alpha_i$ et $\operatorname{Re} \nu_0 = \sum_{\alpha_i \notin \Theta} y_i \alpha_{i|\underline{a}_\Theta}$.

Or $y_i = (\mathrm{Re}\,\lambda_0 , \beta_i)$ est nul si $\alpha_i \in \Theta$. Donc ν_0 est imaginaire pur. Reprenons maintenant les notations du lemme 3. Soit $\lambda \in e\,(\mathcal{P}'_1 , i_{\nu_0}(V))$. Il nous faut montrer que, pour tout $X \in \underline{a}^{\Theta}$ vérifiant $\alpha\,(X) \leq 0$ pour tout $\alpha \in \mathcal{P}'_1$ et X non central dans $\underline{m}_{\Theta}$, on a $\mathrm{Re}\,\lambda\,(X) \leq 0$. Pour cela on remarque que d'après le lemme 3, $\lambda + \nu_0 \in e\,(\mathcal{P}', V)$ et comme V est tempéré $\mathrm{Re}\,(\lambda + \nu_0)_{|\underline{c}} = 0$.

Notons $\Sigma_{\mathcal{P}'} = \{\alpha'_1, \ldots, \alpha'_\ell\}$, $\Theta' = \Sigma_{\mathcal{P}'} \cap \mathcal{P}'_1$ etc... Il nous reste à prouver que pour $\beta'_i \in \Theta'$ on a $(\mathrm{Re}\,\lambda , \beta'_i) > 0$. Comme V est H - tempéré et $\lambda + \nu_0 \in e\,(\mathcal{P}', V)$ on a, pour tout i , $(\mathrm{Re}\,(\lambda + \nu_0), \beta'_i) \geq 0$. Par ailleurs on a vu que $\mathrm{Re}\,\nu_0 = 0$. D'où, pour tout i , $(\mathrm{Re}\,\lambda , \beta'_i) = (\mathrm{Re}\,(\lambda + \nu_0), \beta'_i) \geq 0$. En outre, si $\beta'_i \notin \Theta'$, on a $(\mathrm{Re}\,\lambda , \beta'_i) = 0$ car les $\beta'_i \notin \Theta'$ sont orthogonaux à $(\underline{a}^{\Theta})^*$. Si on avait $(\mathrm{Re}\,\lambda , \beta'_i) = 0$ pour un $\beta'_i \in \Theta'$, cela contredirait la propriété de λ_0 (car $\mathrm{Card}\,\Theta' = \mathrm{Card}\,\Theta$). D'où $(\mathrm{Re}\,\lambda , \beta'_i) > 0$ pour tout $\beta'_i \in \Theta'$. Donc $i_{\lambda_0}(V)$ est dans $L^2\,(G/H)$. Alors on prend pour δ un facteur direct de $i_{\lambda_0}(V)$ (qui est unitarisable) et on applique la réciprocité de Frobenius pour achever la démonstration du théorème 2.

4.3

<u>Théorème 3</u> : Soit V un sous-module de Harish Chandra de $C^{\infty}\,(G/H)$. Alors il existe un sous-groupe parabolique de G contenant $P_{\emptyset}\,(\mathcal{P})$ pour un $\mathcal{P} \in \mathcal{F}$, de la forme $P_{\Theta}\,(\mathcal{P}) = M_{\Theta}\,A_{\Theta}\,N_{\Theta}\,(\mathcal{P})$, τ un sous-module $M_{\Theta} \cap H$ - tempéré de $C^{\infty}\,(M_{\Theta} / M_{\Theta} \cap H)$, $\nu_0 \in (\underline{a}^*_{\Theta})_{\mathbb{C}}$ tel que $\mathrm{Re} < \nu_0, \alpha > \, < 0$ pour tout α racine de $\underline{a}_{\Theta}$ dans $\underline{n}_{\Theta}\,(\mathcal{P})$ et un morphisme non nul de $(\underline{g}, K)$ - modules de V dans

$$\operatorname*{Ind}_{P_{\Theta}(\mathcal{P}) \uparrow G} (\tau \otimes e^{\nu} \otimes 1_{N_{\Theta}(\mathcal{P})}) \; .$$

Démonstration :

Soit $\mathcal{P} \in \mathcal{F}$. Notons $C = \{\lambda \in \underline{a}_{\emptyset}^{*} \mid (\lambda|\alpha) < 0 , \forall \alpha \in \mathcal{P}\}$.

Pour $\lambda \in \underline{a}_{\emptyset}^{*}$, on note $\lambda_{\mathcal{P}}$ sa projection sur le cône convexe fermé $\overline{C}$. Supposons que l'on ait $\lambda \underset{\mathcal{P}}{\leq} \lambda'$. On va voir qu'alors $\|\lambda_{\mathcal{P}}\| \geq \|\lambda'_{\mathcal{P}}\|$ avec égalité seulement si $\lambda_{\mathcal{P}} = \lambda'_{\mathcal{P}}$. En effet, d'après les propriétés de la projection sur un convexe fermé, on a $\lambda - \lambda_{\mathcal{P}} \in C^0 = \{\lambda \in \underline{a}_{\emptyset}^{*} \mid \forall \mu \in C , (\lambda|\mu) \leq 0\}$. Clairement C^0 s'identifie au cône des combinaisons linéaires à coefficients positifs d'éléments de $\mathcal{P}$. En outre, $\lambda - \lambda_{\mathcal{P}}$ et $\lambda_{\mathcal{P}}$ sont orthogonaux. Alors :

$$\|\lambda_{\mathcal{P}}\|^2 = \|\lambda\|^2 - \|\lambda - \lambda_{\mathcal{P}}\|^2 .$$

Par ailleurs, les propriétés de la projection sur $\overline{C}$ impliquent :

$$\|\lambda - \lambda_{\mathcal{P}}\| \leq \|\lambda - \lambda'_{\mathcal{P}}\| ,$$

puisque $\lambda'_{\mathcal{P}} \in \overline{C}$. D'où :

$$\|\lambda_{\mathcal{P}}\|^2 \geq \|\lambda\|^2 - \|\lambda - \lambda'_{\mathcal{P}}\|^2 \qquad (*)$$

Or :

$$\|\lambda\|^2 - \|\lambda - \lambda'_{\mathcal{P}}\|^2 = \|\lambda'_{\mathcal{P}}\|^2 + 2\,(\lambda - \lambda'_{\mathcal{P}} \mid \lambda'_{\mathcal{P}}) .$$

Par ailleurs on a :

$$(\lambda' - \lambda'_{\mathcal{P}} \mid \lambda'_{\mathcal{P}}) = 0 \quad \text{i.e.} \quad (\lambda' \mid \lambda'_{\mathcal{P}}) = (\lambda'_{\mathcal{P}} \mid \lambda'_{\mathcal{P}}) .$$

D'où :

$$\|\lambda\|^2 - \|\lambda - \lambda'_{\mathcal{P}}\|^2 = \|\lambda'_{\mathcal{P}}\|^2 + 2\,(\lambda - \lambda' \mid \lambda'_{\mathcal{P}}) .$$

Mais $\lambda \underset{\mathcal{P}}{\leq} \lambda'$ et $\lambda'_{\mathcal{P}} \in \overline{C}$ impliquent que :

$$(\lambda - \lambda' \mid \lambda'_{\mathcal{P}}) \geq 0 .$$

Donc :

$$\|\lambda\|^2 - \|\lambda - \lambda'_{\mathcal{P}}\|^2 \geq \|\lambda'_{\mathcal{P}}\|^2 .$$

Ceci, joint à (*), donne :

$$\|\lambda_P\|^2 \geq \|\lambda'_P\|^2 \quad .$$

Lorsque $\|\lambda_P\| = \|\lambda'_P\|$ on voit que l'on a nécessairement $\|\lambda - \lambda'_P\| = \|\lambda - \lambda_P\|$, d'où $\lambda_P = \lambda'_P$. Ceci achève de prouver notre assertion. Ceci montre que la fonction sur $e(P,V)$ définie par $\lambda \to \|\mathrm{Re}\,\lambda_P\|$ atteint son maximum en un point de l'ensemble fini $e_\ell(P,V)$. On choisit alors $P \in \mathfrak{F}$ et $\lambda^0 \in e(P,V)$ tel que $\|\mathrm{Re}\,\lambda^0_P\|$ soit maximum (on maximise parmi tous les P et λ possibles).

On note Θ l'ensemble des racines simples de P telles que $(\mathrm{Re}\,\lambda^0 \mid \alpha) = 0$. On notera $\nu^0 = \lambda^0_{|\underline{a}_\Theta} = \lambda^0_P$. Alors on a bien $(\mathrm{Re}\,\nu^0 \mid \alpha) < 0$ pour tout α racine de $\underline{a}_\Theta$ dans $\underline{n}_\Theta(P)$. On va voir également que ν^0 est un élément de $e_\ell(P,\Theta,V)$. Pour cela, soit $\lambda \in e(P,V)$ tel que $\lambda_{|\underline{a}_\Theta} \underset{P,\Theta}{\leq} \nu^0$. Alors on a $\mathrm{Re}\,\lambda_{|\underline{a}_\Theta} \underset{P,\Theta}{\leq} \mathrm{Re}\,\nu^0$. Alors, comme on a $\mathrm{Re}\,\lambda^0_P \in \underline{a}^*_\Theta$, on a

$$(\mathrm{Re}\,\lambda - \mathrm{Re}\,\lambda^0_P \mid \mathrm{Re}\,\lambda^0_P) = (\mathrm{Re}\,\lambda_{|\underline{a}_\Theta} - \mathrm{Re}\,\lambda^0_P \mid \mathrm{Re}\,\lambda^0_P) \quad .$$

Par ailleurs on a $\lambda_{|\underline{a}_\Theta} - \lambda^0_P = \sum_{\alpha \in \Sigma_P - \Theta} m_\alpha\, \alpha_{|\underline{a}_\Theta}$ avec $m_\alpha \in -\mathbb{N}$.

Alors, comme $\mathrm{Re}\,\lambda^0_P = \nu^0 \in \underline{a}^*_\Theta \cap \overline{C}$, on a finalement

$$(\mathrm{Re}\,\lambda - \mathrm{Re}\,\lambda^0_P \mid \mathrm{Re}\,\lambda^0_P) = \sum_{\alpha \in \Sigma_P - \Theta} m_\alpha (\alpha \mid \nu^0) \geq 0 \qquad (**)$$

et l'égalité n'est valable que si les m_α sont tous nuls i.e. $\lambda_{|\underline{a}_\Theta} = \lambda^0_P = \nu_0$. De façon analogue à ce qui précède on voit que (* *) implique $\|\lambda_P\| \geq \|\lambda^0_P\|$ avec inégalité stricte sauf si (* *) est une égalité. D'après l'hypothèse sur λ^0 , on voit que ceci implique que (* *) est une égalité et donc $\lambda_{|\underline{a}_\Theta} = \lambda^0_P = \nu_0$.

Au bout du compte on vient de prouver que $\nu^0 \in e_\ell(\mathcal{P},\Theta,V)$. On construit alors (cf. 3.2) un morphisme non nul de modules de Harish Chandra pour $M_\Theta A_\Theta$, i_{ν_0}, de $V/\underline{n}_\Theta(\mathcal{P}) V$ dans $C^\infty(M_\Theta \mid M_\Theta \cap H) \otimes \mathbb{C}_{\nu_0}$. Alors, grâce à la réciprocité de Frobenius, il suffit, pour achever la démonstration du théorème 3, de prouver que $i_{\nu_0}(V)$ est $M_\Theta \cap H$ - tempéré. On emploie les notations du lemme 3. Soit $\mu \in e(\mathcal{P}'_1, i_{\nu_0}(V))$. Alors on a, d'après ce lemme, $\lambda = \mu + \nu_0 \in e(\mathcal{P}', V)$. Or $(\mathrm{Re}\,\nu_0 \mid \alpha) = 0$ si $\alpha \in \mathcal{P}'_1$ car $\mathrm{Re}\,\nu_0 \in \underline{a}^*_\Theta$ et $\alpha \in (\underline{a}^\Theta)^*$. D'autre part, si $\alpha \in \mathcal{P}'$ et $\alpha_{|\underline{a}_\Theta} \neq 0$, on a $\alpha \in \mathcal{P}$ par définition de $\mathcal{P}'$ et on a alors $(\mathrm{Re}\,\nu_0, \alpha) < 0$. D'où $(\mathrm{Re}\,\nu_0)_{\mathcal{P}'} = \mathrm{Re}\,\nu_0$. Alors l'égalité $(\mathrm{Re}\,\lambda - \mathrm{Re}\,\nu_0 \mid \mathrm{Re}\,\nu_0) = 0$ implique que $\|\mathrm{Re}\,\lambda_{\mathcal{P}'}\| \geq \|\mathrm{Re}\,\nu_0\|$ avec égalité si et seulement si $\lambda_{\mathcal{P}'} = \mathrm{Re}\,\nu_0$. Par définition de ν_0 on doit avoir égalité, donc $\lambda_{\mathcal{P}'} = \mathrm{Re}\,\nu_0$. Alors des propriétés de la projection il résulte que $\mu = \lambda - \lambda_{\mathcal{P}'}$ est de la forme $\sum_{\alpha \in \mathcal{P}'} x_\alpha \alpha$ avec $x_\alpha \geq 0$ pour tout α. Mais μ est nul sur $\underline{a}_\Theta$. Cela implique que $x_\alpha = 0$ dès que $\alpha \in \mathcal{P}'$ et $\alpha_{|\underline{a}_\Theta} \neq 0$. D'où $\mu = \sum_{\alpha \in \mathcal{P}'_1} x_\alpha \alpha$ et l'on a bien $\mu(X) \leq 0$ pour tout X vérifiant $\alpha(X) \leq 0$ pour tout $\alpha \in \mathcal{P}'_1$. Ceci montre que $i_{\nu_0}(V)$ est tempéré et achève de prouver le théorème.

4.4 En combinant les théorèmes 2 et 3 et utilisant l'induction par étage on a :

Corollaire du théorème 3 :

Soit V un sous-module de Harish Chandra de $C^\infty(G/H)$. Alors il existe un sous-groupe parabolique de G contenant $P_\emptyset(\mathcal{P})$ pour un $\mathcal{P} \in \mathcal{F}$, de la forme $P_{\Theta}(\mathcal{P}) = M_{\Theta} A_{\Theta} N_{\Theta}(\mathcal{P})$, une série discrète, δ , pour $M_{\Theta}/M_{\Theta} \cap H$, un élément ν de $(\underline{a}_{\Theta})^{*}_{\mathbb{C}}$ dont la partie réelle est dans la cloture de la chambre de Weyl négative $\underline{a}^{-}_{\Theta}(\mathcal{P})$ et un morphisme non nul de $(\underline{g}, K)$ - modules de V dans $\mathrm{Ind}_{P_{\Theta}(\mathcal{P}) \uparrow G} (\delta \otimes e^{\nu} \otimes 1_{N_{\Theta}(\mathcal{P})})$.

Bibliographie

[v.d.B.] E. van den Ban, Asymptotic behaviour of matrix coefficients related to reductive symmetric spaces (preprint 1984).

[v.d.B.D.] E. van den Ban et P. Delorme (en préparation).

[B.W.] A. Borel et N. Wallach, Continuous cohomology, discrete subgroups and representations of reductive groups, Annals of Math. Studies, 94, (1980), Princeton University Press, Princeton.

[Carm.] J. Carmona, Sur la classification des modules admissibles irréductibles, dans Non Commutative Harmonic Analysis and Lie Groups, Proceedings, Marseille-Luminy 1982, 11-34, L.N. in Mathematics 1020, Springer Verlag 1983.

[Ca.M.] W. Casselman et D. Milicic, Asymptotic behaviour of matrix coefficients of admissible representations. Duke Math. J., 49, 1982, 106-146.

[H.C.] Harish Chandra, Harmonic analysis on real reductive groups I, The theory of the constant term, Journ. of Funct. Anal., 19, 1975, 104-204.

.../...

[H.S.] H. Hecht et W. Schmid, Characters, asymptotics and $\underline{n}$ - homology of Harish Chandra modules, Acta Mathematica, 151, 1983, 49-151.

[O.] T. Oshima, Fourier analysis on semisimple symmetric spaces, dans Non Commutative Harmonic Analysis and Lie Groups, Proceedings, Marseille-Luminy, 1980, 357-369, L.N. in Mathematics 880, Springer Verlag 1981.

[O.S.] T. Oshima et J. Sekiguchi, The restricted root system of a semisimple symmetric pair, dans Group representations and systems of differential equations, 433-497, Advanced in Pure Mathematics, 4, 1984.

(*) Note ajoutée en cours d'épreuve

W. Casselman vient de me communiquer des notes manuscrites rédigées par lui en 1975 sur les équations différentielles satisfaites par les coefficients matriciels des représentations admissibles des groupes réductifs réels. Il est facile d'en extraire une démonstration du fait que $i_{\lambda_o}(F)$ est C^∞. Pour conclure que $i_{\lambda_o}(F)$ est analytique il suffit alors de remarquer qu'elle est annulée par un idéal de codimension finie du centre de $U(\underline{m}_\Theta)$ et d'utiliser un argument standard.

Cette démonstration n'utilise pas les appendices et est donc plus élémentaire. La partie "méthode de descente" de l'appendice A est toutefois commune aux deux preuves.

Je remercie vivement W. Casselman de m'avoir communiqué son manuscrit.

Appendice

par Erik van den BAN et Patrick DELORME.

Cet appendice est divisé en deux parties. La première partie établit un résultat d'extensions de fonctions analytiques de $\underline{a}$ à $\underline{p}$ utilisé dans l'article. Ce résultat contient en particulier le théorème de restriction de Chevalley pour les fonctions analytiques. Notre démonstration originale reposait sur des résultats de Kostant Rallis analysant la structure de $S(\underline{p})^K$ - module de $S(\underline{p})$ ainsi que le théorème de restriction de Chevalley pour les fonctions analytiques. Au lieu de cela nous utilisons le résultat de la deuxième partie (lemme B.1). Nous remercions T. Oshima pour nous avoir fourni une démonstration d'un résultat de même nature. Sa démonstration, qui utilise une propriété classique des polynomes de Tchebishef, pourrait être adaptée pour obtenir ce lemme.

Ce résultat apparaît comme une conséquence immédiate d'un résultat de J. Korevaar et J. Wiegerink (cf. [K.W.], main lemma). Nous le réutiliserons ultérieurement.

A.1. Les notations sont celles de 1.1 et 1.2 . On note $\underline{a}$ un sous-espace abélien maximal de $\underline{p}$, M (resp. M') le centralisateur (resp. normalisateur) de $\underline{a}$ dans K , $W = M'/M$ le groupe de Weyl correspondant. Soit μ une représentation unitaire de K dans un espace de dimension finie E .

<u>Proposition A.1</u> Soit $\psi : S(\underline{p}) \to E$ une fonction μ - sphérique (ici K agit sur $S(\underline{p})$ par représentation adjointe). Soit $\Phi : \underline{a} \to E$ une fonction analytique sur E . On suppose que :

$$\forall D \in S(\underline{a}) \quad , \quad (L_D \Phi)(0) = \psi(D) \quad .$$

Alors il existe une unique fonction μ - sphérique $\Psi : p \to E$ telle que $\Psi_{|\underline{a}} = \Phi$. De plus, Ψ est analytique et pour tout $D \in S(\underline{p})$ on a $(L_D \Psi)(0) = \psi(D)$ (Ici $\underline{p}$ est regardé comme un groupe commutatif et L désigne la différentielle de sa représentation régulière gauche).

Première partie de la démonstration :

L'unicité de Ψ , si elle existe, résulte du fait que $\underline{p} = \mathrm{Ad}\, K \;\mathrm{cl}\; \underline{a}^-$, où $\mathrm{cl}\; \underline{a}^-$ est la fermeture d'une chambre de Weyl $\underline{a}^-$ du système de racines $\Delta(\underline{g}, \underline{a})$ de $\underline{a}$ dans $\underline{g}$.

A.2. On rappelle que $\underline{c}$ est le centre de $\underline{g}$. On remarque que $\underline{c} \cap \underline{p} \subset \underline{a}$. Pour poursuivre la démonstration de la proposition A.1 nous aurons besoin du lemme suivant :

Lemme A.1 :

Soit $X_0 \in \underline{c} \cap \underline{p}$ tel qu'il existe une fonction μ - sphérique linéaire $\psi_{X_0} : S(\underline{p}) \to E$ telle que :

$$\forall D \in S(\underline{a}) \quad , \quad \psi_{X_0}(D) = (L_D \Phi)(X_0) \quad .$$

Alors il existe une fonction analytique, à valeurs dans E , Ψ_{X_0} , définie sur un voisinage V_{X_0} de X_0 dans $\underline{p}$, qui prolonge $\Phi_{|V_{X_0} \cap \underline{a}}$ et vérifiant $(L_D \Psi_{X_0})(X_0) = \psi_{X_0}(D)$ pour tout D dans $S(\underline{p})$.
En outre, si la série de Taylor de Φ en X_0 converge sur la boule $B_{\underline{a}}(X_0, \varepsilon)$ pour un $\varepsilon > 0$, la série de Taylor de Ψ_{X_0} en X_0 converge sur la boule $B_{\underline{p}}(X_0, \varepsilon/B)$ où B est une constante (dépendant seulement de $n = \dim \underline{p}$). Enfin ψ_{X_0} est μ - sphérique sur $B_{\underline{p}}(X_0, \varepsilon/B)$.

<u>Notation.</u> Si F est un espace normé, nous avons employé la notation, pour $x \in F$ et $\varepsilon > 0$, $B_F(x, \varepsilon) = \{y \in F \mid \|x - y\| < \varepsilon\}$.

Démonstration :

On se fixe une base orthonormée $H_1, \ldots, H_n$ de $\underline{p}$. Alors, pour $\omega \in S^{n-1}$ (sphère unité de $\mathbb{R}^n$) et $\omega . H = \omega_1 H_1 + \ldots + \omega_n H_n \in \underline{a}$, on a :

$$\Phi(X_0 + t(\omega . H)) = \sum_{j \geq 0} \frac{1}{j!} \psi_{X_0}((\omega . H)^j)\, t^j ,$$

la somme étant absolument convergente pour $t \in]-\varepsilon, \varepsilon[$. Soit $\eta \in]0, \varepsilon[$. Alors, grâce aux inégalités de Cauchy, il existe une constante $c > 0$ telle que :

(1) $\forall\, \omega \in S^{n-1}$, $\omega . H \in \underline{a} \Rightarrow \|\psi_{X_0}((\omega . H)^j)\| \leq c\, (\frac{1}{\eta})^j\, j!\ (j \in \mathbb{N})$.

Alors, si ω est un élément quelconque de S^{n-1} , il existe un $k \in K$ tel que $(\mathrm{Ad}\, k)(\omega . H) \in \underline{a}$. De plus, Ad k étant une transformation orthogonale, on a $(\mathrm{Ad}\, k)(\omega . H) = \omega' . H$ pour un $\omega' \in S^{n-1}$. Comme ψ_{X_0} est μ - sphérique, on a $\psi_{X_0}(\omega . H)^j) = \mu(k)^{-1}\, \psi_{X_0}((\omega' . H)^j)$. Grâce au fait que μ est unitaire on voit que (1) implique :

(2) $\forall\, \omega \in S^{n-1}$, $\psi_{X_0}((\omega . H)^j) \leq c\, \eta^{-j}\, j!$, $j \in \mathbb{N}$.

Alors, si B est la constante du lemme B.1, on en déduit :

$$\forall\, \ell \in \mathbb{N}^n ,\ \|\psi_{X_0}(H_1^{\ell_1} \ldots H_n^{\ell_n})\| \leq c\, (\frac{B}{\eta})^{|\ell|}\, \ell! ,$$

où $|\ell| = \ell_1 + \ldots + \ell_n$, $\ell! = \ell_1! \ldots \ell_n!$

Il en résulte que la série entière :

$$(3) \qquad \sum_{\ell \in \mathbb{N}^n} \frac{1}{\ell\,!}\, \psi_{X_0}(H_1^{\ell_1} \ldots H_n^{\ell_n})\, t_1^{\ell_1} \ldots t_n^{\ell_n} \quad ,$$

converge absolument pour $|t_j| < \frac{\eta}{B}$, $1 \leq j \leq n$. Par conséquent, il existe une fonction analytique $\Psi_{X_0} : B_{\underline{p}}(X_0 , \eta/B) \to E$ avec $\Psi_{X_0}(X_0 + t.H)$ égale à (3) pour t assez petit dans $\mathbb{R}^n$. Puisque η a été choisi quelconque dans $]0, \varepsilon[$, Ψ_{X_0} s'étend en une fonction analytique sur $B_{\underline{p}}(X_0 , \varepsilon/B)$. Ayant la même série de Taylor en X_0 , $\Psi_{X_0}|B_{\underline{a}}(X_0 , \varepsilon/B)$ doit coïncider avec $\Phi|B_{\underline{a}}(X_0 , \varepsilon/B)$. Finalement, chaque terme de la série de Taylor de Ψ_{X_0} en X_0 est μ - sphérique (car Ad K fixe X_0), donc Ψ_{X_0} est μ - sphérique et ceci achève de prouver le lemme A.1.

A.3. <u>Lemme A.2</u> :

Soit $\mathcal{A}$ l'ensemble des éléments X_0 de $\underline{c} \cap \underline{p}$ tel qu'il existe une application linéaire $\psi_{X_0} : S(\underline{p}) \to E$ vérifiant

a) ψ_{X_0} est μ - sphérique .

b) $\psi_{X_0}(D) = (L_D \Phi)(X_0)$, $\forall\, D \in S(\underline{a}.)$.

Alors $\mathcal{A}$ est égal à $\underline{c} \cap \underline{p}$.

<u>Démonstration</u> :

Par hypothèse $\mathcal{A}$ contient 0 . D'après le lemme A.1, $\mathcal{A}$ est ouvert. En effet, soit $X_0 \in \mathcal{A}$ et soit $\varepsilon > 0$, $\Psi_{X_0} : B_{\underline{p}}(X_0 , \varepsilon/B) \to E$, comme dans le lemme A.1 . Alors, pour $Y_0 \in B_{\underline{p}}(X_0, \epsilon/B) \cap \underline{c}$, on définit, pour $D \in S(\underline{p})$, $\psi_{Y_0}(D) = (L_D \Psi_{X_0})(Y_0)$. Comme Ψ_{X_0} est μ - sphérique et que Ad K centralise Y_0 , (a) est vérifié. De plus, comme Ψ_{X_0} coïncide avec Φ sur $B_{\underline{a}}(X_0, \epsilon/B)$, (b) est vrai. Donc $B_{\underline{a}}(X_0, \epsilon/B) \cap \underline{c} \subset \mathcal{A}$ et $\mathcal{A}$ est ouvert.

Montrons que $\mathcal{A}$ est fermé. Pour cela il suffit de voir que l'intersection de $\mathcal{A}$ avec la fermeture $\mathrm{cl}\, B_{\underline{a}}(0,R)$, de $B_{\underline{a}}(0,R)$, est fermée pour tout $R > 0$. Puisque Φ est analytique dans un voisinage de l'ensemble compact $\mathrm{cl}\,(B_{\underline{a}}(0,R))$, il existe un $\epsilon > 0$ tel que pour tout $X \in \mathrm{cl}\,(B_{\underline{a}}(0,R))$, la série de Taylor de Φ en X converge sur $B_{\underline{a}}(X, \epsilon)$. Soit Y adhérent à $\mathcal{A} \cap \mathrm{cl}\, B_{\underline{a}}(0,R)$. On peut trouver $X \in \mathcal{A} \cap \mathrm{cl}\,(B_{\underline{a}}(0,R))$ tel que $\|X - Y\| < \epsilon/B$. Mais d'après la première partie de la démonstration $B_{\underline{a}}(X, \epsilon/B) \cap \underline{c}$ est contenu dans $\mathcal{A}$. Finalement, $\mathcal{A}$ est fermé dans $\underline{c}$. De plus, il est ouvert et non vide d'après la première partie de la démonstration, donc égal à $\underline{c} \cap \underline{p}$.

A.4 Lemme A.3 :

Il existe une fonction μ - sphérique $\Psi : \underline{p} \to E$ qui prolonge Φ . Cette fonction coïncide avec Ψ_{X_0} (cf. lemme A.1) au voisinage de tout $X_0 \in \underline{c} \cap \underline{p}$. Elle est donc analytique au voisinage de tout point X_0 de $\underline{c} \cap \underline{p}$.

Démonstration :

D'après le lemme A.1 il existe une fonction Ψ_0 analytique définie sur un voisinage Ad K - invariant de 0, U, μ - sphérique et telle que $\Psi_0|_{\underline{a} \cap U} = \Phi|_{\underline{a} \cap U}$. Soit $Y \in \underline{p}$. Alors il existe $k \in K$, $X \in \underline{a}$, tel que $Y = (\mathrm{Ad}\, k)\, X$. Si $k' \in K$, $X' \in \underline{a}$ vérifient $Y = (\mathrm{Ad}\, k')\, X'$, on a $\mu(k)\, \Psi(X) = \mu(k')\, \Psi(X')$. Pour le voir on considère la fonction analytique $f : \mathbb{R} \to E$ définie par $f(t) = \mu(k)\, \Psi(t\, X) - \mu(k')\, \Psi(t\, X')$. Si t est suffisamment petit on a $t\, X$, $t\, X' \in U$ et $f(t) = \Psi_0(t\, (\mathrm{Ad}\, k)\, X) - \Psi_0(t\, (\mathrm{Ad}\, k')\, X') = 0$. Par prolongement analytique on en déduit que f est identiquement nulle. On peut donc définir une fonction $\Psi : \underline{p} \to E$ par $\Psi((\mathrm{Ad}\, k)\, X) = \mu(k)\, \Phi(X)$ pour $k \in K$ et $X \in \underline{a}$. Cette fonction est clairement μ - sphérique et coïncide avec Φ sur $\underline{a}$. Cela implique immédiatement que, pour tout $X_0 \in \underline{c}$, Ψ coïncide avec Ψ_{X_0} sur un voisinage $\mathrm{Ad}\, K$ invariant de X_0 dans $\underline{p}$. Ceci achève de prouver le lemme.

A.5. Fin de la démonstration de la proposition A.1.

On procède par récurrence sur la dimension de G . Si G est de dimension 1, G est commutatif et $\underline{c} \cap \underline{p} = \underline{a}$. Donc Ψ est analytique grâce au lemme A.3. Donc la proposition A.1 est vraie lorsque $\dim G = 1$. On la suppose démontrée pour tous les groupes de dimension strictement plus petite que celle de G . Il reste à prouver que Ψ est analytique en tout point de $\underline{p}$. Comme Ψ est μ - sphérique, il suffit de le voir en tout X_0 de $\underline{a}$. Si $X_0 \in \underline{c}$ cela résulte du lemme A.3. Si $X_0 \notin \underline{c}$, le centralisateur G_{X_0} (resp. $\underline{g}_{X_0}$) de X_0 dans G (resp. $\underline{g}$) est de dimension strictement inférieure à celle de G . Alors on applique l'hypothèse de récurrence à G_{X_0} , qui est clairement dans la classe d'Harish Chandra, pour voir que $\Psi_1 = \Psi_{|\underline{p}_{X_0}}$ est analytique sur $\underline{p}_{X_0} = \underline{p} \cap \underline{g}_{X_0}$. Soit $\underline{s}$ un supplémentaire de $\underline{k}_{X_0} = \underline{k} \cap \underline{g}_{X_0}$ dans $\underline{k}$. Alors l'application f de $\underline{s} \times \underline{p}_{X_0}$ dans $\underline{p}$ définie par :

$\forall\, X \in \underline{s}$, $\forall\, Y \in \underline{p}_{X_0}$, $f(X, Y) = (\mathrm{Ad}\,(\exp X))\, Y$, est un difféomorphisme analytique local au voisinage de $(0, X_0)$. Pour le voir, il suffit de vérifier que $\underline{p} = \underline{p}_{X_0} \oplus [\underline{s}, X_0]$ puisque la différentielle de f en $(0, X_0)$ est donnée par $(X,Y) \in \underline{s} \times \underline{p}_{X_0} \to [X, X_0] + Y$. Or, d'après [War], prop. 1.3.5.4, on a $\underline{g} = \underline{g}_{X_0} \oplus [\underline{g}, X_0]$. En écrivant $\underline{g}_{X_0} = \underline{k}_{X_0} \oplus \underline{p}_{X_0}$, $\underline{g} = \underline{k}_{X_0} \oplus \underline{p} \oplus \underline{s}$ et en utilisant $[\underline{k}, \underline{p}] \subset \underline{p}$, $[\underline{p}, \underline{p}] \subset \underline{k}$, on a facilement l'égalité voulue. Alors, soit V un voisinage de X_0 dans $\underline{p}$ sur lequel f^{-1} est défini et analytique. Notons $j : V \to \underline{s}$ (resp. $h : V \to \underline{p}_{X_0}$) la composée de f^{-1} avec la projection de $\underline{s} \times \underline{p}_{X_0}$ sur $\underline{s}$ (resp. $\underline{p}_{X_0}$) . Alors, pour tout X dans V , on a :

$$\begin{aligned} \Psi(X) &= \Psi[\mathrm{Ad}(\exp(j(X)))(h(X))] \\ &= \mu(\exp(j(X)))\, \Psi_1(h(X)) . \end{aligned}$$

D'où il résulte que Ψ est analytique au voisinage de X_0 . Ceci achève de prouver que Ψ est analytique sur $\underline{p}$. Le fait que, pour $D \in S(\underline{p})$, $(L_D \Psi)(0) = \psi(D)$ est une conséquence du lemme A.1 et du fait que Ψ coïncide avec Ψ_0 au voisinage de 0 (lemme A.3). Ceci achève de prouver la proposition A.1.

B.

Soit $D_1, \ldots, D_n$ la base standard de $\mathbb{R}^n$. Si $\omega \in \mathbb{R}^n$ on pose $\omega . D = \omega_1 D_1 + \ldots + \omega_n D_n$. On utilisera les notations usuelles pour les multiindices.

Lemme B.1.

Soit Ω un ouvert non vide de la sphère unité S^{n-1} dans $\mathbb{R}^n$.
Alors il existe une constante $B > 0$ telle que

(i) Si ψ est une application linéaire de l'algèbre symétrique (complexe) $S(\mathbb{R}^n)$ de $\mathbb{R}^n$ dans un espace normé de dimension finie E , alors :

$$(*) \quad \frac{1}{\ell!} \|\psi(D_1^{\ell_1} \ldots D_n^{\ell_n})\| \leq \sup_{\omega \in \Omega} \|\psi((\omega \cdot D)^{|\ell|})\| \frac{B^{|\ell|}}{|\ell|!} .$$

(ii) Si $p(t) = \sum_{\ell \in \mathbb{N}^n, |\ell| = m} a_\ell t^\ell$ est un polynome homogène de degré m à n variables à valeurs dans E ,

$$\frac{1}{\ell!} \|a_\ell\| \leq \sup_{t \in \Omega} |p(t)| \frac{1}{|\ell|!} B^{|\ell|} .$$

Démonstration :

Il suffit de prouver (i). Grâce à [K.W.], lemme principal, on peut trouver des fonctions intégrables g_ℓ , $\ell \in \mathbb{N}^n$, sur Ω , telles que, pour tout $\ell \in \mathbb{N}^n$:

$$(**) \quad \frac{1}{\ell!} D_1^{\ell_1} \ldots D_n^{\ell_n} = \frac{1}{|\ell|!} \int_\Omega g_\ell(\omega) (\omega \cdot D)^{|\ell|} d\sigma .$$

Ici $d\sigma$ est la mesure euclidienne sur S^{n-1} . De plus on a :

$$(***) \quad \forall \ell \in \mathbb{N}^n , \int_\Omega |g_\ell(\omega)| d\sigma \leq B^{|\ell|} .$$

Comme ψ se restreint à une application linéaire continue de $S(\mathbb{R}^n)_m$ dans E pour tout $m \in \mathbb{N}$, (**) implique

$$\frac{1}{\ell !} \psi (D_1^{\ell_1} \ldots D_n^{\ell_n}) = \frac{1}{|\ell|!} \int_\Omega g_\ell (\omega) \psi ((\omega . D)^{|\ell|}) d\sigma .$$

D'après (***) cela implique (*).

Bibliographie

[K.W.] J. Korevaar et J. Wiegerinck, A representation of mixed derivatives with an application to the edge-of-the-wedge theorem. Indagationes Math. 47 (1985), 77-86.

(*) Département de Mathématique-Informatique
Faculté des Sciences de Luminy
70, route Léon-Lachamp
13288 MARSEILLE CEDEX 9 - France

(**) Rijksuniversiteit Utrecht
Mathematisch Instituut
Budapestlaan 6
Postbus 80.010
3508 TA UTRECHT - Nederlands

On the Cyclicity of Vectors Associated with Duflo Involutions

by

Anthony Joseph

1. Introduction

1.1. Let $\underline{g}$ be a complex semisimple Lie algebra. It was pointed out in ([9],III) that the calculation of the scale factors in the Goldie rank polynomials of primitive quotients of $U(\underline{g})$ requires a deep understanding of the left cells of the Weyl group W. This led to a number of conjectures on left cells. Perhaps the most basic of these, namely C_3 was that the vector in a left cell corresponding to a Duflo involution is cyclic with respect to W module structure on the cell. This was needed to calculate the socle of the Harish-Chandra module associated to linear maps between a pair of simple highest weight modules which in turn gave information on coefficients $z_w : w \in W$ arising in the transformation properties of the Goldie rank polynomials needed to calculate their scale factors.

1.2. The aim of the paper is to prove C_3. This is based on Lusztig's recent proof [18] of conjecture C_4 of ([9],III), his work on cells in affine Weyl groups [17] and on the isomorphism between the Hecke and Weyl group algebras [15]. I should like to thank him for a preview of [18] and drawing my attention to [15] and [17]. In fact given these results and an idea for proving cyclicity which came from [11], the whole analysis became quite straight-forward.

1.3. The present analysis leads to a sharp proliferation of further questions and conjectures about which we are able to say rather little. First an important technical tool in [17] and [18] is a positivity result on the coefficients in the decomposition of products of certain

---

Publisher's Note: For the above paper by A. Joseph, numerous corrections and additions were sent to the publisher after the volume was already in production. As it was not possible to have the paper rewritten appropriately, the corrections were inserted in the text exactly as instructed by the author.

Hecke algebra generators. This had been established via Gabber's decomposition theorem on perverse sheafs. When q = 1 we can show (quite easily) that these positivity constraints are exactly those obtained by decomposing products of projective functors (in the sense of Bernstein-Gelfand [2]). Formally one may further interpret the general positivity constraints in terms of a certain filtration associated to projective modules as developed in ([5], 1.10 and 3.8). Unfortunately this point of view has some difficulties and does not yet lead to a new proof of the positivity constraints, though it gives a little further understanding of projective functors. Secondly Lusztig's analysis involves the structure of leading coefficients in the decomposition of products of Hecke algebra elements. Formally one may interpret these terms as leading terms of the Harish-Chandra module associated to linear maps from an indecomposable projective to an indecomposable injective module (in the Bernstein-Gelfand-Gelfand (BGG) $\mathcal{O}$ category) which should identify with the socle of the Harish-Chandra module associated to linear maps between the corresponding simple modules. This is in fact proved by a very roundabout method (theorem 4.8). Thirdly these points of view lead to a whole new batch of combinatorial conjectures concerning left cells. Here a basic question is to refine the notation of a left cell so that these smaller objects lead to a complete reduction of the group algebra. Our proposal is based on the belief that the Harish-Chandra modules associated to the endomorphism rings of simple highest weight modules are in (some precisely defined sense) as numerous as the involutions in W, whereas of course the primitive quotients of the enveloping algebra are parametrized by the smaller set of Duflo involutions.

Finally we give a conjecture 5.6 for the coefficients $z_w : w \in W$ and show this leads at least for classical groups via recent results of Barbasch and Vogan [1] to a complete (though implicit) description of the scale factors in the Goldie rank polynomials. A main result Cor. 5.5 is that we are able to calculate the product $z_w z_{w^{-1}}$ and to show that the resulting expression is consistent with our conjecture.

2. Projective modules in the $\mathcal{O}$ category

We shall adopt the notation of [9]. (These papers will be designated simply by I, II, III). With regard to labelling certain basic elements of the Hecke algebra, there is some inevitable conflict with the notation of Lusztig [15 - 18].

2.1. Let $\underline{g}$ be a complex semisimple Lie algebra with Cartan subalgebra $\underline{h}$. Given $\mu \in \underline{h}^*$ let $M(\mu)$ denote the corresponding Verma module ([3], Chap. 7) and $L(\mu)$ the unique simple quotient of $M(\mu)$. Fix $\lambda \in \underline{h}^*$ dominant and regular. If we assume the Jantzen conjecture to hold then we have a formula ([5], 4.9) for the multiplicities $[M_j(w\lambda) : L(y\lambda)]$ of the simple modules in each gradation step $M_j(w\lambda)$ in the Jantzen filtration of $M(w\lambda)$ in terms of the Kazhdan-Lusztig polynomials $P_{w,y}(q)$. Indeed

$$(*) \qquad P_{w,y}(q) = \sum_{j=0}^{\infty} q^{\frac{\ell(y)-\ell(w)-j}{2}} [M_j(w\lambda) : L(y\lambda)].$$

Bernstein indicated a proof of the Jantzen conjecture at least for λ integral. Except in this respect our calculations are quite independent of λ being integral and for notational simplicity we just assume that λ is in fact integral taking w_0 to be the longest element of W. Let e denote the identity element of W.

2.2. By (*) above we have $\deg P_{e,y}(q) = \frac{1}{2}(\ell(y) - j)$ where j is the smallest integer such that $L(y\lambda)$ occurs in $M_j(\lambda)$. Label the elements $y \in W$ by the numbers $\ell(y) - 2\deg P_{e,y}(q)$. Fix a left cell $\mathcal{C}$ of W. By (III,4.9) there is a unique $\sigma \in \mathcal{C}$ where the above number takes it minimal value $k_{\mathcal{C}}$ on $\mathcal{C}$ and moreover σ is exactly the Duflo involution associated to $\mathcal{C}$ (see I, 3.4). We may conveniently write $\mathcal{C} = \mathcal{C}(\sigma)$. We showed in (III, 4.17) that $k_{\mathcal{C}}$ is constant on double cells and conjectured (III, C_4) that

$$(**) \qquad k_{\mathcal{C}(\sigma)} = \text{card } R^+ - d(L(\sigma w_0\lambda))$$

where d denotes Gelfand-Kirillov dimension. This was already clear in type A_n and has now been proved in general by Lusztig [18]. For the moment we observe the following natural but not quite obvious corollary. Let B' be a subset of simple roots B and $W_{B'}$ the subgroup of W generated by the reflections $s_\alpha : \alpha \in B'$. Let $\sum_{B'}^0$ denote the set of Duflo involutions of $W_{B'}$.

Corollary. - $\sum_{B'}^0 \subset \sum_B^0$.

Let $C_{B'}$ be a left cell of $W_{B'}$. As an immediate consequence of the definition of left cells, $C_{B'}$ is contained in a left cell C of W, an inclusion which is in general strict. However we shall show that the numbers we have assigned to the elements of $W_{B'}$ as well as the minimal value on each left cell $C_{B'}$ of $W_{B'}$ are unchanged if we replace $W_{B'}$ by W and $C_{B'}$ by C. This will prove the corollary.

Obviously $\ell(y)$ is independent of whether we regard $y \in W_{B'}$ as an element of $W_{B'}$ or of W. Taking account of the inductive definition of the Kazhdan-Lusztig polynomials one easily sees this to be also true of $\deg P_{e,y}(q)$.

It remains to establish that $k_{C_{B'}} = k_C$. By (II, 5.1) the right hand side of (**) is just the degree of the Goldie rank polynomial $p_{\sigma w_0}$. Again by (II, 5.1) the degree of p_w is just the smallest integer m such that

$$\sum_{w' \in W} a(w,w')w'^{-1}\delta^m \neq 0$$

where the $a(w,w')$ the coefficients of the inverse of the Jantzen matrix $\{b(w,w') := [M(w\lambda) : L(w'\lambda)]\}_{w,w' \in W}$. Here we also remark that δ can be any regular element of $\underline{h}^*$ (c.f. [12], 5.1). Take any $y \in C_{B'}$ and consider

$$\sum_{w' \in W} a(yw_0,w'w_0)w'^{-1}\delta^m$$

which determines $\deg p_{yw_0}$. One has

$$a(yw_0,w'w_0) = (-1)^{\ell(y)+\ell(w')}b(w',y) = (-1)^{\ell(y)+\ell(w')}P_{w',y}(1).$$

In particular the left hand side is zero unless $w' < y$ and this in turn implies $w' \in W_{B'}$. From the inductive definition of the Kazhdan-Lusztig polynomials it follows that $P_{w',y}(1)$ is independent of whether we view $w',y \in W_{B'}$ as elements of $W_{B'}$ or of W. Finally deg P_{yw_0} is constant on the left cells and so we can also take $y \in C$. This proves the required independence.

2.3. In discussing simple highest weight modules, it is natural to identify the Grothendieck group element $[M(ww_0\lambda)]$ with $w \in W$ and then $[L(yw_0\lambda)]$ is identified with the element

$$a(y) := \sum_{w\in W} a(yw_0, ww_0)w$$

in the group ring $\mathbb{Z}W$. This was the point of view introduced in [8] and which gave rise to the notion of cells in W. Moreover replacing $\mathbb{Z}W$ by the Hecke algebra $\underline{H}$ (whose generators as a $\mathbb{Z}[q,q^{-1}]$ module we will denote as usual by T_y : $y \in W$) Kazhdan and Lusztig were then led to their conjectures [14] for the matrix elements $a(w,y)$. Set $t = q^{1/2}$, $\tilde{\underline{H}} = \underline{H}[t]$. To interpret the more refined data given by the Hecke algebra, we identify $[M(ww_0\lambda)]$ with $\tilde{T}_w := t^{-\ell(w)}T_w$ and view the appearance of $[L(yw_0\lambda)]$ in the jth step of the Jantzen filtration of $M(ww_0\lambda)$ as providing the term $t^j a_y$ in $\tilde{T}_w$.

Explicitly

$$\tilde{T}_w = \sum_{y\in W} \sum_{j=0}^{\infty} t^j [M_j(ww_0\lambda) : L(yw_0\lambda)] a_y$$

$$= \sum_{y\in W} t^{\ell(w)-\ell(y)} P_{ww_0,yw_0}(q^{-1}) a_y$$

by (*). Inversion formulae then give

$$a_w = \sum_{y\in W} (-1)^{\ell(w)-\ell(y)} q^{1/2\ell(w)-\ell(y)} P_{y,w}(q^{-1}) T_y$$

which are exactly the self-dual elements of Kazhdan-Lusztig identified with the simple (self-dual) highest weight modules $L(ww_0\lambda)$.

2.4. The multiplication formulae

$$a_x a_y = \sum_{z\in W} \alpha_{x,y,z} a_z$$

exhibit remarkable positivity constraints. One has [17]

$$\alpha_{x,y,z} = \sum_{i\in\mathbb{Z}} c^{(i)}_{x,y,z}(-t)^i$$

where the coefficients $c^{(i)}_{x,y,z}$ are non-negative integers. This is proved using deep results in the theory of perverse sheaves. We interpret this result in terms of Verma modules by extending the point of view of 2.3 though by making some modifications. This will involve the projective covers $P(w\lambda)$ of the $L(w\lambda)$. Here we identify the Grothendeck group element $[M(y\lambda)]$ with $y \in W$ and then $[P(w\lambda)]$ is identified with the element $b(w) \in \mathbb{Z}\, W$ given by

$$b(w) := \sum_{y\in W} [P(w\lambda) : M(y\lambda)]y \quad ,$$

$$= \sum_{y\in W} [M(y\lambda) : L(w\lambda)]y \quad , \quad \text{by BGG duality}$$

$$= \sum_{y\in W} b(y,w)y.$$

Let $\underline{K}$ denote the category of all finitely generated $U(\underline{g})$ modules satisfying $\dim(Z(\underline{g})/\mathrm{Ann}_{Z(\underline{g})}M) < \infty$, $\forall\, M \in \mathrm{Ob}\,\underline{K}$. Then each $M \in \mathrm{Ob}\,\underline{K}$ admits primary decomposition with respect to $Z(\underline{g})$ and we let $\underline{K}_{\hat{\lambda}}$ denote the subcategory of objects in $\underline{K}$ annihilated by a power of $\mathrm{Ann}_{Z(\underline{g})}M(\lambda)$. Fix $w \in W_\lambda$. After Bernstein and Gelfand [2] there is a uniquely determined projective $\hat{\lambda}$ functor θ_w satisfying $\theta_w M(\lambda) = P(w\lambda)$. Since the product of projective functors is again a projective functor they were able to conclude that

$$b(x)b(y) = \sum_{z\in W} \beta^{\circ}_{x,y,z} b(z)$$

where the $\beta^{\circ}_{x,y,z}$ are non-negative integers. In order to introduce the Hecke algebra we take up the point of view of ([5], 3.8). Here we introduced a linear map $\tilde{\theta}_w : w \in W$ on $\mathbb{Z}[t,t^{-1}]W$ and conjectured that this gave the symbol of $P(w\lambda)$ associated with a contravariant form constructed on $\theta_w M(\lambda)$ when the base field is replaced by $k[t]_{(t)}$. From the formulae in ([5], 3.8) one easily checks that if we replace $ww_0 \in W$ by $\tilde{T}^{-1}_{w^{-1}} \in \tilde{\underline{H}}$ then the action of θ_{s_α} is just right multiplication by $\tilde{T}^{-1}_{s_\alpha} + t$. If we make this replacement for the conjectured symbol on $P(w\lambda)$ given in ([5], 3.9) we obtain the element

$$b_w := \sum_{y\in W} t^{\ell(w)-\ell(y)} P_{y,w}(q^{-1})\tilde{T}^{-1}_{y^{-1}} \in \tilde{\underline{H}}$$

which specializes to $b(w)$ taking $t = 1$ and $\tilde{T}^{-1}_{y^{-1}} = y$.

Now use the formula

$$\sum_{y\in W} t^{\ell(w)-\ell(y)} P_{y,w}(q^{-1})\tilde{T}^{-1}_{y^{-1}} = \sum_{y\in W} t^{-(\ell(w)-\ell(y))} P_{y,w}(q)\tilde{T}_y$$

which expresses the self-duality of the $a_w : w \in W$. This shows that b_w can be obtained from a_w by replacing t by $\sigma(t) := -t^{-1}$. Since $(\tilde{T}_\alpha - t)(\tilde{T}_\alpha + t^{-1}) = 0$, then taking $\sigma(\tilde{T}_\alpha) = \tilde{T}_\alpha$ defined an involution σ on $\tilde{\underline{H}}$ and we conclude that

$$b_x b_y = \sum_{z\in W} \beta_{x,y,z} b_z$$

where

$$\beta_{x,y,z} = \sigma(\alpha_{x,y,z}) = \sum_{i\in\mathbb{Z}} c^{(i)}_{x,y,z} t^{-i}.$$

Here the $c^{(i)}_{x,y,z}$ are the non-negative integers introduced previously. Comparison with the multiplication formulae for the $b(x) : x \in W$ shows that the Bernstein-Gelfand results corresponds to the weaker positivity constants, namely $\sum c^{(i)}_{x,y,z}, \in \mathbb{N}$, obtained by specialization at $t = 1$.

2.5. The positivity of the individual coefficients, $c^{(i)}_{x,y,z}$ indicates the existence of a level structure in the Bernstein-Gelfand projective functions $\theta_w : w \in W$. Let $\delta_{\mathcal{O}}$ be the $\mathcal{O}$ dual (see 2.9). Here we shall develop a formula for $L(P(x\lambda), \delta_{\mathcal{O}}P(y\lambda))$ in terms of a product of projective functions and then the idea is that the submodule $L(L(x\lambda),L(y\lambda))$ should be described by the "lowest level term" in this product. This is proved in 4.8 by a different and rather roundabout method.

2.6. Let $a \longmapsto \check{a}$ (resp. $a \longmapsto {}^t a$) denote the principal (resp. a fixed Chevalley) antiautomorphism of $\underline{g}$. Since t and v commute their product τ defines an involutive automorphism of $\underline{g}$. If M is a $U(\underline{g})$ module we let M^{τ} denote the $U(\underline{g})$ module with the same underlying space; but with the action $m \longmapsto a.m$ defined by $a.m = \tau(a)m$. We shall always assume the dual M^* of M to have a $U(\underline{g})$ module structure defined via v. The same remarks apply to $\underline{g} \times \underline{g}$ modules.

If M, N are $U(\underline{g})$ modules we define $\mathrm{Hom}_{\mathbb{C}}(M,N)$ to be the $U(\underline{g}) \otimes U(\underline{g})$ module with the action $(a \otimes b)x = {}^t\check{a}x\check{b}$. Identify $U(\underline{g}) \otimes U(\underline{g})$ canonically with $U(\underline{g} \times \underline{g})$ and let j be the map of $\underline{g}$ into $\underline{g} \times \underline{g}$ defined by $j(X) = (X, {}^t\check{X})$. Set $\underline{k} = j(\underline{g})$ and let $L(M,N)$ be the $U(\underline{g} \times \underline{g})$ submodule of $\mathrm{Hom}_{\mathbb{C}}(M,N)$ of locally $\underline{k}$ finite elements.

2.7. Recall that we have fixed $\lambda \in \underline{h}^*$ dominant. Our result uses a generalization of an equivalence of categories theorem ([4], 1.16 or [2], 5.9) due to Soergel ([19], 2.5.3.9). Let ker λ denote the annihilator of the one dimensional $U(\underline{b})$ module $\mathbb{C}_{\lambda}$ and set $M^i(\lambda) := U(\underline{g}) \otimes_{U(\underline{b})} (U(\underline{b})/\mathrm{Ker}\ \lambda)^i$. It is clear that we have a sequence of maps

$$0 \longleftarrow M^1(\lambda) \longleftarrow M^2(\lambda) \longleftarrow \ldots.$$

each with kernel $\oplus$ M(λ). In particular $M^i(\lambda)$ belongs to the BGG $\tilde{\mathcal{O}}$ category.

To be more precise set $\Lambda = \lambda + P(R)$ and for each $i \in \mathbb{N}^+$ let $\mathcal{O}^i_\Lambda$ denote the subcategory of all $M \in \mathrm{Ob}\ \tilde{\mathcal{O}}$ satisfying

$$M = \bigoplus_{\mu \in \Lambda} M^i_\mu$$

where

$$M^i_\mu = \{m \in M \mid (H - (\mu,H))^i m = 0\}.$$

Recall that $\mathrm{Ann}_{Z(\underline{g})} M(\lambda)$ is a maximal ideal of $Z(\underline{g})$ depending only the orbit $\hat{\lambda}$ of λ under W. For each $j \in \mathbb{N}^+$ let ${}^j_{\hat{\lambda}}\mathcal{O}^i_\Lambda$ denote the subcategory of all $M \in \mathrm{Ob}\ \mathcal{O}^i_\Lambda$ satisfying $\mathrm{Ann}_{Z(\underline{g})} M \subset \left(\mathrm{Ann}_{Z(\underline{g})} M(\lambda)\right)^i$. One has $M^i(\lambda) \in \mathrm{Ob}\ {}^i_\lambda\mathcal{O}^i_\Lambda$ ([19], 2.3.12).

Let $\mathcal{H}$ denote the category of all Harish-Chandra modules for the pair $(\underline{g} \times \underline{g}, \underline{k})$. Each $V \in \mathrm{Ob}\ \mathcal{H}$ may be viewed as a $U(\underline{g})$ - $U(\underline{g})$ bimodule via $avb = {}^t\check{v}a\check{b}$ and this leads for each $i, j \in \mathbb{N}^+$ to an analogous definition of the subcategories ${}^j_{\hat{\lambda}}\mathcal{H}^i_{\hat{\lambda}}$. Soergel ([19], 2.5.3.9 and 2.3.2.2) establishes that the functors $L^\infty_\lambda: {}^j_\lambda\mathcal{O}^i_\Lambda \longrightarrow {}^j_{\hat{\lambda}}\mathcal{H}^i_{\hat{\lambda}}$ and $T^\infty_\lambda: {}^j_{\hat{\lambda}}\mathcal{H}^i_{\hat{\lambda}} \longrightarrow {}^j_{\hat{\lambda}}\mathcal{O}^i_\Lambda$ defined by

$$L^\infty_\lambda M := \varinjlim L(M^i(\lambda), M)$$

$$T^\infty_\lambda V := \varprojlim V \otimes_{U(\underline{g})} M^i(\lambda)$$

are mutually inverse.

2.8. Through the symmetry inherent in $\mathcal{H}$ one has ([19], 2.6.1) an involutive functor s on $\mathcal{H}$ taking ${}^{i}_{\hat\lambda}\mathcal{H}^{j}_{\hat\lambda}$ onto ${}^{j}_{\hat\lambda}\mathcal{H}^{i}_{\hat\lambda}$. Then $S := T^{\infty}_{\lambda} s L^{\infty}_{\lambda}$ is an involutive functor taking ${}^{i}_{\hat\lambda}\mathcal{O}^{j}_{\Lambda}$ onto ${}^{j}_{\hat\lambda}\mathcal{O}^{i}_{\Lambda}$ ([19], 2.6.2.2).

Fix $w \in W_{\lambda}$. The Bernstein-Gelfand projective functor θ_{w} restricts to an exact functor on ${}^{\infty}_{\hat\lambda}\mathcal{O}^{\infty}_{\Lambda}$ and on ${}^{\infty}_{\hat\lambda}\mathcal{O}^{1}_{\Lambda}$. Set $\theta'_{w} = S\theta_{w}S$ viewed as an exact functor on ${}^{\infty}_{\hat\lambda}\mathcal{O}^{\infty}_{\Lambda}$ or on ${}^{1}_{\hat\lambda}\mathcal{O}^{\infty}_{\Lambda}$.

2.9. Recall that the $\mathcal{O}$ dual $\delta_{\mathcal{O}}$ is the exact involutive functor which to $M \in \text{Ob}\ \mathcal{O}$ associates the submodule of $\underline{h}$ finite elements in the dual $M^{*\tau}$. Similarly one may define an $\mathcal{H}$ dual $\delta_{\mathcal{H}}$ (see [10], 2.8) which associates to each $V \in \text{Ob}\ \mathcal{H}$ the submodule of $\underline{k}$ finite elements in $V^{*\tau}$. Again $\delta_{\mathcal{H}}$ is exact and involutive. By ([10], 2.8) we have $L^{\infty}_{\lambda}\delta_{\mathcal{O}} = \delta_{\mathcal{H}}L^{\infty}_{\lambda}$. Since $\delta_{\mathcal{H}}s = s\delta_{\mathcal{H}}$ we obtain that $S\delta_{\mathcal{O}} = \delta_{\mathcal{O}}S$. Again by ([13], 3.11) we have $\delta_{\mathcal{O}}\theta_{w} = \theta_{w}\delta_{\mathcal{O}}$. Consequently $\delta_{\mathcal{O}}\theta'_{w} = \theta'_{w}\delta_{\mathcal{O}}$. Finally view $V \in \text{Ob}\ \mathcal{H}$ as a left $U(\underline{g})$ module as in 2.7. From say ([4], 1.13) one checks quite easily that $\theta_{w}L^{\infty}_{\lambda}M = L^{\infty}_{\lambda}\theta_{w}M$ for all $M \in \text{Ob}\ \tilde{\mathcal{O}}$, that is $\theta_{w}L^{\infty}_{\lambda} = L^{\infty}_{\lambda}\theta_{w}$. Consequently $\theta_{w}\delta_{\mathcal{H}} = \delta_{\mathcal{H}}\theta_{w}$.

2.10. Let E be a finite dimensional simple $U(\underline{g})$ module and take $V \in \text{Ob}\ \mathcal{H}$. Then $V \otimes E := V \otimes (\mathbb{C} \otimes E)$ is a Harish-Chandra module for the diagonal action of $\underline{g} \times \underline{g}$. View V and $V \otimes E$ as right $U(\underline{g})$ modules as in 2.7. If M is a left $U(\underline{g})$ module it is easily checked that the identity map on $V \otimes E \otimes M$ factors to an isomorphism of $(V \otimes E) \otimes_{U(\underline{g})} M$ onto $V \otimes_{U(\underline{g})} (E \otimes M)$. Take $\alpha \in B_{\lambda}$ and recall that $\theta_{\alpha} := \theta_{s_{\alpha}}$ is a direct summand of $E^{*} \otimes (E \otimes -)$ for an appropriate choice of E and appropriate primary decomposition with respect to $Z(\underline{g})$. We conclude that $V\theta_{\alpha} \otimes_{U(\underline{g})} M \cong V \otimes_{U(\underline{g})} \theta_{\alpha}M$ and then by say ([13], 3.8) that $V\theta_{w} \otimes_{U(\underline{g})} M \cong V \otimes_{U(\underline{g})} \theta_{w}M$ for all $w \in W$. Similar reasoning gives that $L(\theta_{\alpha}M,N) \cong L(M,N)\theta_{\alpha}$ and then that $L(\theta_{w}M,N) \cong L(M,N)\theta_{w^{-1}}$.

2.11. Fix $x \in W_{\lambda}$. One easily checks from say ([4], 1.13) that $\theta_{x}L(M,N) \cong L(M,\theta_{x}N)$. Again by say ([11], 2.4) we have $s(L(M,N) \cong L(\delta_{\mathcal{O}}N,\delta_{\mathcal{O}}M)$ and so $(s\theta_{x}s)L(M,N) \cong L(\theta_{x}M,N) \cong L(M,N)\theta_{x^{-1}}$. Then $L^{\infty}_{\lambda}\theta'_{x}M \cong$ $\cong (s\theta_{x}s)L^{\infty}_{\lambda}M \cong \varinjlim L(\theta_{x}M^{i}(\lambda),M)$. We conclude that θ'_{x} commutes with

$E \otimes -$ and in particular with θ_y for all $y \in W_\lambda$. Since θ'_x is exact it follows that it is completely determined by its action on $M(\lambda)$. Set $J(\lambda) = \mathrm{Ann}_{U(\underline{g})} M(\lambda)$. Recall ([4], 3.4) that by Kostant's theorem $L(M(\lambda), M(\lambda)) = U(\underline{g})/J(\lambda)$.

Lemma. - *For all* $x \in W_\lambda$ *one has*

$$\theta'_x M(\lambda) = \varprojlim \theta_{x^{-1}} M^i(\lambda)/J(\lambda)\theta_{x^{-1}} M^i(\lambda).$$

For any left $U(\underline{g})$ module M one has

$$(*) \qquad U(\underline{g})/J(\lambda) \otimes_{U(\underline{g})} M \xrightarrow{\sim} M/J(\lambda)M.$$

The remarks above give

$$L^\infty_\lambda \theta'_x M(\lambda) \cong \varinjlim L(M^i(\lambda), M(\lambda))\theta_{x^{-1}}$$

$$\cong L(M(\lambda), M(\lambda))\theta_{x^{-1}}, \text{ by ([19], 2.3.2.2).}$$

Hence $\theta'_x M(\lambda) \cong \varprojlim (U(\underline{g})/J(\lambda) \otimes_{U(\underline{g})} \theta_{x^{-1}} M^j(\lambda))$ and then the result obtains from (*).

Remark: One has $\theta_x M(\lambda) \in \mathrm{Ob}\ {}^j_\lambda\mathcal{O}^1_\Lambda$ for j sufficiently large. The above limit is reached for such a value of j.

2.12. Take $M \in \mathrm{Ob}\ \underline{K}_{\hat{\lambda}}$ and let E be a finite dimensional simple $U(\underline{g})$ module. Then $(E \otimes L(M(\lambda), M(\lambda)) \otimes_{U(\underline{g})} M \cong E \otimes (L(M(\lambda), M(\lambda)) \otimes_{U(\underline{g})} M) \cong$ $\cong E \otimes (M/J(\lambda)M)$ by (*) above. As in 2.10 we conclude that

$$(**) \qquad \theta_y L(M(\lambda), M(\lambda)) \otimes_{U(\underline{g})} M \cong \theta_y (M/J(\lambda)M)$$

for all $y \in W_\lambda$.

Lemma. - *For all* $x, y \in W_\lambda$ *one has*

$$L(P(x\lambda), \delta_{\mathcal{O}} P(y\lambda)) \cong \underset{\longrightarrow}{\ell im}\, L(M^i(\lambda), \delta_{\mathcal{O}} \theta_y \theta_x' M(\lambda)).$$

It is enough to show that

$$\delta_{\mathcal{O}} T_\lambda^\infty L(P(x\lambda), \delta_{\mathcal{O}} P(y\lambda)) \cong \theta_y \theta_x' M(\lambda).$$

Now by 2.9 the left hand side is isomorphic to

$$\begin{aligned}
T_\lambda^\infty \delta_{\mathcal{H}} L(P(x\lambda), \delta_{\mathcal{O}} P(y\lambda)) &\cong T_\lambda^\infty \delta_{\mathcal{H}} s\theta_x sL(M(\lambda), \delta_{\mathcal{O}} P(y\lambda)), \text{ by 2.11,}\\
&\cong T_\lambda^\infty s\theta_x s\delta_{\mathcal{H}} L(M(\lambda), \delta_{\mathcal{O}} P(y\lambda)), \text{ by 2.9,}\\
&\cong T_\lambda^\infty s\theta_x sL(M(\lambda), P(y\lambda)), \quad \text{by ([10], 2.8),}\\
&\cong T_\lambda^\infty L(M(\lambda), P(y\lambda))\theta_{x^{-1}}, \quad \text{by 2.10, 2.11,}\\
&\cong \underset{\longleftarrow}{\ell im}\, \theta_y L(M(\lambda), M(\lambda)) \otimes_{U(\underline{g})} \theta_{x^{-1}} M^j(\lambda),\\
&\cong \underset{\longleftarrow}{\ell im}\, \theta_y (\theta_{x^{-1}} M^j(\lambda)/J(\lambda)\theta_{x^{-1}} M^j(\lambda)), \text{ by (**),}\\
&\cong \theta_y \theta_x' M(\lambda), \quad \text{by 2.11.}
\end{aligned}$$

Remarks. As before the limit is reached for j sufficiently large. By the commutativity properties established above it is clear that $\delta_{\mathcal{O}} \theta_y \theta_x' M(\lambda) \cong \theta_x' \delta_{\mathcal{O}} P(y\lambda)$. One easily checks from 2.11 using S that

$[\theta_x' M(\lambda)] = [\theta_{x^{-1}} M(\lambda)]$. Thus in the Grothendieck group the expression $L(P(x\lambda),\delta_0 P(y\lambda))$ is determined by the product $\theta_y\theta_{x^{-1}}$. Since $L(L(x\lambda),L(y\lambda)) \in \text{Ob}\ {}_{\lambda}^{1}H_{\lambda}^{1}$ and is a submodule of $L(P(x\lambda),\delta_0 P(\lambda))$ we conclude that it is a submodule of $L(M(\lambda),\theta_y(\theta_{x^{-1}}M(\lambda)/J(\lambda)\theta_{x^{-1}}M(\lambda)))$. As pointed out in ([5], 3.8) the form defined on $\theta_y\theta_{x^{-1}}M(\lambda)$ is not quite what would be obtained from the product $\tilde{\theta}_y\tilde{\theta}_{x^{-1}}$ (equivalently from $b_y b_{x^{-1}}$). As we see this is not in any case quite what is needed to compute $L(P(x\lambda),\delta_0 P(y\lambda))$. Nevertheless one should expect that in the sense of the t expansion the latter is given by lowest order term in the product $b_y b_{x^{-1}}$ or equivalently the lowest order term in the product $a_y a_{x^{-1}}$. This is proved by a different approach in 4.8. (To compare this with the above one must use the involution * introduced in 4.2).

3. The Cyclicity Property.

3.1. To establish cyclicity we shall need to determine the lowest order terms in the product $a_x a_y$ both in the sense of cells (cf. [11], Sect. 5) and in the sense of the t expansion. This truncated or circle product takes a relatively simple form and establishes cyclicity at the level of the Hecke algebra. Finally ([15]) is applied to give cyclicity in $\mathbb{Q}W$. For all this we recall in the nect sections several results of Lusztig. (A sign factor is missing from the formulae in lemmas 3.3 and 3.4. See A.1).

3.2. (Notation 2.4). Fix $z \in W$ and set

$$m(z) = \sup_{x,y \in W} \inf \{m \in \mathbb{Z} \mid t^m \alpha_{x,y,z} \in \mathbb{Z}[t]\}.$$

We can write

$$\alpha_{x,y,z} = t^{-m(z)}(c_{x,y,z^{-1}} + 0(t))$$

where $c_{x,y,z^{-1}} \in \mathbb{Z}$.

After Lusztig one has the

Theorem. - ([17], Prop. 6.4, Thm. 6.1, Cor.6.3)

(i) $m(z) = \deg p_{zw_0}$ ($= \operatorname{card} R^+ - d(L(zw_0\lambda))$).

(ii) $c_{x,y,z}$ *is cyclically symmetric.*

(iii) *If* $c_{x,y,z} \neq 0$ *then* x,y^{-1} *are in the same left cell of* W.

3.3. Following Lusztig we compute $a_x a_y$ in two ways retaining only the coefficient of $\tilde{T}_e$. Indeed define a $\mathbb{Z}[t,t^{-1}]$ linear map $\varphi : \underline{\tilde{H}} \to \mathbb{Z}[t,t^{-1}]$ by $\varphi(\tilde{T}_e) = 1$, $\varphi(\tilde{T}_x) = 0 : x \neq e$. Then

$$\varphi(\tilde{T}_x\tilde{T}_y) = \begin{cases} 0 : x \neq y^{-1} \\ 1 : x = y^{-1} \end{cases}$$

and a brief calculation gives

$$\varphi(a_x a_y) = \sum_{w\in W} (-1)^{\ell(x)+\ell(y)} t^{\ell(x)+\ell(y)-2\ell(w)} P_{w,x}(q^{-1}) P_{w^{-1},y}(q^{-1}).$$

Since $P_{w,x}(q) \neq 0$ implies $w \leq x$; $P_{w,w}(q) = 1$ and $\deg P_{w,x}(q) \leq$ $\leq \frac{1}{2}(\ell(x)-\ell(w)-1)$ if $w < x$ we conclude that the above expression is a polynomial in t and admits a constant term (equal to one) if and only if $x = y^{-1}$.
On the other hand

$$\varphi(a_x a_y) = \sum_{z\in W} \alpha_{x,y,z}\varphi(a_z) = \sum_{z\in W}\sum_{i\in\mathbb{Z}} c^{(i)}_{x,y,z}(-t)^{\ell(z)+i} P_{e,z}((-t)^{-2}).$$

Now the positivity of the $c^{(i)}_{x,y,z}$ and of the coefficients in the $P_{e,z}(q)$ means that we can have no cancellations between terms corresponding to different $z \in W$. Thus for each $z \in W$

$$\sum_{i\in\mathbb{Z}} c^{(i)}_{x,y,z}(-t)^{\ell(z)+i} P_{e,z}((-t)^{-2})$$

is a polynomial in t. Hence taking account of 3.2(i) we obtain

$$(*)\qquad \ell(z) - 2\deg P_{e,z}(q) \geq m(z) = \operatorname{card} R^+ - d(L(zw_0\lambda)).$$

Yet since the constant term does occur when $x = y^{-1}$ equality holds for some $z_0 \in W$. Moreover by (*) and 3.2(ii),(iii) we see that $z_0 \in C$ where C denotes the left cell of W containing x^{-1}. That is for every left cell C of W there exists $z_0 \in W$ such that equality holds in (*). Now in 2.2 we remarked that the left hand side of (*) takes its minimal value k_C on just one $\sigma \in C$ and moreover the latter is the Duflo involution. This gives (**) of 2.2 and is Lusztig's proof of C_4. Finally since the constant term occurs exactly when $x = y^{-1}$ and then equals one, we have also found the

Lemma. - *Take* $\sigma \in \Sigma^o$. *Then*

(i) $c_{x,y,\sigma} \neq 0$ *implies* $x = y^{-1}$ *and* $y \in C(\sigma)$.

(ii) $c_{x,x^{-1},\sigma} = 1$ *for all* $x^{-1} \in C(\sigma)$.

3.4. Consider the double cell $\mathcal{DC}$ containing C. (We recall that a double cell is a minimal union of left cells stable under $x \longmapsto x^{-1}$). For each $x \in \mathcal{DC}$, the integer $m(x)$ takes a fixed value which we denote simply by m. Set $\tilde{a}_x = a_x t^m$, for all $x \in \mathcal{DC}$.

The expression $a_x a_y$: $x,y \in \mathcal{DC}$ involves only terms lying in the same left cone (called a block in (III,4.6)) as y and the same right cone as x. Lets use $a_x \cdot a_y$ to denote the truncated product (dot product) in which only the terms lying in $\mathcal{DC}$ are retained. In view of the function (constant on double cells) which assigns to a_x the degree of p_{xw_0} it is easy to show that the dot product is associative (see also [11], Sect. 5). Let us use $a_x \circ a_y$ to denote the truncated product (circle product) defined by

$$a_x \circ a_y = \sum_{z \in W} c_{x,y,z} a_{z^{-1}}.$$

From the definition of $c_{x,y,z}$ and 3.2 the circle product is associative. By 3.2 again it is a truncation of the dot product.

Lemma. - *Take* $\sigma \in \Sigma^o$ *and let* $C(\sigma)$ *denote the left cell containing* σ. *Then for all* $x \in C(\sigma)$

(i) $\tilde{a}_x \circ \tilde{a}_\sigma = \tilde{a}_x$.

(ii) $\tilde{a}_\sigma \circ \tilde{a}_{x^{-1}} = \tilde{a}_{x^{-1}}$.

These follow from 3.2 and 3.3.

3.5. Let S be a subset of W. We denote by [S] the $\mathbb{Q}(t)$ module with basis $a_x : x \in S$, and by $[S]_t$ (resp. $[S]_1$) the $\mathbb{Q}(t)$ (resp. $\mathbb{Q}$) module with basis $a(x) : x \in S$. If C is a left cell then left multiplication by the dot product gives $[C]$ the structure of a left $\underline{H}$ module. Actually at the level of the group algebra this was a key result of [8] and lay behind the notion of a left cell. Its generalization to the Hecke algebra is immediate. Again at the level of the group algebra relations analogous to 3.4 for the dot product were conjectured in ([11], Sect. 5) and would of course imply the cyclicity of $a(\sigma) \in [C(\sigma)]_1$ considered as a left W module. We shall use 3.4 to obtain this result.

Recalling the definition of the dot product we define a card $C(\sigma)$x card $C(\sigma)$ matrix $M(t)$ whose entries in the polynomial ring $\mathbb{Z}[t]$ are defined by

$$(*) \qquad \tilde{a}_x \cdot \tilde{a}_\sigma = \sum_{y \in C(\sigma)} M_{x,y}(t)\tilde{a}_y \quad .$$

By 3.4(i) one has $M(0) = \mathrm{Id}$. Set $d(t) = \det M(t)$. Then $d(0) = 1$ and so certainly $d(t) \neq 0$. Thus $M(t)$ admits an inverse with coefficients in the localized ring $\mathbb{Z}[t,d(t)^{-1}]$. Multiplying (*) by this inverse matrix we obtain the

<u>*Corollary*</u>. - *For each* $x \in C(\sigma)$ *there exists a unique* $\tilde{a}'_x \in [C(\sigma)]$ *such that*

$$\tilde{a}'_x \cdot \tilde{a}_\sigma = \tilde{a}_x \quad .$$

In particular a_σ *is a cyclic vector for* $[C(\sigma)]$.

3.6. We cannot immediately conclude from 3.5 the corresponding result for the group ring, because specialization may cause difficulties. Explicitly one may have $d(1) = 0$. This difficulty may be circumvented by a further result of Lusztig. Fix a Duflo involution $\sigma \in \sum^o$ and let

$\mathcal{DC}(\sigma)$ be the double cell containing σ. Consider the $\mathbb{Q}(t)$ module E with basis $e_x : x \in \mathcal{DC}(\sigma)$. Following Lusztig, we define a left (resp. right) action of $\mathbb{Q}(t)\underline{H}$ on E by $e_x \longmapsto T_w . e_x$ (resp. $e_x \longmapsto e_x . T_w$) taking $e_x = a_x \in \underline{H}$ where the dot designates (as before) that only the terms in the double cell are retained. These actions commute trivially. Again we define a left (resp. right) action of $\mathbb{Q}(t)W$ on E by $e_x \longmapsto w \cdot e_x$ (resp. $e_x \longmapsto e_x . w$) taking $e_x = a(x) \in \mathbb{Z}W$ where the dot designates that only terms in the double cell are retained. By ([15], lemma 2.3) the actions $e_x \longmapsto T_s . e_x$ and $e_x \longmapsto e_x . s$ commute.

Observe that setting $e_x = a_x$ (resp. $e_x = a(x)$) identifies E with $[\mathcal{DC}(\sigma)]$ (resp. $[\mathcal{DC}(\sigma)]_t$). These are rings under the dot product and furthermore $[\mathcal{DC}(\sigma)]_t \simeq e\mathbb{Q}(t)W$ for some central idempotent $e \in \mathbb{Q}(t)W$ and p_0 is semisimple artinian. Consider E as a left $[\mathcal{DC}(\sigma)]$ module and as a right $[\mathcal{DC}(\sigma)]_t$ module. By construction E is isomorphic to $[\mathcal{DC}(\sigma)]_t$ as a right $[\mathcal{DC}(\sigma)]_t$ module. Yet $[\mathcal{DC}(\sigma)]_t$ is a semisimple artinian ring so the left action of $[\mathcal{DC}(\sigma)]_t$ on E defines an isomorphism $a \longmapsto (e_y \longmapsto ae_y)$ of $[\mathcal{DC}(\sigma)]_t$ onto $\mathrm{End}_{[\mathcal{DC}(\sigma)]_t} E$. By Lusztig's commutativity result we also have a ring homomorphism $a \overset{\varphi}{\longmapsto} (e_y \longmapsto ae_y)$ of $[\mathcal{DC}(\sigma)]$ into $\mathrm{End}_{[\mathcal{DC}(\sigma)]_t} E$. Actually it is easy to see that φ is injective hence an isomorphism by comparison of dimension and this is the core of Luszitg's proof ([15], 3.2) of the Benson-Curtis theorem. All we need to know here is that $\mathrm{Im}\varphi$ identifies with a subring of $[\mathcal{DC}(\sigma)]_t$ and consequently from 3.5, which gives $[\mathcal{DC}(\sigma)] \, . \, a = [\mathcal{C}(\sigma)]$, we obtain $[\mathcal{DC}(\sigma)]_t \, . \, a(\sigma) = [\mathcal{C}(\sigma)]_t$. Thus to each $x \in \mathcal{C}(\sigma)$ we can find $a'(x) \in \mathbb{Q}(t)W$ such that $a'(x) \, . \, a(\sigma) = a(x)$. Choose $0 \neq d'(t) \in \mathbb{Q}[t]$ such that $d'(t)a'(x) \in \mathbb{Q}[t]W$ for all $x \in \mathcal{C}(\sigma)$ and specialize at t_0 where $d'(t_0) \neq 0$. This gives $a''(x) \in \mathbb{Q}W$ such that $a''(x) \, . \, a(\sigma) = a(x)$ and implies the required cyclicity. Note that we do not need to take $t_0 = 1$. We have proved the

<u>*Theorem.*</u> - *Fix a Duflo involution* $\sigma \in \mathcal{I}^0$. *Then* a(σ) *is a cyclic vector for the left* $\mathbb{Q}W$ *module* $[\mathcal{C}(\sigma)]_1$.

4. A Formula for $L(L(x\lambda),L(y\lambda))$.

4.1. The main result of this section is to show that the circle product determines completely $\mathrm{Soc}L(L(x\lambda),L(y\lambda))$, a result suggested by the calculations of Section 2.

4.2. Define an involutory antiautomorphism $*$ on $\mathbb{Q}W$ by $w^* = w^{-1}$, for all $w \in W$. A key fact pointed out in ([11], 5.9) is that $a(w)^* = a(w^{-1})$. This follows from the identity $a(w,w') = a(w^{-1},w'^{-1})$, for all $w,w' \in W$ which obtains from the corresponding property $P_{y,w}(q) = P_{y^{-1}w^{-1}}(q)$ for the Kazhdan-Lusztig polynomials; but which had also been known earlier ([7], 3.3). As usual one may define a W invariant non-degenerate inner product on $\mathbb{Q}W$ through the formula $(y,w) = \mathrm{Tr}\, y^{-1}w$, for all $y,w \in W$, where the trace is taken with respect to the regular representation. The elements of W form an orthogonal basis with respect to this scalar product. If we view $a \in \mathbb{Q}W$ as a matrix with matrix elements $a_{y,w} :=$ (y,aw), then $(a^*)_{y,w} = a_{w,y}$, so a^* is the adjoint matrix of a.

Let I be a left ideal of $\mathbb{Q}W$. We claim that there exists a unique self-adjoint projection $e \in \mathbb{Q}W$ such that $I = \mathbb{Q}We$. Indeed set $J = \{a \in \mathbb{Q}W \mid (a,b) = 0$, for all $b \in I\}$. Then J is a left ideal of $\mathbb{Q}W$ and $I \oplus J = \mathbb{Q}W$. The projection p of $\mathbb{Q}W$ onto I defined by the above decomposition is self-adjoint and satisfies $p \in \mathrm{End}_W \mathbb{Q}W$. Since the map $a \longmapsto (b \longmapsto ba)$ of $\mathbb{Q}W$ into $\mathrm{End}_W \mathbb{Q}W$ is bijective we have $I = p(\mathbb{Q}W) = \mathbb{Q}We$ for some self-adjoint projection $e \in \mathbb{Q}W$. For uniqueness observe that if $\mathbb{Q}We = \mathbb{Q}We'$, then $e'(1-e) = 0$ so $e' = e'e$. Taking adjoints gives $e = e'$.

4.3. The simple considerations above lead to a number of results. We start with a result of Lusztig ([16], 12.15). Let $\mathcal{C}_1$, $\mathcal{C}_2$ be left cells of W. The dot product defines a map $\varphi: a \longmapsto (b \longmapsto b\cdot a)$ of $[\mathcal{C}_1 \cap \mathcal{C}_2^{-1}]_1$ into $\mathrm{Hom}_W([\mathcal{C}_2]_1, [\mathcal{C}_1]_1)$. We can assume that $\mathcal{C}_1$, $\mathcal{C}_2$ belong to the same double call $\mathcal{DC}$, as otherwise both spaces are zero. Let F denote the subspace of $\mathbb{Q}W$ generated by the a(w) where w lies in the double cone containing $\mathcal{DC}$ but not in $\mathcal{DC}$. Then F is a two-sided ideal of $\mathbb{Q}W$ and so can be written in the form $F = \mathbb{Q}Wf$ for some self-adjoint central projection f. Writing $e = 1 - f$, we can regard $[\mathcal{DC}]_1$ with its dot product as a subring of the semisimple artinian ring $\mathbb{Q}We$.

<u>Lemma</u>. - *φ is bijective.*

Suppose $\varphi(a) = 0$. Since $a \in [C_2^{-1}]$ we have $a^* \in [C_2]_1$ and then by definition of φ that $a^* \cdot a = 0$. The latter implies $a = 0$, as required. Then surjectivity follows by an easy dimension counting argument (as in Lusztig [16], 12.15) using the fact that $\oplus [C]_1$ is just the left regular representation of W.

Remark. The analogous result for the Hecke algebra can be proved similarly.

4.4. Let C be a left cell of W. We denote by $\ell(C)$ the length of $[C]_1$ as a left W module. Lusztig shows that C is nearly always multiplicity free (for example in the classical groups). In this special case $\ell(C) = \text{card}\,(C \cap \sum)$. Actually

Lemma. - For any left cell C one has $\ell(C) = \text{card}\,(C \cap \sum)$.

Take $C_1 = C_2 = C$ in 4.3. Then by 4.3 $[C \cap C^{-1}]_1$ with its dot product identifies with a semisimple subring $\text{End}_W[C]_1$ of the semisimple artinian ring $\mathbb{Q}We$.

As is well-known every irreducible W module can be realized over $\mathbb{Q}$. Consequently if $n_1, n_2, \ldots, n_r$ are the multiplicities of the irreducible representations of W occurring $[C]_1$ then $[C \cap C^{-1}]_1$ is a direct sum of $n_i \times n_i$ matrix rings over $\mathbb{Q}$. Hence $\text{card}\,(C \cap C^{-1}) = \dim_{\mathbb{Q}}[C \cap C^{-1}]_1 = \sum_{i=1}^{r} n_i^2$. Again over $\mathbb{Q}$ the set V of self-adjoint elements of $[C \cap C^{-1}]_1$ form a subspace whose dimension is given by

$$\dim_{\mathbb{Q}} V = \frac{1}{2} \sum_{i=1}^{r} n_i(n_i + 1).$$

On the other hand we saw in 4.2 that $a(w)^* = a(w^{-1})$ and so the subspace V' of all elements $a \in [C \cap C^{-1}]_1$ satisfying $a = a^*$ satisfies

$$\dim_{\mathbb{Q}} V' = \text{card}\,(C \cap C^{-1} \cap \textstyle\sum) + \frac{1}{2}\,\text{card}\,(C \cap C^{-1} \setminus C \cap C^{-1} \cap \textstyle\sum).$$

Finally by restricting the scalar product in 4.2 to the left ideal $[C]_1$ of $\mathbb{Q}W$ we may identify V with V' and then the above three formulae give

$$\text{card } (C \cap \Sigma) = \text{card } (C \cap C^{-1} \cap \Sigma) = \sum_{i=1}^{r} n_i = \ell(C),$$

as required.

4.5. Lemma. - *For all* $w \in W$ *one has* $\mathbb{Q}\,W\,a(w) = \mathbb{Q}\,W\,a(w^{-1})a(w)$.

By 4.2 we can write $\mathbb{Q}\,W\,a(w^{-1}) = \mathbb{Q}\,W\,e$ for some unique self-adjoint projection e. Then $a(w^{-1})(1-e) = 0$ and taking adjoints gives $ea(w) = a(w)$. Then $\mathbb{Q}\,W\,a(w^{-1})a(w) = \mathbb{Q}\,W\,ea(w) = \mathbb{Q}\,W\,a(w)$ as required.

4.6. Fix a double cell $\mathcal{DC}$ of W. Recall the definition of the circle and dot products in $\underline{\tilde{H}}$.

Proposition. - *For all* $a,b,c \in [\mathcal{DC}]$ *one has*

(i) $(a \cdot b) \circ c = a \cdot (b \circ c)$

(ii) $(a \circ b) \cdot c = a \circ (b \cdot c)$.

Both parts are similar and we consider only (i). The proof is similar to ([15], Lemma 2.3). First observe that it suffices to prove that

$$(*) \qquad (T_\alpha \cdot \tilde{a}_x) \circ \tilde{a}_y = T_\alpha \cdot (\tilde{a}_x \circ \tilde{a}_y)$$

for all $\alpha \in B$ and $x,y \in \mathcal{DC}$ where $T_\alpha \cdot a : a \in [\mathcal{DC}]$ designates that in the expansion of $T_\alpha a$ as a sum of terms of the form $a_w : w \in W$ we retain only those belonging to the double cell $\mathcal{DC}$. This holds because the a_x form a $\mathbb{Z}[t,t^{-1}]$ basis for $[\mathcal{DC}]_1$, because the $T_\alpha : \alpha \in B$ generate $\underline{H}$ as an algebra, because the dot product corresponds (c.f. 4.3) to multiplication in a quotient algebra of $\underline{H}$ and because the circle is a truncation of the dot product.

For each $x \in W$ we define

$$\tau(x) = \{\alpha \in B \mid s_\alpha x < x\} .$$

A key point in the proof of (*) is that if a_y and a_z are in the same right cell of W, then $\tau(y) = \tau(z)$. This is just the Borho-Jantzen-Duflo τ-invariance translated via the Kazhdan-Lusztig conjectures and the theory of cells to the Hecke algebra. It may also be proved directly ([14], 2.4).

Suppose in (*) that $\alpha \in \tau(x)$. Then $a_x \circ a_y$ is a sum of terms a(z) with z in the same right cell as x. In particular $s_\alpha z < z$. Yet if $\alpha \in \tau(z)$ we have $T_\alpha \tilde{a}_z = -\tilde{a}_z$ and so the assertion follows easily in this case.

Now suppose $\alpha \notin \tau(x)$. Then $\alpha \notin \tau(z)$ for every term a_z in the expansion of $a_x \circ a_y$. A second key fact is that when $\alpha \notin \tau(z)$ we have (c.f. [15], 2.2)

$$T_\alpha \cdot \tilde{a}_z = t^2 \tilde{a}_z + t \sum_{s_\alpha w < w} \tilde{\mu}(w,z)\tilde{a}_w$$

where the $\tilde{\mu}(w,z)$ are certain non-negative integers. Thus we obtain

$$(T_\alpha \cdot \tilde{a}_x) \circ \tilde{a}_y - T_\alpha \cdot (\tilde{a}_x \circ \tilde{a}_y)$$

$$= t \{ \sum_v \sum_{s_\alpha w < w} \tilde{\mu}(w,x) c_{w,x,v^{-1}} \tilde{a}_v$$

$$- \sum_v \sum_{s_\alpha w < w} c_{x,y,v^{-1}} \tilde{\mu}(w,v)\tilde{a}_w \}.$$

However the term in brackets on the right hand side is just the leading term in the expansion in powers of t of the expression $(T_\alpha \cdot \tilde{a}_x) \cdot \tilde{a}_y - \tilde{T}_\alpha \cdot (\tilde{a}_x \cdot \tilde{a}_y)$ which is identically zero by associativity of the dot product. This proves (*) and the proposition.

4.7. *Corollary. - The rings* $[\mathcal{DC}]$ *under dot and circle product are isomorphic (and both isomorphic to the semisimple Artinian ring* $[\mathcal{DC}]_1$*).*

This follows from 4.6 as in the proof of Lusztig's result described in 3.6. Here we use for injectivity the fact that (up to a power of $-t$)

$$\sum_{\sigma\in\sum^{\circ}\cap\mathcal{DC}} a_\sigma$$

is an identity in $[\mathcal{DC}]$ under the circle product. This follows from 3.2(iii) and 3.4(ii).

4.8. To obtain our main result we shall first need to refine 3.6. We claim that the obvious inclusion $[\mathcal{C}(\sigma)]_1 \cdot a(\sigma) \subset [\mathcal{C}(\sigma)]_1$ is an equality. By comparison of dimension it is enough to show that $b \cdot a(\sigma) = 0$ for $b \in [\mathcal{C}(\sigma)]_1$ implies $b = 0$. Taking adjoints in 3.6 we have $a(\sigma) \cdot [\mathcal{DC}(\sigma)]_1 = [\mathcal{C}(\sigma)^{-1}]_1$. Yet $b^* \in [\mathcal{C}(\sigma)^{-1}]_1$ and so we should have $b\cdot b^* = 0$. This implies that $b = 0$, as required. It thus makes sense to define $a(x)' := a(x)\cdot a(\sigma)^{-1}$ for each $x \in \mathcal{C}(\sigma)$. Trivially $a(x)'\cdot a(\sigma) = a(x)$.

Now take $x,y \in \mathcal{C}(\sigma)$. The rule to compute $\operatorname{Soc} L(L(x^{-1}\lambda),L(y^{-1}\lambda))$ as a direct sum of the simple Harish-Chandra modules $L(M(\lambda),L(z\lambda))$ established in (III, 4.14) can be described as follows. Take an element $a(x)' \in [\mathcal{DC}]_1$ such that $a(x)' \cdot a(\sigma) = a(x)$ and write $a(x)' \cdot a(y^{-1})$ as a sum of the form $\sum n(z)a(z)$. Then

$$\operatorname{Soc} L(L(x^{-1}\lambda),L(y^{-1}\lambda)) = \oplus\, L(M(\lambda),L(z\lambda))^{n(z)} .$$

By the above we can choose $a(x)' \in [\mathcal{C}(\sigma)]_1$. Correspondingly in the Hecke algebra we have by 3.5 a unique element $\tilde{a}_x' \in [\mathcal{C}(\sigma)]$ such that $\tilde{a}_x' \cdot \tilde{a}_\sigma = \tilde{a}_x$. Moreover by 4.6 and 3.4(ii) we obtain that $\tilde{a}_x' \cdot \tilde{a}_{y^{-1}} = \tilde{a}_x' \cdot (\tilde{a}_\sigma \circ \tilde{a}_{y^{-1}}) = (\tilde{a}_x' \cdot \tilde{a}_\sigma) \circ \tilde{a}_{y^{-1}} = \tilde{a}_x \circ \tilde{a}_{y^{-1}}$. As before we write

$$(*) \qquad \tilde{a}_x \circ \tilde{a}_{y^{-1}} = \sum_{z\in W} c_{x,y^{-1},z^{-1}}\, \tilde{a}_z .$$

We show that $n(z) = c_{x,y^{-1},z^{-1}}$. For this we must show that specialization at $t = 1$ behaves well. As in 4.3 we may identify $[C(\sigma) \cap C(\sigma)^{-1}]$ with the semisimple, artinian ring $\mathrm{End}_{\mathbb{Q}(t)\underline{H}}[C(\sigma)]$. By 3.5, a_σ is a regular hence invertible element of this ring. Set $\Gamma = C(\sigma) \cap C(\sigma)^{-1}$. Let us write a_σ^{-1} in the form

$$a_\sigma^{-1} = \sum_{\sigma' \in \Gamma} f_{\sigma'}(t) a_{\sigma'} \quad .$$

Suppose some $f_{\sigma'}(t)$ has a denominator having a non-trivial factor of $(t-1)$. Then there exists a strictly positive integer k such that $(t-1)^k a_\sigma^{-1}$ specializes at $t = 1$ to a non-zero element $b \in [\Gamma]_1$ satisfying $b\, a(\sigma) = 0$. This contradicts the fact established above that $a(\sigma)$ is an invertible element in $[\Gamma]_1$. We conclude that the specialization of a_σ^{-1} at $t = 1$ is defined and equals $a(\sigma)^{-1}$. This is just what we require.

Taking account of ([4], 3.8) and 3.2(iii) we obtain the following result (see 4.10).

<u>Theorem</u>. - *Assume $\lambda \in \underline{h}^*$ dominant, regular and integral. Then for all* $x,y \in W$,

$$\mathrm{Soc}\ L(L(x^{-1}\lambda), L(y\lambda)) = \bigoplus_{z \in W} L(M(\lambda), L(z^{-1}\lambda))^{c_{x,y,z}}$$

where the $c_{x,y,z}$ *are defined by (*) above.*

<u>Remarks</u>. Replacing W by W_λ we can drop as usual the assumption that λ is integral. Again by ([11], 5.2) taking $\sigma \in \overset{\circ}{\sum}$ we also have

$$L(L(x^{-1}\lambda), L(y\lambda)) \cong \bigoplus_{z \in C(\sigma)} L(\sigma\lambda), L(z^{-1}\lambda))^{c_{x,y,z}}$$

for all $x, y^{-1} \in C(\sigma)$.

4.9. Take $\sigma \in \sum^{\circ}$. In ([11], 5.11) we conjected that $a(\sigma) \cdot a(\sigma) = n_\sigma a(\sigma)$ for some positive integer n_σ. In virtue of Lusztig's positivity result described in 2.4 this would imply and hence be equivalent to the existence of a polynomial $n_\sigma(q)$ in $q = t^2$ such that $a_\sigma \cdot a_\sigma = n_\sigma(q) a_\sigma$. Observe that by 4.6 we should then also have that $a_x \cdot a_{y^{-1}} = n_\sigma(q) a_x \circ a_{y^{-1}}$ for all $x,y \in C(\sigma)$. (Although this holds in type A_n, A. Melnikov (dissertation, Weizmann Institute) has shown it to fail in type B_3).

4.10. Here there is an error - see Section A. In the formulae below $c_{x,y,z}$ should be replaced by $(-1)^{m(x)} c_{x_*,y_*,z_*}$ - see A.3.6.).

5. The Scale Factors (see 5.10).

5.1. Our interest in 4.8 arose from a desire to calculate the scale factors in the Goldie rank polynomials. Let us see how its conclusion might help. Set $\Lambda = \lambda + P(R)$ and let Λ^{++} denote the dominant regular elements of Λ. For each $w \in W_\lambda$ we define functions on Λ^{++} through

$$p_w(\mu) = \mathrm{rk}\ (U(\underline{g})/\mathrm{Ann}\ L(w\mu)),\ q_w(\mu) = \mathrm{rk}\mathcal{L}(L(w\mu),L(w\mu)).$$

In I, II it was shown that p_w, q_w extend to polynomials on $\underline{h}^*$ which are both proportional to $a(w^{-1})\rho^m$ where $m = \mathrm{card}\ R^+ - d(L(w\mu))$. (The latter only depends on the double cell containing w which will be assumed fixed. To see that one may use ρ in this formula apply ([12], 5.1(*))). As before there is no loss of generality in assuming $W_\lambda = W$.

Through the Goldie rank additivity principle one may determine how q_w behaves under the action of W. This was discussed in (II, 5.5, Remark 1). Moreover it was pointed out in (III, 4.12) that as a further consequence of conjecture $\mathcal{C}_3$ (established in 3.6) one may precisely interrelate the q_w as w runs through a right cell. (Right cells arise because the tensor product functor occuring in the additivity principle implements right multiplication at the level of the Grothendieck group).

Lemma. - *Fix* $\sigma \in \mathring{\Sigma}$. *There exists a positive rational number* c_σ *such that for all* $w \in \mathcal{C}(\sigma)^{-1}$ *one has*

$$q_w = c_\sigma a(w^{-1})\rho^m.$$

By 3.6 there exists $a(w)' \in [\mathcal{DC}(\sigma)]_1$ such that $a(\sigma) \cdot a(w)' = a(w)$. From the formula in (III, 5.5, Remark 1) one checks that $(a(w)'^*)q_\sigma = q_w$. Yet $q_\sigma = c_\sigma a(\sigma)\rho^m$ for some positive rational number c_σ and so

$$q_w = c_\sigma(a(w)'^*)a(\sigma)\rho^m = c_\sigma a(w^{-1})\rho^m.$$

5.2. (Notation 5.1). Let us write $z_w = q_w/p_w$ which we recall (I, 5.12(iii) is a positive integer. A basic problem is to compute these coefficients. Recall that $z_\sigma = 1$, furthermore if $w \in \mathcal{C}(\sigma)$ then q_w and q_σ are proportional. This gives the

<u>*Corollary*</u>. - *For all* $\tau \in \mathcal{C}(\sigma) \cap \mathcal{C}(\sigma)^{-1}$ *one has*

$$z_\tau = \frac{a(\tau^{-1})\rho^m}{a(\sigma)\rho^m} \quad .$$

5.3. It is entirely natural to conjecture that there exists a positive rational number c such that

$$(*) \qquad q_w = c\, a(w^{-1})\rho^m$$

for all $w \in \mathcal{DC}(\sigma)$. We remark that this formula is consistent with the formula in (II, 5.5 , Remark 1) and it holds in rank < 2. To establish such a result we would at least need to analogously interrelate the q_w when $w \in \mathcal{DC}(\sigma)$ - though even this is not enough because $a(\sigma)$ can fail to be a cyclic vector in its <u>double cell</u>.

We note that conjecture (*) determines the z_w. Indeed it implies that

$$(**) \qquad z_w = \frac{a(w^{-1})\rho^m}{a(\sigma)\rho^m}\ , \text{ for all } w \in \mathcal{C}(\sigma).$$

Conversely

<u>*Lemma*</u>. - *(**) implies (*).*

Take $\sigma' \in \sum^\circ \cap \mathcal{DC}(\sigma)$. By ([16], Cor. 12.16) there exists $w \in \mathcal{C}(\sigma)$ such that $w^{-1} \in \mathcal{C}(\sigma')$. Then by (**) and 5.1 we have

$$p_w = z_w^{-1} q_w = \left(\frac{a(w^{-1})\rho^m}{a(\sigma)\rho^m}\right)^{-1} c_{\sigma'} a(w^{-1})\rho^m = c_{\sigma'} a(\sigma)\rho^m.$$

Yet $p_w = p_\sigma = q_\sigma = c_\sigma a(\sigma)\rho^m$ and so $c_\sigma = c_{\sigma'}$ as required.

5.4. According to (III, 4.15) we can now compute the products $z_w z_{w^{-1}}$. Actually we can obtain directly from 5.1 the following

<u>Lemma</u>. - *For all* $w \in C(\sigma) \cap C(\sigma')^{-1}$,

$$z_w z_{w^{-1}} = \frac{a(w^{-1})\rho^m}{a(\sigma)\rho^m} \, \frac{a(w)\rho^m}{a(\sigma')\rho^m} .$$

5.5. Although conjecture (*) (or (**)) seems very natural, it is not so trivial as it looks and indeed we don't even know that the right hand side of (**) is an integer! Again the choice of ρ is quite crucial. Suppose we take $w \in C(\sigma)$, $\delta \in P(R)^{++}$ and set $z_w(\delta) = a(w^{-1})\delta^m / a(\sigma)\delta^m$. By ([12], 5.1(*)) this rational function on $\underline{h}^*$ is actually a constant; but is does depend on δ. Remarkably the product $z_w(\delta) z_{w^{-1}}(\delta)$ is independent of δ. This again follows from ([12], 5.1(*)).

5.6. Take $\sigma' \in \Sigma^\circ \cap \mathcal{D}C(\sigma)$ and set

$$\bar{n}_{\sigma'} = \frac{a(\sigma')^2 \rho^m}{a(\sigma')\rho^m}$$

which is a positive rational number.

<u>Lemma</u>. - *For all* $x, y \in C(\sigma')^{-1}$ *one has*

$$a(x^{-1}) a(y) \rho^m = \bar{n}_{\sigma'} \, \frac{(a(x^{-1})\rho^m)(a(y)\rho^m)}{a(\sigma')\rho^m} .$$

By 3.6 we can choose $b \in [\mathcal{D}C(\sigma)]_1$ such that $b \cdot a(\sigma') = a(x^{-1})$. Then (noting that dots can be omitted) we obtain

$$a(x^{-1})a(\sigma')\rho^m = b\ a(\sigma')a(\sigma')\rho^m \quad ,$$

$$= \bar{n}_{\sigma'}\ b\ a(\sigma')\rho^m \quad ,$$

$$= \bar{n}_{\sigma'}\ a(x^{-1})\rho^m \quad .$$

Yet $a(y)\rho^m$ is proportional to $a(\sigma')\rho^m$ and so

$$a(x^{-1})a(y)\rho^m = \frac{a(y)\rho^m}{a(\sigma')\rho^m}\ a(x^{-1})a(\sigma')\rho^m$$

$$= \bar{n}_{\sigma'}\frac{(a(x^{-1})\rho^m)(a(y)\rho^m)}{a(\sigma')\rho^m} \quad ,$$

as required.

5.7. (Notation 5.6).

<u>*Corollary*</u>. - $z_x/z_{x^{-1}}$ *is independent of the choice of* $x \in C(\sigma) \cap C(\sigma')^{-1}$ *and equals one if* $\sigma = \sigma'$.

Take $x,y \in C(\sigma) \cap C(\sigma')^{-1}$. By 5.1 we have

$$\frac{z_x}{z_y} = \frac{a(x^{-1})\rho^m}{a(y^{-1})\rho^m} \ , \ \frac{z_{x^{-1}}}{z_{y^{-1}}} = \frac{a(x)\rho^m}{a(y)\rho^m} \quad .$$

Taking account of 5.6 we see that it is enough to show that $a(x^{-1})a(y)\rho^m = a(y^{-1})a(x)\rho^m$. Yet $a(x^{-1})a(y) \in C(\sigma) \cap C(\sigma)^{-1}$ and $a(y^{-1})a(x) = (a(x^{-1})a(y))^*$, so by 5.2 it is enough to show that $z_\tau = z_{\tau^{-1}}$ for all $\tau \in C(\sigma) \cap C(\sigma)^{-1}$. This follows from (III, 3.4, 4.12 (ii)).

5.8. Take $w \in C(\sigma)$. From 4.8 and (III, 3.4, 4.12, 4.15) we obtain

$$z_w z_{w^{-1}} = \sum_{\tau \in C(\sigma) \cap C(\sigma)^{-1}} c_{w^{-1},w,\tau} z_\tau \ .$$

Actually, this is just a special case of the following general result.

Lemma. - *Take* $x,y^{-1} \in C(\sigma)$. *Choose* $\sigma',\sigma'' \in \Sigma^\circ$ *such that* $x^{-1} \in C(\sigma')$, $y \in C(\sigma'')$ *and set* $\Gamma = C(\sigma') \cap C(\sigma'')^{-1}$. *Then (see 5.10)*

$$z_{x^{-1}} z_{y^{-1}} = \sum_{w \in \Gamma} c_{x,y,w} z_w \ .$$

We may conveniently define the circle product on $[\mathcal{D}C(\sigma)]_1$ through

$$a(x) \circ a(y) = \sum_{w \in \Gamma} c_{x,y,w} a(w^{-1}) \ ,$$

for all $x,y \in \mathcal{D}C(\sigma)$. The analogue of 4.6 carries over to $\mathbb{Q}W$ by specialization at $t = 1$. Recalling 4.3 and 4.8 we may consider $a(\sigma)$ as a regular, hence invertible, element of the semisimple artinian ring $[C(\sigma) \cap C(\sigma)^{-1}]_1$. Since $y \in C(\sigma)^{-1}$ the product $a(\sigma)^{-1} \cdot a(y)$ is defined and is an element of $[C(\sigma)^{-1}]_1$. Now for any $w \in C(\sigma)^{-1}$ one has $a(w)\rho^m \in \mathbb{Q}a(\sigma)\rho^m$ and so

$$(a(\sigma)^{-1} \cdot a(y))\rho^m = c\ a\ (\sigma)\rho^m$$

for some $c \in \mathbb{Q}$. Applying $a(\sigma)$ to both sides (and recalling as before that the dot may be omitted) we obtain (notation 5.6) that

$$a(y)\rho^m = c\, \bar{n}_\sigma a(\sigma)\rho^m = \bar{n}_\sigma (a(\sigma)^{-1} \cdot a(y))\rho^m \ .$$

Since $x,y^{-1} \in C(\sigma)$ we have $a(x) \circ a(\sigma) = a(x)$ and $a(\sigma) \circ a(y)$. From the analogue of 4.6 we obtain that

$$a(x) \cdot a(\sigma)^{-1} \cdot a(y) = a(x) \circ a(y) \ .$$

Substitution from the above using 5.6 gives

$$\frac{(a(x)\rho^m)(a(y)\rho^m)}{a(\sigma)\rho^m} = \sum_{w\in\Gamma} c_{x,y,w}a(w^{-1})\rho^m .$$

By 5.1 the left hand side equals $q_{x^{-1}}q_{y^{-1}}/p_\sigma c_{\sigma''} = z_{x^{-1}}z_{y^{-1}}p_{\sigma'}/c_{\sigma''}$. Yet for all $w \in \Gamma$ we have $a(w^{-1})\rho^m = c^{-1}_{\sigma''}q_w = c^{-1}_{\sigma''}p_{\sigma'}z_w$, and so the required conclusion is obtained.

5.9. We remark that due to recent work of Barbasch and Vogan [1] one can assume conjecture C_1 of III to hold for the classical groups. Consequently the equivalent conjectures of 5.3 determine (at least implicitly) the scale factors in the Goldie rank polynomials. For the special case of a double cell associated to a Richardson orbit (i.e. of the form $\mathcal{DC}(w_B w_{B'})$ for some $B' \subset B$) this would give the following result. Set $R^+_{B'} = \mathbb{N}B' \cap R^+$ and $r_{B'} = \text{card } R^+_{B'}$. Then assuming 5.3(*) to hold we have

$$p_\tau = \frac{1}{r_{B'}!} \frac{a(\tau)\rho^{r_{B'}}}{\prod_{\alpha\in R^+_{B'}}(\rho,\alpha)}$$

Even though this would implicity determine all such primitive ideals which are completely prime, this formula does not even permit one to say that all such ideals are induced.

5.10. An error occurs throughout this section. In the formulae a(x) must be replaced by a'(x) defined in Section A.1, and the $c_{x,y,z}$ by $(-1)^{m(x)}c_{x_*,y_*,z_*}$ - see A.3.6.

6. Further Conjectures.

6.1. An obvious interpretation of 4.4 is that $[C \cap \Sigma]_1$ is a maximal commutative subring of $[C \cap C^{-1}]_1$. It would be enough to show it is a subring or that the $a(\sigma') : \sigma' \in C \cap \Sigma$ commute for the dot product. It is not even obvious that they commute for the circle product. Of course Lusztig ([16], 12.17) has shown that $[C]_1$ is multiplicity free for the classical groups and so by 4.3 it holds in these cases.

6.2. A question related to 6.1 is to show that $a(w^{-1}) \circ a(w) \in [C(w) \cap \Sigma]_1$. This is strongly motivated by 4.8 and the self-duality of $L(L(w\lambda), L(w\lambda))$. Since we already have $a(w^{-1}) \cdot a(w) \in [C(w) \cap C(w)^{-1}]_1$, this is true for the classical groups. Actually we further conjecture that

$$[C \cap \Sigma]_1 = \sum_{w \in C} \mathbb{Q}\, a(w^{-1}) \circ a(w)$$

for any left cell C. It is a way of expressing that there are enough modules of the form $L(L(w\lambda), L(w\lambda))$ to exhaust Σ (whereas the Ann $L(w\lambda)$ only exhaust $\Sigma°$).

6.3. Outside type A_n the partition of W into left cells is not fine enough to completely reduce the regular representation. This fact motivates a finer decomposition of cells. Fix a left cell C. For each $w \in W$, we write $\mathbb{Q}W \cdot a(w)$ for $\mathbb{Q}Wa(w)$ considered as a W submodule of $[C]_1$ (by appropriate projection - that is ignoring terms corresponding to smaller Gelfand-Kirillov dimension). We may define an equivalence relation on C by $w \sim y$ if and only if $\mathbb{Q}W \cdot a(w) = \mathbb{Q}W \cdot a(y)$. We conjecture that $L(L(w\lambda), L(w\lambda)) \cong L(L(y\lambda), L(y\lambda))$ if and only if $w \sim y$. This is motivated by 4.5 and 4.8. For example if $a(\sigma) \cdot a(\sigma)$ is multiple of $a(\sigma)$ (that is $a(\sigma)$ in a multiple of a projection for the dot product - this condition may be relaxed, see 6.5) then for all $w \in C(\sigma)^{-1}$ one has $a(w^{-1}) \cdot a(w) = a(w^{-1}) \circ a(\sigma) \cdot a(\sigma) \circ a(w) = n_\sigma a(w^{-1}) \circ a(\sigma) \circ a(w) = n_\sigma a(w^{-1}) \circ a(w)$. If we further suppose that $a(w^{-1}) \circ a(w)$ is a multiple of a (necessarily self-adjoint) projection under the dot product then $\mathbb{Q}W \cdot a(w^{-1}) \cdot a(w) = \mathbb{Q}W \cdot a(y^{-1}) \cdot a(y)$ if and only if $a(w^{-1}) \cdot a(w) = a(y^{-1}) \cdot a(y)$ which in view of 4.8 is what we require. Furthermore in this case conjecture 6.2 would imply that the submodules $\mathbb{Q}W \cdot a(w)$ reduce

$[C]_1$ - that is they generate the Grothendieck group of $[C]_1$. In fact already in type G_2 it can fail that $a(w^{-1}) \circ a(w)$ (or even $a(w^{-1}) \cdot a(w)$) is a multiple of a projection. Nevertheless it is still possible for our conjecture, that the $\mathbb{Q}W \cdot a(w)$ generate the Grothendieck group of $[C]_1$, to be true.

6.4. Take $\sigma \in \overset{\circ}{\sum}$ and $w \in C(\sigma)$ and $m \in \mathbb{N}^+$ as in 5.1. The truth of conjecture (**) of 5.3 would imply that $a(w^{-1})\rho^m/a(\sigma)\rho^m$ is an integer. As in (II, 5.4) this further implies that $a(\sigma)y\rho^m/a(\sigma)\rho^m \in \mathbb{Z}$ for all $y \in W$. More generally one may conjecture that $a(w)y\rho^m/a(w)\rho^m \in \mathbb{Z}$ for all $y \in W$. Conversely suppose the latter holds can one deduce the former assertion? This leads to a further conjecture. We say that $a(\tau)$ is <u>arithmetically</u> cyclic if $\mathbb{Z}W \cdot a(\tau) = \oplus\{\mathbb{Z}\, a(w) : w \in C(\tau)\}$. One may ask if cyclicity implies arithmetic cyclicity. If this and the former conjecture hold then $a(w^{-1})\rho^m(a(\sigma)\rho^m)$ is a positive integer and equal to one if $a(w)$ is cyclic. This is what we would expect from 5.3(**) and 6.3.

6.5. Fix a double cell $\mathcal{DC}$ and consider the ring $A := [\mathcal{DC}]_1$. Let us clarify the relationship between the dot and circle product (c.f. 5.8) on A. Specialization at $t = 1$ of 4.6 gives

(*) $$(a \cdot b) \circ c = a \cdot (b \circ c), \quad (a \circ b) \cdot c = a \circ (b \cdot c)$$

for all $a,b,c \in A$. We claim that

(**) $$(a \circ b)^* = b^* \circ a^*$$

for all $a,b \in A$. This is equivalent to the relation $c_{x,y,z} = c_{y^{-1},x^{-1},z^{-1}}$, for all $x,y,z \in \mathcal{DC}$. This can be proved by defining an involutory anti-automorphism on $\underline{H}$ through $T^*_\alpha = T_\alpha$ for all $\alpha \in B$. Then $T^*_w = T_{w^{-1}}$ and so $a^*_w = a_{w^{-1}}$ by virtue of the relations $P_{w,y}(q) = P_{w^{-1},y^{-1}}(q)$, for all $w,y \in W$. Then $(a_x \cdot a_y)^* = a_{y^{-1}} \cdot a_{x^{-1}}$, expansion of which gives the required relations. Finally it follows easily from 3.3 that

(***) $$a \circ a^* = 0 \Rightarrow a = 0.$$

Consider A as a ring under the dot product. It is isomorphic to a direct summand of $\mathbb{Q}W$, hence is a semisimple ring. In particular A admits an identity. It follows that if we consider A as a right A module then the map $\theta \longmapsto \theta(1)$ defines an isomorphism of $\mathrm{End}_A A$ onto A. By (*) the map $b \mapsto a_\circ b$ of A into A is an element of $\mathrm{End}_A A \xrightarrow{\sim} A$ and so there exists a unique element of $\varphi(a) \in A$ satisfying $\varphi(a).b = a_\circ b$ for all $b \in A$. By (***) the map $a \longmapsto \varphi(a)$ is injective, hence bijective and we let ψ denote its inverse. By definition we have $\psi(a)\,{}_\circ b = a \cdot b$. Now $\varphi(a_\circ b) \cdot c = a \circ (\varphi(b) \cdot c) = (a \,{}_\circ\varphi(b)) \cdot c = (\varphi(a) \cdot \varphi(b)) \cdot c$. Hence $\varphi(a_\circ b) = \varphi(a) \cdot \varphi(b)$ and $\psi(a) \circ \psi(b) = \psi(a \cdot b)$. Set

$$d = \sum_{\sigma,\tau \in \overset{\circ}{\Sigma} \cap \mathcal{D}C} a(\sigma) \cdot a(\tau).$$

From (*) and 3.2, 3.3 one easily checks that $\psi(a) = a_\circ d$. Moreover d is self-adjoint and

$$0 = a_\circ d = \sum_{\sigma \in \overset{\circ}{\Sigma} \cap \mathcal{D}C} a \cdot a(\sigma)$$

implies $a \cdot a(\sigma) = 0$ for all $\sigma \in \overset{\circ}{\Sigma} \cap \mathcal{D}C$ (since each term lies in a distinct direct summand $[C(\sigma)]_1$). By the remark in 4.8 this further implies that $a = 0$ and so d is a regular, hence invertible element of A. Finally by (**) we have $\psi(a)^* = d_\circ a^* = d \circ \psi(a^*) \circ d^{-1}$.

<u>Lemma</u> . - *Fix* $\sigma \in \overset{\circ}{\Sigma} \cap \mathcal{D}C$ *and* $w \in C(\sigma)$. *Then*

(i) $A_\circ a(w^{-1}) \circ a(w) = A \circ a(w)$.

(ii) *Take* $\lambda \in P(R)^{++}$. *If* $U(\underline{g})/\mathrm{Ann}\, L(w\lambda) \to L(L(w\lambda), L(w\lambda))$: *extends to an isomorphism of rings of fractions, then* a(w) *is cyclic in its left cell.*

(i) Consider A as a ring for the circle product. By φ above, A is a semisimple ring which by (**) admits an involutory antiautomorphism $a \longmapsto a^*$. We may view A as a direct sum of matrix rings and so define a non-degenerate scalar product on A through $(x,y) \longmapsto \mathrm{Tr}\, x^* \,{}_\circ y$. Just as in 4.2 it follows that every left ideal J of A can be written $J = A \circ e$ where e is a self-adjoint projection. Just as in 4.5 we conclude that $A \circ a^* \circ a = A \circ a$ for all $a \in A$. In particular $A \circ a(w^{-1}) \circ a(w) = A \circ a(w)$.

(ii) By 4.8 the hypothesis of the lemma is equivalent to $a(w^{-1}) \circ a(w) = a(\sigma)$. Thus $A \cdot a(w) = \varphi(A) \cdot a(w) = A \circ a(w) = A \circ a(\sigma) = [C(\sigma)]_1$, as required.

Remarks. As usual we can drop the assumption that λ is integral by replacing W by W_λ. By ([4], 4.4) we see for any $B' \subset B_\lambda$ that $a(w_{B'} w_\lambda)$ is a cyclic vector in its left cell. The truth of the conjecture in 6.3 would imply in particular the converse to (ii). By (i) it is enough to show that $a(w^{-1}) \circ a(w)$ is a multiple of a projection under the circle product.

References

[1] D. Barbasch and D.A. Vogan (to appear)

[2] J.N. Bernstein and S.I. Gelfand (1980) Tensor products of finite and infinite dimensional representations of semisimple Lie algebras. Compos Math 41:245-285

[3] J. Dixmier (1974) Algèbres enveloppantes. Cahiers Scientifiques, XXXVII, Gauthier-Villars, Paris

[4] O. Gabber and A. Joseph (1981) The Bernstein-Gelfand-Gelfand resolution and the Duflo sum formula. Compos Math 43:107-131

[5] O. Gabber and A. Joseph (1981) Towards the Kazhdan-Lusztig conjecture. Ann Ec Norm Sup 14:261-302

[6] J.C. Jantzen (1979) Moduln mit einem höchsten Gewicht. Springer, Berlin Heidelberg New York, LN 750

[7] A. Joseph (1979) Dixmier's problem for Verma and principal series submodules. J Lond Math Soc 20:193-204

[8] A. Joseph (1979) W-module structure in the primitive spectrum of the enveloping algebra of a semi-simple Lie algebra. Springer, Berlin Heidelberg New York, LN 728, pp 116-135

[9] A. Joseph (1980) Goldie rank in the enveloping algebra of a semi-simple Lie algebra, I, II, III. J Algebra 65:269-283, 284-306; J Algebra 73 (1981):295-326

[10] A. Joseph (1982) The Enright functor in the Bernstein-Gelfand-Gelfand category $\mathcal{O}$. Invent Math 67:423-445

[11] A. Joseph (1981) Completion functors in the $\mathcal{O}$ category. Springer, Berlin Heidelberg New York, LN 1020, pp

[12] A. Joseph (1984) On the variety of a highest weight module. J Algebra 88:238-278

[13] A. Joseph, Three topics in enveloping algebras. In: Proceedings Durham Symposium 1983 (to appear)

[14] D. Kazhdan and G. Lusztig (1979) Representations of Coxeter groups and Hecke algebras. Invent Math 53:165-184

[15] G. Lusztig (1981) On a theorem of Benson and Curtis. J Algebra 71:490-498

[16] G. Lusztig (1984) Characters of reductive groups over a finite field. Princeton University Press, New Jersey

[17] G. Lusztig (1985) Cells in affine Weyl groups. Advanced Studies in Pure Math., Vol. 6, Kinokuniya and North Holland

[18] G. Lusztig (1986) Cells in affine Weyl groups II. Preprint, M.I.T.

[19] W. Soergel (1985) Über den "Erweiterungs-Abschluß" der Kategorie $\mathcal{O}$ in der Kategorie aller Moduln über einer halbeinfachen Liealgebra. Diplomarbeit im Fach Mathematik an der Universität Bonn

Corrigenda and addenda to "On the cyclicity of vectors associated with Duflo involutions" by A. Joseph

A.1. There is a significant error in this paper arising from the difference in the definition of a(y) given in 2.3 and the quantity defined in say III, 4.5 which we shall denote by a'(y). In fact we have that $a'(w) = a(ww_0)w_0$ for all $w \in W$. There is a further, somewhat freudian, sign error in the definition of the $c_{x,y,z}$ which in order to be positive quantities need to be multiplied by a factor of $(-1)^{m(z)}$. Apart from sign changes, these errors do not alter any of the combinatorial analysis and so only concern their applications to enveloping algebras. Thus we have not shown conjecture $\underline{C}_3$ of III, that $a'(\sigma)$ is a cyclic vector in its cell for any Duflo involution σ, but rather that $a'(\sigma w_0)$ is cyclic. Again the formulae of 4.8 imply that Soc $L(L(x\lambda), L(y\lambda))$ can be a virtual module!

Here we correct these errors. First we use the remarkable inversion formulae for the Kazhdan-Lusztig polynomials to show that the $a'(\sigma):\sigma \in \sum^0$ are also cyclic. Secondly we prove an old conjecture of Lusztig, namely that at the level of left cells, left multiplication by w_0 permutes (up to a factor of the form $(-1)^{m(z)}$) the a(w). This corrects 4.8 which is modified in only a very small way.

A.2. Fix a left cell $\mathcal{C}$ and let $\mathcal{DC}$ be the double cell containing $\mathcal{C}$. Set $A = [\mathcal{DC}]_1$. We shall say that for $w \in \mathcal{C}$ the element a(w) is cocyclic in $\mathcal{C}$ if for any non-zero $a \in [\mathcal{C}]_1$ some ya has a non-zero coefficient of a(w), equivalently if $\langle A\cdot a, \xi(w)\rangle \neq 0$ with respect to a dual basis $\xi(w)$ for $[\mathcal{C}]_1$ or again if $\xi(w)$ is a cyclic vector with respect to the transposed action of A in $[\mathcal{C}]_1^*$. In the notation of 6.5 we have that $A\cdot a = \varphi(A)\bullet a = A\circ a$ for all $a \in [\mathcal{C}]_1$ and so it is enough to prove cyclicity or co-cyclicity with respect to the circle product.

Given $w \in \mathcal{C}$ let $*\mathcal{C}$ be the unique left cell containing ww_0. Let sg denote the sign representation of W. One has the

Lemma. -

(i) There is a natural isomorphism $[\mathcal{C}]_1^* \xrightarrow{\sim} [*\mathcal{C}]_1 \otimes$ sg *of left* $\mathbb{Q}W$ *modules. In particular* a(w) *is cyclic if and only of* $a(ww_0)$ *is co-cyclic.*

(ii) Each $a(\sigma) : \sigma \in \sum^{\circ}$ *is cocyclic. Thus each* $a'(\sigma) : \sigma \in \sum^{0}$ *is cyclic.*

(i) is an old observation of Lusztig though we shall give a few details for completion. The inversion formulae for the Kazhdan-Lusztig polynomials implies via say the formula in III, 4.7 that the map $\xi(w) \longmapsto (-1)^{\ell(w)} a(ww_0)$ defines an isomorphism of $\left[\overline{C(w)}\right]_1^* \xrightarrow{\sim} \left[\overline{C(w)}w_0\right]_1 \otimes sg$ of left $\mathbb{Q}W$ modules. (Here $\overline{C(w)}$ denotes the left cone containing w and $\overline{C(w)}w_0 = \{yw_0 | y \in \overline{C(w)}\}$). Its restriction gives the required isomorphism, where we note by III, 4.6, 4.7 that $Cw_0 = {}^*C$.

(ii) Let σ denote the unique Duflo involution in C. Take a non-zero element $a \in [C]_1$. It is enough to show that $a^* \circ a$ has a non-zero coefficient of $a(\sigma)$. Here we note that 3.3 (ii) should have read $c_{x,x^{-1},\sigma} = (-1)^{m(\sigma)}$ but we can ignore this sign factor which is constant on double cells and so does not affect our computation. We write

$$a = \sum_{w \in C} c_w a(w) \ : \ c_w \in \mathbb{Q}.$$

Then by 3.3 the coefficient of $a(\sigma)$ in $a^* \circ a$ is just

$$\sum_{w \in C} c_w^2$$

and hence non-zero.

<u>Remark</u>. This result means that 5.1 - 5.7 which only use cyclicity are correct if we replace a(w) by a'(w).

A.3. We shall now prove a deeper result which allows one to correct 4.8 and 5.8.

A.3.1. Observe that $w \longmapsto w_0 w w_0^{-1}$, $s\alpha \longmapsto s_{w_0\alpha}$ is an involution of $(W,S) : S = \{s_\alpha : \alpha \in B\}$ considered as Coxeter group. Consequently $P_{w_0w,w_0y}(q) = P_{ww_0,yw_0}(q)$ for all $w,y \in W$. This gives $w_0 a(y) w_0^{-1} = a(w_0 y w_0^{-1})$ for all $y \in W$. Thus if C is a left cell with Duflo involution

σ, then $w_0 \mathcal{C} w_0^{-1}$ is a left cell with Duflo involution $\tilde{\sigma} := w_0 \sigma w_0^{-1}$. We have already observed (A.2) that $*\mathcal{C} = \mathcal{C}w_0$ is a left cell with say Duflo involution ${}_*\sigma$ and so $\mathcal{C}* = w_0\mathcal{C}$ is also a left cell with Duflo involution $\sigma_* := {}_*\tilde{\sigma}$. From $[\mathcal{DC}(\sigma)]_1 \ni w_0 \cdot a(\sigma) = a(\tilde{\sigma}) \cdot w_0 \in [\mathcal{DC}(\tilde{\sigma})]_1 = [w_0 \mathcal{DC}(\sigma) w_0^{-1}]_1$ we conclude that $w_0 \mathcal{DC}(\sigma) w_0^{-1} = \mathcal{DC}(\sigma)$.

A.3.2. Define an involutory antiautomorphism $*$ on $\underline{\tilde{H}}$ by $T_\alpha^* = T_\alpha$ and $\mathbb{Z}[t,t^{-1}]$ linearity. The relations $P_{w,y}(q) = P_{w^{-1},y^{-1}}(q)$ imply that $a_w^* = a_{w^{-1}}$ for all $w \in W$. Since $\ell(yw_0) + \ell(w_0 y^{-1} w_0) = \ell(w_0)$ we have $T_{w_0} = T_{yw_0} T_{w_0 y^{-1} w_0}$ and so $T_{w_0} T_{w_0 y} = T_{yw_0} T_{w_0 y^{-1} w_0} T_{w_0 y} = T_{yw_0} T_{w_0}$ from which it easily follows that $T_{w_0} a_{w_0 w} = a_{ww_0} T_{w_0}$ for all $w \in W$. Thus $(T_{w_0} a_{w_0 w})^* = a_{w^{-1} w_0} T_{w_0} = T_{w_0} a_{w_0 w^{-1}}$ for all $w \in W$. Thus if $w \in \sum$, then $T_{w_0} \cdot a_{w_0 w}$ is self-adjoint and lies in $[\mathcal{C}(w_0 w)]$. In type A_n this would imply that $T_{w_0} \cdot a_{w_0 \sigma} : \sigma \in \mathring{\sum}$ is proportional to a_{σ_*}. We shall prove that this holds in general.

A.3.3. Fix $\sigma \in \mathring{\sum}$ and recall the notation of A.3.1.

<u>Lemma</u>. - *For each* $y \in \mathcal{C}(\sigma)$ *there exists a unique* $y' \in \mathcal{C}(\sigma)$ *such that* $w_0 \cdot a(y) = (-1)^{m(\tau)} a(y')$.

Fix $y \in \mathcal{C}(\sigma)$ and choose $\tau \in \sum^0$ such that $y \in \mathcal{C}(\tilde{\tau})^{-1}$. One has $m(\tau) = m(\sigma)$. Set $a = w_0 \cdot a(\tau)$. By 3.4 and 4.6 we have $a \circ a^* = w_0 \cdot a(\tau) \circ a(\tau) . w_0 = (-1)^{m(\tau)} w_0 \cdot a(\tau) . w_0 = (-1)^{m(\tau)} a(\tilde{\tau})$, where we have used that in 3.4 $a(\tau) \circ a(\tau)$ should have been correctly given by $(-1)^{m(\tau)} a(\tau)$.

On the other hand $w_0 . a(\tau)$ is contained in the $\mathbb{Z}$ module generated by $a(z) : z \in \mathcal{C}(\tau)$. Moreover $w_0 . a(\tilde{\tau}) = a(\tau) . w_0$ and taking adjoints we conclude that each z occuring in this decomposition also satisfies $z^{-1} \in \mathcal{C}(\tilde{\tau})$. Writing

$$a = \sum \lambda_i a(z_i) : z_i \in \mathcal{C}(\tau) \cap \mathcal{C}(\tilde{\tau})^{-1} ,$$

we obtain from 3.4 that the coefficient of $a(\tau)$ in $a \circ a^*$ is just

$(-1)^{m(\tau)}\sum\lambda_i^2$. Since $\lambda_i \in \mathbb{Z}$ we conclude that $a = \pm\, a(z)$ for some $z \in C(\tau)$.

Similarly consider $a(y^{-1})\circ a$. We have $y \in C(\sigma) \cap C(\tilde{\tau})^{-1}$ and so $(a(y^{-1})\circ a)\circ(a(y^{-1})\circ a)^* = a(y^{-1})\circ a\circ a^*\circ a(y) = (-1)^{m(\tau)}a(y^{-1})\circ a(\tilde{\tau})\circ a(y) = (-1)^{m(\tau)}a(y^{-1})\circ a(y)$. We conclude from 3.4 that the coefficient of $a(\sigma)$ in this product is exactly $(-1)^{m(\tau)}$. Yet $(a(y^{-1})\circ a)^* = a^*\circ a(y)$ is a $\mathbb{Z}$ linear combination of the $a(z) : z \in C(\sigma)$ and so as before we must have $a^*\circ a(y) = \pm\, a(y')$ for some $y' \in C(\sigma)$. Finally $a^*\circ a(y) = (w_0\cdot a(\tau))^*\circ a(y) = (a(\tilde{\tau}).w_0)^*\circ a(y) = w_0\cdot a(\tilde{\tau})\circ a(y) = \pm\, w_0\cdot a(y)$. This gives the assertion of the lemma where we note that the overall sign can be determined by examining the behaviour of $a(y)\rho^{m(\tau)}$ under w_0 which by 3.2(i) (and making correct identifications, that is $a(y)\rho^{m(\tau)} = a'(yw_0)w_0\rho^{m(\tau)}$) is a negative multiple of the Goldie rank polynomial $p_{w_0y^{-1}}$ (which has the same degree as p_{yw_0} and hence as $p_{\sigma w_0}$).

Although w_0 does not always quite act by -1 (and in this case y is certainly different to y') it does change a polynomial taking positive values on $P(R)^{++}$ to one taking $(-1)^{m(\tau)}$ times a positive value. (We also remark that it can happen that $y \neq y'$ even if w_0 acts by -1, for example in type B_2).

A.3.4. Take $y \in C$. By definition $a'(y) = a(yw_0)w_0$ and this is the correct object that should be used in 4.8. Yet by A.3.3 we have up to cells (that is calculating $a(yw_0).w_0$) and an appropriate sign factor that the latter is of the form $a(y')$ for some $y' \in w_0C = C^*$. However we still have to show that $a(\sigma w_0).w_0 : \sigma \in \mathring{\sum}$ is proportional to σ_*. This is a delicate point.

It is natural that we should resolve this question through the characterization of $\mathring{\sum}$ described in Section 3 which involves the Hecke algebra. Take $\sigma \in \mathring{\sum}$ and set $\tau = \sigma_*$. Consider the expansion of $\tilde{T}_{w_0}.a_{w_0y} : y \in \mathcal{D}C(\sigma)$ as a $\mathbb{Z}[t,t^{-1}]$ linear combination of the $a_z : z \in w_0\mathcal{D}C(\sigma)$.

Lemma. –
(i) For all $y \in \mathcal{D}C(\sigma)$ *one has*

$$\tilde{T}_{w_0}\cdot a_{w_0y} = a_{yw_0}\cdot\tilde{T}_{w_0} = t^{m(\sigma)-m(\tau)}(d_y + O(t)),$$

where d_y *is a* $\mathbb{Z}$ *linear combination of the* $a_z : z \in w_0\mathcal{D}C(\sigma)$.

(ii) If $y \in C(\sigma)$*, then* $d_y = t^{m(\tau)} a_{yw_0} \circ a_{w_0\sigma}$*. Moreover in this case* a_τ *occurs in the expansion of* d_y *if and only if* $y = \sigma$.

We have

$$\tilde{T}_{w_0} = \sum_{w \in W} P_{e,w}(t^{-2}) t^{\ell(w)} a_{ww_0} = \sum_{w \in W} P_{e,w}(t^{-2}) t^{\ell(w)} a_{w_0 w} .$$

Now in computing $T_{w_0} \cdot a_{w_0 y}$ it is possible to omit from this sum all those $w \in W$ for which $a_{ww_0} a_{w_0 y}$ makes no contribution to $[w_0 \mathcal{DC}(\sigma)]$. From the positivity constraints of Section 3 (see A.4) it is enough that $a(ww_0)a(w_0y)$ makes no contribution to $[w_0 \mathcal{DC}(\sigma)]_1$. Take $z \in \mathcal{DC}(\sigma)$. If $w \notin \overline{\mathcal{DC}(\sigma)}$ (the double cone containing σ) then $\xi(z)$ cannot occur in $\mathbb{Z}\, \xi(w) W$. Consequently by the map defined in A.2(i) we conclude that $a(zw_0)$ cannot occur in $\mathbb{Z}\, a(ww_0)W$ and a fortiori in $a(ww_0)a(w_0y)$. Since $w_0\, \mathcal{DC}(\sigma) w_0^{-1} = \mathcal{DC}(\sigma)$ we conclude that $a(w_0 z)$ cannot occur in $a(ww_0)a(w_0y)$. It is thus enough to take $w \in \overline{\mathcal{DC}(\sigma)}$.

Take $\sigma' \in \overset{\circ}{\sum} \cap \overline{\mathcal{DC}(\sigma)}$, $w \in \mathcal{DC}(\sigma')$. Then $\deg(P_{e,w}(t^{-2}) t^{\ell(w)}) = \ell(w) - 2 \deg P_{e,w}(q) > m(\sigma') \geqslant m(\sigma)$. Again if $\sigma' \in \overset{\circ}{\sum} \cap \mathcal{DC}(\sigma)$, $w \in C(\sigma')$ then $\deg(P_{e,w}(t^{-2}) t^{\ell(w)}) > m(\sigma') = m(\sigma)$ with equality if and only if $w \in \overset{\circ}{\sum} \cap C(\sigma') = \{\sigma'\}$. On the other hand by 3.2 the worst pole (namely $t^{-m(\tau)}$) occurs in $a_{yw_0} a_{w_0 w}$ only when $w^{-1} \in C(\dot{y})$. Recalling 3.4 this proves the lemma. Moreover to obtain the non-zero contribution of $t^{m(\sigma)-m(\tau)} a_\tau$ we need simultaneously that $(w_0 w)^{-1} = yw_0$ and $w \in \overset{\circ}{\sum} \cap \mathcal{DC}(\sigma)$.

Remark. It should be noted that one cannot further restrict the summation to $\mathcal{DC}(\sigma)$, nor may we merely compute the dot product $a_{ww_0} \cdot a_{w_0 y}$.

A.3.5. We conclude from A.3.4 that a_τ occurs in $T_{w_0} \cdot a_{w_0\sigma}$ with a non-zero coefficient $\beta \in \mathbb{Z}[t,t^{-1}]$. Unfortunately we cannot conclude that $a(\tau)$ occurs in $w_0 \cdot a(w_0\sigma)$ (and hence that $w_0 \cdot a(\tau) = a(w_0\sigma)$ up to sign) because β could have a zero at $t = 1$, a problem we have met before. We therefore need the following stronger result.

Theorem. - *Take* $\sigma \in \sum^{\circ}$ *and set* $\tau = \sigma_*$, $m = m(\tau)$, $n = \ell(w_0) + m(\sigma)$. *Then* $T_{w_0} \cdot a_{w_0 y} = a_{yw_0} \cdot T_{w_0} = t^n a_{yw_0} \circ a_{w_0\sigma}$, *for all* $y \in C(\sigma)$). *Moreover* $a_{\sigma w_0} \circ a'_{w_0\sigma} = (-t)^{-m} a_\tau$.

Consider the linear transformation ψ_0 defined on $[C(\sigma)w_0]$ by $\psi_0(a) = a \circ a_{w_0\sigma}$. In A.2 we showed that $a_{\sigma w_0}$ is a cyclic vector in $[C(\sigma)w_0]$ for the circle product under left multiplication. Applying * it follows that $a_{w_0\sigma}$ is a cyclic vector in $[w_0 C(\sigma)^{-1}]$ for the circle product under right multiplication. Yet $a^* \in [w_0 C(\sigma)^{-1}]$ and $a \circ a^* = 0$ implies that $a = 0$. We conclude that ψ is an invertible linear transformation. Now by A.3.4 we have for all $y \in C(\sigma)$ that

$$(*) \quad a_{yw_0} \cdot T_{w_0} = t^{n-m}(t^m a_{yw_0} \circ a_{w_0\sigma} + 0(t)) .$$

We conclude that the linear map $\psi(a) = t^{-(n-m)} a \cdot T_{w_0}$ on $[C(\sigma)w_0]$ satisfies $\det \psi \in \mathbb{Z}[t]$ and $\det \psi\big|_{t=0} \in \mathbb{Z} \setminus \{0\}$. Now apply the antiautomorphism $t \longmapsto t^{-1}$, $T_y \longmapsto T^{-1}_{y^{-1}}$ of $\underline{\tilde{H}}$. By definition the $a_y : y \in W$ are invariant under this map, so A.3.4(i) gives $a_{w_0 y} \cdot T^{-1}_{w_0} = T^{-1}_{w_0} \cdot a_{yw_0} = t^{-(n-m)}(d_y + 0(t^{-1}))$, for all $y \in \mathcal{D}C(\sigma)$. Replacing y by $w_0 y w_0^{-1}$ we conclude that ψ and hence $\det \psi$ is invertible in $\mathbb{Z}[t,t^{-1}]$. By our previous observation it follows that $\det \psi \in \mathbb{Z} \setminus \{0\}$ and consequently ψ is invertible in $\mathbb{Z}[t]$. This implies the vanishing of the coefficients in $t^{-\ell} : \ell > 0$ in the expansion of $t^{n-m} T^{-1}_{w_0} \cdot a_{yw_0}$ and hence the vanishing of the coefficients of $t^\ell : \ell > 0$ in the expansion of $t^{(n-m)} T_{w_0} \cdot a_{w_0 y}$ as required. The last part follows from A.3.3 and 3.3 on specialization at $t = 1$ and comparison of both sides in (*).

Remark. Up to sign and t factors the action $a \longrightarrow T_{w_0} \cdot a$ on $[C(\sigma)]$ is by permuting the basis elements $a_y : y \in C(\sigma)$.

A.3.6. Take $\sigma \in \sum^{\circ}$ and set $\tau = \sigma_*$. We have shown that $a(\sigma w_0) \cdot w_0 = (-1)^{m(\tau)} a(\tau)$. Thus $a'(\sigma)$ equals $a(\tau)$ up to cells. This allows one to correct 4.8 and 5.8. The modifications are quite minor; but we describe these for 4.8. In fact it is easy to see that the multiplicity of the simple module $L(M(\lambda),L(z\lambda))$ in Soc $L(L(x\lambda),L(y\lambda))$ is precisely the coefficient of $a'(z)$ in the decomposition of the product $a'(x^{-1})\circ a'(y)$.

Let us observe that up to cells (which is all we need) one may compute the product $a'(x^{-1})\circ a'(y)$ in two different ways. Take $y \in C(\sigma)$. By A.3.3 there is a unique $y_* \in C(\sigma_*)$ such that $a(yw_0)\cdot w_0 = (-1)^{m(\sigma)}a(y_*)$. By A.5, $y_{**} = y$ equivalently that $(yw_0)_* = y_*w_0$. Since $a((y_*)^{-1}) = a(y_*)^* = (-1)^{m(\sigma)}\, w_0\cdot a(w_0y^{-1}) = (-1)^{m(\sigma)}a(y^{-1}w_0)\cdot w_0 = a((y^{-1})_*)$ we have $(y_*)^{-1} = (y^{-1})_*$. On the one hand for all $x,y \in \mathcal{DC}(\sigma)$,

$$a'(x^{-1})\circ a'(y) = a(x_*^{-1})\circ a(y_*)$$

$$= \sum_{z\in\mathcal{DC}(\sigma)} c_{x_*^{-1},y_*\,z_*^{-1}}(-1)^{m(\sigma)}a'(z).$$

This leads to the stated modification of 4.8 and 5.8.

On the other hand

$$a'(x^{-1})\circ a'(y) = w_0\cdot a(w_0x^{-1})\circ a(yw_0)\cdot w_0$$

$$= \sum_{z\in\mathcal{DC}(\sigma)} c_{w_0x^{-1},yw_0,w_0z_*^{-1}w_0}(-1)^{m(\sigma)}a'(z).$$

This implies a symmetry within each double cell. Also the (-1) factor cancels with that in $c_{x,y,z}$ giving an overall positive contribution to each coefficient.

A.4. (notation 2.4). Since $t^{-\ell(w)}a_w \in \underline{H}$ we conclude that $c^{(i)}_{x,y,z} = 0$ unless $i + \ell(x) + \ell(y) + \ell(z) = 0 \bmod 2$. This should have been noted in discussing cancellations in 3.3 and at specialization. Also in deriving 3.3(*) it is helpful to note that one may take $x,y \in W$ such that $c_{x,y,z} \neq 0$.

A.5. Lemma.- *One has* $y_{**} = y$, *for all* $y \in W$.

Define a linear map φ on $\mathbb{Q}W$ by $\varphi(a(y)) = (-1)^{\ell(y)} a(w_0 y)$. Obviously $\varphi^2 = (-1)^{\ell(w_0)} \mathrm{Id}$. As in (III, 4.6) we can write

$$w_0 a(y) = \sum_{w \in W} c(y,w) a(w).$$

Since $w_0 \cdot a(y)$ has only one term (A.3.3) and w_0 is an involution, we have $c(y,w) = c(w,y)$ if $w \in \mathcal{DC}(y)$. Then

$$\varphi(w_0 . a(y)) = \sum_{w \in \mathcal{DC}(y)} c(w,y)(-1)^{\ell(w)} a(w_0 w),$$

$$= (-1)^{\ell(y)+\ell(w_0)} \sum_{w \in \mathcal{DC}(y)} c(w_0 y, w_0 w) a(w_0 w) ,$$

as in (III,4.7) ,

$$= (-1)^{\ell(y)+\ell(w_0)} w_0 . a(w_0 y) ,$$

since $w_0 \mathcal{DC}(y) = \mathcal{DC}(w_0 y)$,

$$= (-1)^{\ell(w_0)} w_0 . \varphi(a(y)) .$$

Now $\ell(y) = m(y)$ mod2, so $a(y_*) = w_0 . \varphi(a(y))$. Then $a(y_{**}) = w_0 . \varphi(a(y_*)) = (-1)^{\ell(w_0)} \varphi(w_0 . a(y_*)) = (-1)^{\ell(w_0)} \varphi^2(a(y)) = a(y)$, as required.

Department of Theoretical Mathematics, The Weizmann Institute of Science
Rehovot 76100, Israel

and

Laboratoire de mathématiques fondamentales (Equipe de recherche associée au C.N.R.S.), Université Pierre et Marie Curie, France.

Atomic Hardy Spaces on Semisimple Lie Groups*

by Takeshi Kawazoe

§1. Introduction.

We shall recall a relation between maximal operators and the Hardy space $H^1(\mathbf{R})$ on the 1 dimensional Euclidean space $\mathbf{R}$. For f in $L^1_{loc}(\mathbf{R})$ Hardy-Littlewood's maximal operator M_{HL} is defined by

$$M_{HL}f(x) = \sup_{r>0} \frac{1}{2r} \int_{x-r}^{x+r} |f(y)|dy. \tag{1}$$

If we use the characteristic function χ on [-1,1] and its dilation:

$$\chi_\varepsilon(x) = \frac{1}{\varepsilon}\chi\left(\frac{x}{\varepsilon}\right) \qquad (\varepsilon > 0), \tag{2}$$

we can rewrite the operator as follows.

$$M_{HL}f(x) = \frac{1}{2}\sup_{\varepsilon>0} \chi_\varepsilon * |f|(x). \tag{3}$$

Then this operator is of type (L^p, L^p) $(p > 1)$ and of weak (L^1, L^1) (cf. [S], p.5). Therefore, it is worth while considering the problem:

"when p=1, what is a reasonable way to get a strong boundedness ?"

Obviously, the operator must be changed to a strict one and the domain $L^1(\mathbf{R})$ must be replaced by a small subspace of it. As is generally known, we can obtain an answer according to the following steps.

(a) First we replace the characteristic function χ in (3) by a smooth function ϕ. Namely, we shall consider a radial maximal operator defined by

$$M_A f(x) = \sup_{\phi\varepsilon A,\ \varepsilon>0} |\phi_\varepsilon * f(x)|, \tag{4}$$

where A is a class of smooth functions on $\mathbf{R}$ with suitable decreasing order (cf. [FS], Chap. 2).

*This paper is a shortened version of a paper containing complete proofs, which is to appear later.

(b) Next we replace the domain $L^1(\mathbf{R})$ by the subspace $H^1(\mathbf{R})$, which is the space of real parts of boundary values of all functions in the classical Hardy space H^1 on the upper half plane. The norm of $\mathrm{Re}(f(x))$ in $H^1(\mathbf{R})$ is given by the H^1-norm of the Poisson transform of it.

Then we can obtain the strong boundedness of M_A on $H^1(\mathbf{R})$, that is,

<u>Theorem</u> *M_A is a bounded mapping from $H^1(R)$ to $L^1(R)$.*

Our aim is to obtain the similar results on the other metric spaces X with a measure. At that time, generally, it is impossible to expect a complex structure like the classical Hardy space H^1 for $H^1(\mathbf{R})$. However, fortunately, we can define the Hardy space $H^1(\mathbf{R})$ without using the upper half plane. As shown in [C], this space coincides with the atomic Hardy space $H^1_{\infty,0}(\mathbf{R})$, which is defined by

$$H^1_{\infty,0}(\mathbf{R}) = \{f = \sum_{i=0}^{\infty} \lambda_i a_i \ ; \ \sum_{i=0}^{\infty} |\lambda_i| < \infty, \text{ each } a_i \text{ is a } (1,\infty,0)\text{-atom on } \mathbf{R}\}, \tag{5}$$

where a $(1,\infty,0)$-atom a means that a is supported in an open ball (interval) with radius r (length 2r) and satisfies:

$$\text{(i)} \qquad \|a\|_\infty < \frac{1}{2r}, \tag{6}$$

$$\text{(ii)} \qquad \int_{\mathbf{R}} a(t)dt=0. \tag{7}$$

The norm of f in $H^1_{\infty,0}(\mathbf{R})$ is given by

$$\rho^1_{\infty,0}(f) = \inf \sum_{i=0}^{\infty} |\lambda_i|, \tag{8}$$

where the infimum is taken over all atomic decomposition of f, and it is equivalent with the norm on $H^1(\mathbf{R})$. Thus, our problem can be stated just as:

"define a maximal operator on X and an atomic Hardy space $H^1_{\infty,0}(X)$ on which the operator is bounded to $L^1(X)$."

§2. <u>Homogeneous Type</u>.

As mentioned by Coifman-Weiss (see [CW]), when the space X has a family of dilations and is of homogeneous type, in the sense that each open ball satisfies

$$|Q(x,2r)| \leqq C|Q(x,r)| \qquad (r > 0), \tag{9}$$

where $Q(x,r)$ means the open ball with radius r and centered at x, $|S|$ the volume of the set S and C a constant which does not depend on x, r, these concepts: maximal operators, atomic Hardy spaces, ... in the case of $\mathbf{R}$ can be extended to the case of X. For example, when X is a Heisenberg group with a family of dilations, the theory of maximal operators and atomic Hardy spaces on X are constructed by Folland-Stein (see [FS]). This case includes the case of the n dimensional Eucledian space $\mathbf{R}^n$ as a special case of X.

§3. Non-compact Symmetric Spaces.

Now we shall consider the case of $X=G/K$, where G is a non-compact connected real rank one semisimple Lie group with finite center and K a maximal compact subgroup of G. Let $G=KAN$ be an Iwasawa decomposition of G. Especially, we shall handle only K-biinvariant (spherical) functions on G, that is, right K-invariant functions on X. In what follows any "functions" on G mean "K-biinvariant functions" on G. Then by the Cartan decomposition $G=KCL(A^+)K$, where $CL(A^+)$ is the closure of the positive Weyl chamber A^+ of A and $CL(A^+) \cong [0,\infty)$, we can identify such functions f on G with even functions on $\mathbf{R}$, which we denote by the same letter f, that is, $f(x)=f(a_{t(x)})=f(t(x))$, where $x \in Ka_{t(x)}K$ and $t(x) \geqq 0$. According to this identification, a Haar measure dg on G corresponds to a weighted measure

$$\Delta(x)dx= c(\mathrm{sh}\ x)^p(\mathrm{sh}\ 2x)^q\ dx, \tag{10}$$

where p and q are the multiplicities of a simple root γ of (G,A) and 2γ respectively, and c a constant. For simplicity, we put $\rho=(p+2q)/2$ and define α,β as $p=2(\alpha-\beta)$, $q=2\beta+1$. Here we note the following

Lemma 1. *For each* $c_0 > 0$,

$$\begin{aligned} &(a) \qquad |Q(e,r)| \sim e^{2\rho r} \qquad (r \geqq c_0), \\ &(b) \qquad |Q(e,r)| \sim r^{2(\alpha+1)} \qquad (0 \leqq r \leqq c_0). \end{aligned} \tag{11}$$

Therefore, it is easy to see that the space X=G/K is not of homogeneous type.

3.1. For f in $L^1_{loc}(G//K)$ Hardy-Littelwood's maximal operator M_{HL} on G is defined by

$$M_{HL}f(x) = |Q(x,r)|^{-1} \int_{Q(x,r)} |f(g)|dg. \qquad (12)$$

Then without the assumption of real rank one, Clerc-Stein show that this operator is of type $(L^p(G//K), L^p(G//K))$ $(p > 1)$ (see [CS]) and Strömberg shows that it is of weak type when p=1 (see [St]). Now we shall consider the problem which was stated in the end of §1.

3.2. To define a radial maximal operator on G like the case of **R**, first we must define dilations of functions on G. Taking notice that the dilation (2) on **R** preserves the L^1-norm on **R**, we shall define a dilation of a K-biinvariant function f on G, by identifying f with an even function on **R**, as follows.

$$f_\varepsilon(t) = \frac{1}{\varepsilon}\,\Delta(t)^{-1}\Delta(\frac{t}{\varepsilon})\ f(\frac{t}{\varepsilon}) \quad (\varepsilon > 0). \qquad (13)$$

Then we can obtain an answer according to the following steps.

(a) First we shall define a radial maximal operator M_A on G by

$$M_A f(x) = \sup_{\phi\varepsilon A,\ \varepsilon>0} (1+\varepsilon)^{-1}|\phi_\varepsilon * f(x)| \quad (x \in G), \qquad (14)$$

where A is a class of K-biinvariant functions on G satisfying

(i) $\phi(t),\ \phi'(t) \in L^1(\mathbf{R},\Delta(t)dt)$,

(ii) $|\phi(t)| \leqq \Delta(t)^{-1}t^{-1}$ for $0<t<1$ and $\Delta(t)^{-1}t^{-2}$ for $t \geqq 1$,

(iii) $|\phi'(t)| \leqq \Delta(t)^{-1}t^{-2}$ for $t \in \mathbf{R}^+$, (15)

and "*" the convolution on G.

(b) Next we shall define an atomic Hardy space $H^1_{\infty,0}(G//K)$ on G, which is a subspace of $L^1(G//K)$, as follows.

$$H^1_{\infty,0}(G//K) = \{f = \sum_{i=0}^{\infty} \lambda_i a_i\ ;\ \sum_{i=0}^{\infty} |\lambda_i| < \infty, \text{ each } a_i \text{ is a } (1,\infty,0)\text{-atom on G}\}, \qquad (16)$$

where a $(1,\infty,0)$-atom a on G means that a is a spherical function on G supported in an open ball $Q(r)$ with radius r and centered at the origin e of G and satisfies:

$$\text{(i)} \qquad \|a\|_\infty \leq |Q(r)|^{-1} = \left(\int_0^r \Delta(t)dt\right)^{-1}, \tag{17}$$

$$\text{(ii)} \qquad \int_{\mathbf{R}} a(t)\Delta(t)dt = 0. \tag{18}$$

The norm on $H^1_{\infty,0}(G//K)$ is given by (8). Then we can obtain the following

Theorem 2. *M_A is a bounded mapping from $H^1_{\infty,0}(G//K)$ to $L^1(G//K)$.*

Proof. Because of the definition (16) of $H^1_{\infty,0}(G//K)$ and its norm (8), it is enough to show that there exists a constant $C > 0$ such that $\|M_A a\|_1 < C$ for all $(1,\infty,0)$-atoms a on G. Now let us suppose that the support of a is contained in an open ball $Q(r)$.

Case 1: $r \leq 1$. We put

$$\int_G M_A a(x)dx = \int_{Q(2r)} M_A a(x)dx + \int_{Q(2r)_c} M_A a(x)dx$$

$$= I_{11} + I_{12}.$$

Since $\|a\|_\infty \leq |Q(r)|^{-1}$ and $\|\phi_\varepsilon\|_1 = \|\phi\|_1 \leq C$ for all $\phi \in A$, we see that

$$\|M_A a\|_\infty \leq \sup_{\phi \in A,\ \varepsilon>0} \|\phi_\varepsilon * a\|_\infty \leq C|Q(r)|^{-1}.$$

Therefore, under the assumption that $r \leq 1$,

$$I_{11} \leq C|Q(r)|^{-1} \int_{Q(2r)} 1\, dx \leq C \quad \text{by Lemma 1.}$$

Next we shall estimate I_{12}. Let ϕ be in A. Then by the vanishing moment condition (18), we see that

$$\begin{aligned} a*\phi_\varepsilon(x) &= \int_G \phi_\varepsilon(a_x g^{-1})a(g)dg \\ &= \int_0^r \int_K \phi_\varepsilon(a_x k a_y{}^{-1})dk\, a(y)\Delta(y)dy \\ &= \int_0^r \int_K \frac{d}{dy}(\phi_\varepsilon(a_x k a_y{}^{-1}))\, dk \int_0^y a(s)\Delta(s)ds\, dy \end{aligned} \tag{19}$$

Here we note that

$$\left|\frac{d}{dt}\phi_\varepsilon(t)\right| \leqq C(1+\varepsilon)\Delta(t)^{-1}t^{-2} \quad (0<t<\infty) \quad \text{by (13) and (15),}$$

and

$$\left|\int_0^y a(s)\Delta(s)ds\right| \leqq \|a\|_1 \leqq 1 \quad \text{by (17).}$$

Then it is easy to see that

$$|a*\phi_\varepsilon(x)| \leqq C(1+\varepsilon)r(x-r)^{-2}\Delta(x)^{-1} \quad (x \geqq 2r).$$

In particular, $M_A a(x) \leqq Cr(x-r)^{-2}\Delta(x)^{-1}$ $(x \geqq 2r)$, and thus,

$$I_{21} \leqq Cr \int_{2r}^{\infty} (x-r)^{-2}\Delta(x)^{-1}\Delta(x)dx \leqq C.$$

This completes the proof of Case 1.

<u>Case 2</u>: $r > 1$. We put

$$\int_G M_A a(x)dx = \int_{Q(r+1)} M_A a(x)dx + \int_{Q(r+1)^c} M_A a(x)dx$$

$$= I_{21} + I_{22}.$$

As in the case of I_{11}, by Lemma 1, we see that $I_{21} \leqq C$. Moreover, by using (13) and (15), but without using integration by parts in (19), we see that

$$|a*\phi_\varepsilon(x)| \leqq C\varepsilon(x-r)^{-2}\Delta(x)^{-1} \quad (x \geqq r+1).$$

In particular, $M_A a(x) \leqq C(x-r)^{-2}\Delta(x)^{-1}$ $(x \geqq r+1)$, and thus,

$$I_{22} \leqq C \int_{r+1}^{\infty} (x-r)^{-2}\Delta(x)^{-1}\Delta(x)dx \leqq C.$$

This completes the proof of Case 2.

Q.E.D.

Here we shall state some remarks.

(1) In the definition (14) of the radial maximal operator M_A on G the term "$(1+\varepsilon)^{-1}$" appears, which does not exist in (4) on R. When X=R, by differentiating $\phi_\varepsilon(t)$ with respect to t, we can obtain the order "ε^{-1}".

However, when X=G/K, we cannot obtain the order, because ϕ_ε (see (13)) contains the term "$\Delta(t)^{-1}$" which does not depend on ε. Therefore, to modify this fact we need the term "$(1+\varepsilon)^{-1}$".

(2) We consider only atoms on G supported in open balls centered at the origin e of G.

(3) The reverse of the theorem is not obtained.

Conjecture. "if $M_A f$ belongs to $L^1(G//K)$, then f belongs to $H^1_{\infty,0}(G//K)$."

(4) As is generally known (cf. [Wa2], Theorem 9.2.2.12), Abel transform:

$$f \longrightarrow F_f(t) = e^{\rho t} \int_N f(a_t n)dn \quad (f \in \mathcal{C}(G//K)) \qquad (20)$$

sets up a topological isomorphism between $\mathcal{C}(G//K)$ (Schwartz space on G) and $\mathcal{S}_e(\mathbf{R})$ (the space of all even functions in $\mathcal{S}(\mathbf{R})$). Therefore, we hope that $(1,\infty,0)$-atoms on G are mapped into $(1,\infty,0)$-atoms on **R** by Abel transform. However, this is not true, because for a $(1,\infty,0)$-atom $\mathfrak{a}$ on G $F_{\mathfrak{a}}$ does not satisfy the condition (6). If we define $(1,\infty,0)$-atoms on G by using the strong condition (17)' instead of (17):

$$(17)' \qquad \|\mathfrak{a}\|_\infty \leq (r^{\alpha+3/2} e^{2\rho r})^{-1},$$

we see that Abel transform maps such atoms on G to $(1,\infty,0)$-atoms on **R**.

3.3. The boundedness of an operator T on the atomic Hardy space can usually be proved by estimating $T\mathfrak{a}$ uniformly for each atom $\mathfrak{a}$. At that time, it is generally not true that T maps the atom $\mathfrak{a}$ to an atom $T\mathfrak{a}$. However, as mentioned in [CW] and [TW], sometimes $T\mathfrak{a}$ becames a function enjoying many of good properties of atoms, which are called molecules. This situation is same for the case of X=G/K. We shall define a $(1,\infty,0,\varepsilon)$-molecule $\mathfrak{m}$ on G as follows. For $\varepsilon > 0$, we put $a=\varepsilon$ and $b=1+\varepsilon$. Then we say a spherical function $\mathfrak{m}$ in $L^\infty(G)$ is a $(1,\infty,0,\varepsilon)$-molecule on G if $\mathfrak{m}$ satisfies the following conditions:

$$\text{(i)} \qquad M(\mathfrak{m}) = \|\mathfrak{m}\|_\infty^{a/b} \, \|\mathfrak{m}Q^b\|_\infty^{1-a/b} < \infty, \qquad (21)$$

$$\text{where } Q(x) = \int_0^{t(x)} \Delta(s)ds,$$

$$\text{(ii)} \qquad \int_{\mathbf{R}} \mathfrak{m}(t)\Delta(t)dt = 0. \qquad (22)$$

Of course, if we put $\Delta(t)=1$, that is, p=q=0, this definition coincides with one in the case of **R**. As in the case of **R**, we can obtain the following

Theorem 3.

(1) If $\mathfrak{a}$ *is a* $(1,\infty,0)$*-atom on* G*, then* $\mathfrak{a}$ *is also a* $(1,\infty,0,\varepsilon)$*-molecule for all* $\varepsilon > 0$ *and* $M(\mathfrak{a}) \leqq C$*, where* C *is independent of the atom* $\mathfrak{a}$.

(2) If $\mathfrak{m}$ *is a* $(1,\infty,0,\varepsilon)$*-molecule on* G*, then there exist* $(1,\infty,0)$*-atoms* $\mathfrak{a}_i$ *on* G *and* $\lambda_i \geqq 0$ $(i \in \mathbf{N})$ *such that*

$$(i) \qquad \mathfrak{m} = \Sigma\, \lambda_i \mathfrak{a}_i,$$

$$(ii) \qquad \Sigma\, \lambda_i \leqq CM(\mathfrak{m}),$$

where C *is independent of the molecule* $\mathfrak{m}$*. Namely,* $\mathfrak{m} \in H^1_{\infty,0}(G//K)$ *and* $\rho^1_{\infty,0}(\mathfrak{m}) \leqq CM(\mathfrak{m})$.

Sketch of the proof. (1) Let us suppose that $\mathfrak{a}$ is supported in $Q(r)$. Since $\|\mathfrak{a}\|_\infty \leqq |Q(r)|^{-1} = |Q(r)|^{a-b}$ and $\|\mathfrak{a}Q^b\|_\infty \leqq \|\mathfrak{a}\|_\infty |Q(r)|^b \leqq |Q(r)|^a$, (1) is easily obtained. (2) Let $\mathfrak{m}$ be a $(1,\infty,0,\varepsilon)$-molecule on G. Without loss of generality, we may assume that $M(\mathfrak{m})=1$. Now we divide $\mathbf{R}^+=[0,\infty)$ into a sum of

$$\mathbf{R}^+ = \bigcup_{k=0}^{\infty} I_k, \text{ where } I_k = \begin{cases} [0,N) & (k=0) \\ [2^{k-1}N, 2^k N) & (0<k\leqq k_0) \\ [N_0+k-k_0-1, N_0+k-k_0) & (k > k_0) \end{cases}$$

and N is determined by $\|\mathfrak{m}\|_\infty = |Q(N)|^{a-b}$, k_0 the smallest integer satisfying $2^k N \geqq 1$, $N_0 = 2^{k_0}N$. For simplicity, we denote each I_k as $I_k = [r_k, r_k')$. Here we choose functions h_k on $\mathbf{R}^+$ satisfying the following conditions:

(1) $\quad \mathrm{supp}(h_k) \subset I_k$,

(2) $\quad \int_{\mathbf{R}} h_k(t)\Delta(t)dt = 1$,

(3) $\quad \|h_k\|_\infty \leqq c|Q(r_k')|^{-1}$.

Now we denote the restriction of $\mathfrak{m}$ to I_k by $\mathfrak{m}_k$ and put $\mathfrak{m}_k^0 = \int \mathfrak{m}_k(t)\Delta(t)dt$. Then, as in the case of $\mathbf{R}$ (see [C]), the desired atomic decomposition of $\mathfrak{m}$ is given as follows.

$$\begin{aligned}
m &= \sum_{k=0}^{\infty} m_k \\
&= \sum_{k=0}^{\infty} (m_k - m_k^0 h_k) + \sum_{k=0}^{\infty} m_k^0 h_k \\
&= \sum_{k=0}^{\infty} (m_k - m_k^0 h_k) + \sum_{\ell=1}^{\infty} m_\ell^0 (h_1 - h_0) + \sum_{\ell=2}^{\infty} m_\ell^0 (h_2 - h_1) + \cdots \\
&= \sum_{k=0}^{\infty} \alpha_k + \gamma_1 + \gamma_2 + \cdots \\
&= \sum_{k=0}^{\infty} \lambda_k \beta_k + \sum_{k=1}^{\infty} \mu_k \delta_k,
\end{aligned}$$

where $\lambda_k = |Q(r_k')|^{b-a} \| \alpha_k \|_\infty$ and $\mu_k = |Q(r_k')|^{b-a} \| \gamma_k \|_\infty$.

To complete the proof we must show that each β_k and δ_k is a $(1,\infty,0)$-atom on G, and $\sum_k |\lambda_k| + \sum_k |\mu_k| \leq C$ (see [K]).

References

[C] R.R. Coifman: A real variable characterization of H^p, Studia Math., 51 (1974), 269-274.

[CW] R.R. Coifman and G. Weiss: Extensions of Hardy space and their use in analysis, Bull. Amer. Math. Soc., 83 (1977), 569-645.

[CS] J.L. Clerc and E.M. Stein: L^p-multipliers for non-compact symmetric spaces, Proc. Nat. Acad. Sci. U.S.A., 71 (1974), 3911-3912.

[FS] G.B. Folland and E.M. Stein: Hardy spaces on homogeneous groups, Math. Notes, Princeton Univ. Press, 28 (1982).

[K] T. Kawazoe: Atomic Hardy spaces on semisimple Lie groups, to appear.

[S] E.M. Stein: Singular integrals and differentiability properties of functions, Princeton Univ. Press (1970).

[St] Jan-Olov Strömberg: Weak type L^1 estimates for maximal functions on non-compact symmetric spaces, Ann. of Math., 114 (1981), 115-126.

[TW] M. Taibleson and G. Weiss: The molecule characterization of certain Hardy spaces, Astérisque, 77 (1980), 67-149.

[Wa^2] G. Warner: Harmonic analysis on semi-simple Lie groups II. Springer-Verlag (1972).

Orbital Integrals on Symmetric Spaces

Jeremy Orloff
Mathematics Department
Norhteastern University
360 Huntington Ave.
Boston, MA 02115
USA

Introduction

A main step in Harish-Chandra's proof of the Plancherel Formula for semisimple Lie groups is a certain limit formula. This formula can be described as follows. Let G be a semisimple symmetric space. Let J be a Cartan subgroup, and J' the regular elements in J. For $j \in J'$ and $f \in C_c^\infty(G)$ let

$$Mf(j) = \int_{G/J} f(gjg^{-1})d\dot{g}.$$

Here $d\dot{g}$ is a G invariant measure on G/J. Note, it has to be shown that this integral converges. Note also, in Harish-Chandra's treatment he actually uses a normalized orbital integral

$$\Phi_f(j) = \Delta(j)Mf(j).$$

The orbital integral is smooth on J'. The normalized orbital integral is smooth on the closure of any connected component of J'.

The limit formula is then

$$\lim_{j \to e} D\Delta(j)Mf(j) = cf(e), \tag{0.1}$$

where D is a certain differential operator on J', c is a constant independent of f and the limit is taken inside some component of J'. Furthermore, $c \neq 0$ if and only if J is a fundamental Cartan subgroup. As a relationship between distributions, the right hand side is the delta function and the orbital integrals in the left hand side are invariant under conjugation by G. Thus equation 0.1 represents a kind of primitive Plancherel formula.

The proof of this formula proceeds in two steps. First a similar formula is proved on the Lie algebra level and then this formula is "exponentiated" to the group.

Roughly speaking a semisimple symmetric space is a homogeneous space G/H, where G is a semisimple Lie group and H is the fixed point of some involution of G. We can consider the group G itself to be a semisimple symmetric space in the following way. Let $\tilde{G} = G \times G$. Let

σ be the involution of $\tilde{G}$ such that $\sigma(g, g_1) = (g_1, g)$. The fixed point group of σ is the diagonal $\Delta(G)$ in $\tilde{G}$. The map $(g, g_1) \mapsto gg_1^{-1}$ induces a diffeomorphism $G/\Delta(G) \approx G$.

A natural question is to find formulas similar to equation 0.1 on semisimple symmetric spaces. First of all, this requires a decent formulation of the question. As a part of this problem one should try to find a normalizing factor that gives the orbital integral as pleasing a nature as possible.

Harish-Chandra's proof of the limit formula and the niceness of the normalized orbital integral relies heavily on a very beautiful formula for the radial parts of certain differential operators. Because we don't have such nice formulas in general we can not expect as congenial a normalization.

Once the limit formula problem has been formulated, we must decide on a technique for solving it. A major tool we will use is something we call generalized Riesz potentials. These generalize M. Riesz's generalization of the classical Mellin transform (or Riemann-Liouville integral), which is defined by

$$I^s f = \int_0^\infty f(t) t^{s-1} dt.$$

In the 30's and 40's Riesz (see Riesz [1949]) introduced this generalization on $\mathbf{R}^n$ (in fact manifolds) with a Lorentzian metric. His "Riesz potential" is integration against complex powers of the Lorentzian inner product over the region where the inner product is positive. Analogously to the Mellin transform, this integral converges if the real part of s is large enough. The resulting function in s extends meromorphically to the whole complex plane.

Also, analogously to the Mellin transform, a certain residue is $cf(0)$. (c a nonzero constant, independent of f.) Finally, these potentials satisfy a certain functional equation analogous to the equation

$$I^s \frac{df}{dt} = -(s-1) I^{s-1} f$$

satisfied by the Mellin transform.

We generalize this by integrating against powers of some relevant functions. The hope in doing this is to find the delta function (the right hand side of eq. 0.1) at one value of s , (something related to) the orbital integral at another value of s, and use a functional equation to relate the two. We then hope to use this data to solve the limit formula problem. Actually carrying this out requires a detailed knowledge of the integrals involved. This includes several integral formulas corresponding to various decompositions of the space. This program is carried out in Gelfand-Graev[1955] for the complex classical groups, and in Helgason[1959] for the Lorentzian isotropic spaces of even dimension.

In response to a question of Gelfand, Atiyah[1970] and Bernstein[1972] showed that integrating against complex powers of a polynomial always leads to a meromorphic function (in the complex power). Atiyah's method leads to some general statements about the nature of the poles and Bernstein's to some statements about the functional equation. However, we need more specific information than either general method can supply. Atiyah's theorem relies on the existence of a resolution (à là Hironaka) of the polynomial that gives it as (essentially) a monomial. This reduces the problem to the case of a Mellin transform. Bernstein uses a beautiful algebraic set-up to prove the existence of operators that produce a functional equation. We would like to know both of these things as explicitly as possible, so the general existence theorems will not suffice.

For the symmetric space given by $G = SO(n,1)$ and $H = SO(n-1,1)$, the action of H on $T_{eH}G/H$ corresponds to the action of H on $\mathbf{R}^n$. The only H invariant polynomial is the Lorentzian product of signature $(n,1)$. Thus, this is the case considered by Riesz[1949]. Given the special nature of symmetric spaces it seems reasonable to expect that we can find detailed information about the residues when H invariant polynomials are used as the potential function. This indicates that the definition and analysis of generalized Riesz potentials on semisimple symmetric spaces is an interesting problem in its own right.

The organization of this paper is the following. In section 1 we set up our notation and give a clear statement of the limit formula problem. This requires a number of propositions that are either well known or obvious extensions of well known propositions.

In section 2 we carry out the program described above in the case of rank one symmetric spaces. Interesting deviations from the group case can occur. It can happen that more than one Cartan subspace gives a (nonzero) limit formula, and it can happen that one needs more than one Cartan subspace, working in concert, to have a limit formula. We define the relevant Riesz potentials and use them to find limit formulas for all rank one spaces. Section 2.1 describes the classical Mellin transform. The theorems in this section are useful for motivating and for carrying out the analysis of generalized Riesz potentials. In section 2.2 we define and find special values for three types of generalized Riesz potentials. The first two are discussed in Gelfand-Graev [1955] and Gelfand-Shilov [1964]. Our analysis follows Riesz [1949] (where only the Lorentzian case is discussed). The third, I_0^s (eq. 2.45), is a new Riesz potential (essentially a clever combination of the first two). The analysis leading up to the definition of I_0^s is patterned after Helgason[1959] and Tengstrand[1960]. This analysis plays a central role in proving the limit formulas in section 2.3. Here we note that the limit formulas for $\mathbf{R}^n$ in Theorem 2.22 follow readily from Tengstrand's results. However, our method demonstrates the program for using generalized Riesz potentials to solve the limit formula problem. For section 2.4 we reduce the problem from the tangent space (of a rank one symmetric space) to the previously handled case of $\mathbf{R}^n$. This gives limit formulas for this case. Section 2.5 is given to exponentiating these formulas to the symmetric space itself. Rather than simply using Lemma 1.2 in section 1 to lift these formulas, we lift the Riesz potentials and use them to derive limit formulas. This has the advantage of producing a differential operator for these formulas that is better related to the symmetric space.

As a final note, we apologize in advance for the profusion of uses of the symbol "M". All of our uses are standard. We think the meaning will always be clear from the context.

1.

Notation

We will use $\mathbf{R}$ to denote the real numbers and $\mathbf{C}$ to denote the complex numbers.

On $\mathbf{R}^n$, $x_1 \cdots x_n$ will be the standard coordinates and $e_1 \cdots e_n$ will be the standard basis. On $\mathbf{C}^n$ these will be denoted $z_1 \cdots z_n$ and $e_1 \cdots e_n$.

For a vector space V (over any field) we will let V^* be its dual space.

For a vector space V over $\mathbf{R}$, $V_{\mathbf{C}} = V \otimes_{\mathbf{R}} \mathbf{C}$ will be its complexification. Similarly we will use the subscript $\mathbf{C}$ to denote the complexification of other structures. For example if $\langle\, ,\, \rangle$ is a symmetric bilinear form on V then $\langle\, ,\, \rangle_{\mathbf{C}}$ is its (unique) extension to a symmetric bilinear (over $\mathbf{C}$) form on $V_{\mathbf{C}}$.

The classical gamma function will have its usual notation Γ.

The support of a continuous function f will be denoted by $\operatorname{supp} f$. The restriction of a function f on a space A to a subspace B will be denoted $f|_B$.

Let G be a Lie group and let $\mathbf{g}$ be its Lie algebra. In general we will denote Lie groups by upper case Roman letters such as G, H, M and their Lie algebras by the corresponding lower case boldface (in lieu of German because of the capabilities of this type setting system) letter such as $\mathbf{g}$, $\mathbf{h}$, $\mathbf{m}$. We will also use Lie(G) to denote the Lie algebra of G.

The Killing form will be denoted $\quad B(X,Y) = \langle X,Y \rangle = tr(adX \cdot adY) \quad$ for any $X, Y \in \mathbf{g}$.

We will often write $Ad(g)(X) = g \cdot X$.

More generally, if the group H acts on a space M we will let $(h, m) \mapsto h \cdot m$ denote the action ($h \in H$ and $m \in M$). We also let τ_h be the map of M to itself given by the action of h. If $S \subset M$ and $A \subset H$ we will put $Z_A(S)$ and $N_A(S)$ for the centralizer and normalizer respectively of S in A.

All other notation regarding Lie groups can be found in Helgason [1978].

Symmetric Spaces

We take G to be a connected noncompact semisimple Lie group over $\mathbf{R}$. We also assume G has finite center. We let θ be a Cartan involution of G and σ any involution of G commuting with θ. We let K be the fixed point set of θ (K is necessarily a connected and maximal compact subgroup of G) and we let H be the identity component of the fixed point group of σ. The homogeneous space G/H is denoted X, it is called a *a semisimple symmetric space.* (In general, we can take $(G_\sigma)_0 \subset H \subset G_\sigma$, where G_σ is the fixed point group for σ and $(G_\sigma)_0$ is its identity component.)

We note, it is well known (Berger [1957]) that for any involution τ there is a Cartan involution commuting with it. Thus we have imposed no restriction on σ.

We let the differentials of θ and σ also be denoted θ and σ. At the Lie algebra level we let $\mathbf{k}$ and $\mathbf{p}$ be the +1 and -1 eigenspaces of θ and $\mathbf{h}$ and $\mathbf{q}$ be the +1 and -1 eigenspaces of σ. Of course $\mathbf{k} = Lie(K)$ and $\mathbf{h} = Lie(H)$, and

$$\mathbf{g} = \mathbf{k} \oplus \mathbf{p} = \mathbf{h} \oplus \mathbf{q}$$

We let $\pi : G \to G/H = X$ and we have $\mathbf{q} \approx T_o X$.

The exponential map

$$\mathrm{Exp} : \mathbf{q} \to X$$

is defined by $\mathrm{Exp}(Y) = \pi \circ \exp(Y)$ for $Y \in \mathbf{q}$.

Suppose $\mathbf{a} \subset \mathbf{q}$ is an abelian subspace such that adX is semisimple for all $X \in \mathbf{a}$. Then over $\mathbf{C}$ we can simultaneously diagonalize all the maps adX. For $\alpha \in \mathbf{a_C}^*$ we define

$$\mathbf{g}_{\mathbf{C}}^{\alpha} = \{Y \in \mathbf{g_C} \mid [X,Y] = \alpha(X) \text{ for all } X \in \mathbf{a_C}\}.$$

If $\mathbf{g}_{\mathbf{C}}^{\alpha} \neq \{0\}$ then we call α a *root* of $\mathbf{a_C}$ in $\mathbf{g_C}$ and we call $\mathbf{g}_{\mathbf{C}}^{\alpha}$ a *root space*. The *multiplicity* , m_α of α is $\dim \mathbf{g}_{\mathbf{C}}^{\alpha}$. The set of all nonzero roots will be denoted by $\Delta(\mathbf{g_C}, \mathbf{a_C})$. We can put a lexicographic order on $\mathbf{a}$. If the roots are all real on $\mathbf{a}$ then those that are positive with respect to this order will be called the positive roots and generally denoted Δ^+.

We will often use the following notations to indicated a symmetric space, (G, σ), $(\mathbf{g}, \sigma)$ and $\mathbf{g} = \mathbf{h} \oplus \mathbf{q}$.

Exponentiation Theorem

Harish-Chandra's proof of the limit formula for groups is done in stages. First a limit formula is proved on the Lie algebra and then it is lifted (via the exponential map) to the group. This requires a theorem that says (locally) we can lift G invariant analysis from $\mathbf{g}$ to G. Hence (see Harish-Chandra [1957a] Lemma 11)

Lemma 1.1: *There is a* G *invariant neighborhood* U *of* 0 *in* $\mathbf{g}$ *such that*

$$\exp : \mathrm{U} \to \mathrm{G}$$

is a diffeomorphism onto its image.

We generalize this lemma to symmetric spaces.

Lemma 1.2: *There is an* H *invariant neighborhood* W *of* 0 *in* q *such that* Exp:W $\rightarrow$ X *is a diffeomorphism onto its image.*

Proof: Choose U as in Lemma 1.1 . Let $W_1 = (U/2) \cap \mathbf{q}$. Clearly W_1 is H invariant. First we show Exp is one to one on W_1 . Suppose $X_1, X_2 \in W_1$ and $\text{Exp}X_1 = \text{Exp}X_2$. Then, $\exp X_1 = \exp X_2 \cdot h$ for some $h \in H$. Applying σ we get

$$\exp(-X_1) = \exp(-X_2) \cdot h.$$

Thus

$$\exp 2X_1 = \exp 2X_2.$$

Since $2X_1$ and $2X_2$ are in U we have $2X_1 = 2X_2$ so $X_1 = X_2$. This shows Exp is one to one on W_1. Since $d\text{Exp}_0$ is nonsingular we can choose a neighborhood $W_2 \subset W_1$ of 0 on which Exp is a diffeomorphism. Let $W = H \cdot W_2$ since $W \subset W_1$ Exp is one to one on W. If $X \in W_2, h \in H$ then $d\text{Exp}_{h \cdot X} = d\tau_h \circ d\text{Exp}_X \circ Ad(h^{-1})$. Hence Exp is nonsingular on W. ∎

Structure

Given G, σ, H, $\mathbf{q}$, etc. as above, we want to have on hand a reasonably detailed knowledge of the structure involved. Essentially, we will see that at the Lie algebra level the structure of a semisimple symmetric space is much the same as the familiar cases of a semisimple group or a Riemannian (semisimple) symmetric space (or its complexification).

We have the following well known and mostly obvious lemma.

Lemma 1.3: *We have the following bracket relations,*

$$[\mathbf{h},\mathbf{h}] \subset \mathbf{h}, \quad [\mathbf{h},\mathbf{q}] \subset \mathbf{q}, \quad [\mathbf{q},\mathbf{q}] \subset \mathbf{h} \qquad \textit{and}$$
$$[\mathbf{k},\mathbf{k}] \subset \mathbf{k}, \quad [\mathbf{k},\mathbf{p}] \subset \mathbf{p}, \quad [\mathbf{p},\mathbf{p}] \subset \mathbf{k}.$$

Also, h *and* q *are orthogonal with respect to the Killing form on* g *as are* k *and* p. *Finally the Killing form of* g *is negative definite on* k *and positive definite on* p.

For $Y \in \mathbf{q}$ these bracket relations allow us to define $T_Y : \mathbf{q} \rightarrow \mathbf{q}$ by $T_Y(X) = (adY)^2(X)$. We have the following very important formula due to Helgason[1958]. A proof is in Helgason[1978] Chapter 4.

Lemma 1.4: *The exponential map* Exp : q $\rightarrow$ X *has differential*

$$\text{dExp}_{\text{Y}} = \text{d}\tau_{\text{ExpY}} \circ \sum_{0}^{\infty} \frac{\text{T}_{\text{Y}}^{\text{n}}}{(2\text{n}+1)!}.$$

The following definition is a natural generalization from the case of a Riemannian symmetric space.

Definition 1.5: A *Cartan subspace* of $\mathbf{q}$ is a subspace $\mathbf{a} \subset \mathbf{q}$ such that

1. $\mathbf{a}$ is maximal abelian,
2. $\mathbf{a}$ is composed of semisimple elements.

In the following proposition parts 1 and 2 can be found in Lepowsky-McCollum [1976] but were certainly known previously. Part 3 is not difficult and proofs can be found in Matsuki [1979] or Kosters [1983].

Proposition 1.6: *Let* $\mathrm{G/H}$ *be a semisimple symmetric space,* H *connected. Then*

1. *Cartan subspaces exist.*
2. *They all have the same dimension.*
3. *There are finitely many* H *conjugacy classes of Cartan subspaces.*

From 3, let $\mathbf{a}_1 \ldots \mathbf{a}_r$ be representatives from each of the conjugacy classes.

Definition 1.7: If $\mathbf{a}$ is a Cartan subspace then we put

$$\mathbf{a}' = \{X \in \mathbf{a} \mid z_{\mathbf{q}}(X) = \mathbf{a}\}.$$

Likewise we define

$$\mathbf{q}' = \{Y \in \mathbf{q} \mid z_{\mathbf{q}}(Y) \text{ is a Cartan subspace } \}.$$

Proposition 1.8: 1. $\mathbf{a}'$ *is open in* $\mathbf{a}$.

2. $\mathrm{H} \cdot \mathbf{a}'$ *is open in* $\mathbf{q}$.
3. $\bigcup \mathrm{H} \cdot \mathbf{a}'_i = \mathbf{q}'$ *is open and dense in* $\mathbf{q}$.
4. $\bigcup \mathrm{H} \cdot \mathbf{a}_i$ *is the set of all semisimple elements in* $\mathbf{q}$. *(i.e. every semisimple element is in some Cartan subspace)*

In Sano[1984] parts 3 and 4 are attributed to Oshima-Matsuki[1980]. Part 1 is easy and follows from Corollary 1.12. Part 2 is also easy and follows from Lemma 1.16.

In Matsuki[1979] we also have the following result.

Lemma 1.9: *Every Cartan subspace is* H *conjugate to a* θ *stable Cartan subspace.*

Because σ and θ commute and $\mathbf{h}$, $\mathbf{q}$ and $\mathbf{k}$, $\mathbf{p}$ are the eigenspaces of σ and θ respectively we have

$$\begin{aligned}
\mathbf{g} &= \mathbf{h} \oplus \mathbf{q} = \mathbf{k} \oplus \mathbf{p}. \\
\mathbf{h} &= \mathbf{h} \cap \mathbf{k} \ \oplus \ \mathbf{h} \cap \mathbf{p}. \\
\mathbf{q} &= \mathbf{q} \cap \mathbf{k} \ \oplus \ \mathbf{q} \cap \mathbf{p}. \\
\mathbf{k} &= \mathbf{k} \cap \mathbf{h} \ \oplus \ \mathbf{k} \cap \mathbf{q}. \\
\mathbf{p} &= \mathbf{p} \cap \mathbf{h} \ \oplus \ \mathbf{p} \cap \mathbf{q}.
\end{aligned} \tag{1.1}$$

We define the following subspaces of $\mathbf{g_C}$;

$$\begin{aligned} \mathbf{k}^d &= (\mathbf{h}\cap\mathbf{k})\oplus i(\mathbf{h}\cap\mathbf{p}) \\ \mathbf{p}^d &= (\mathbf{q}\cap\mathbf{p})\oplus i(\mathbf{q}\cap\mathbf{k}) \\ \mathbf{g}^d &= \mathbf{k}^d \overset{\theta^d}{\oplus} \mathbf{p}^d. \end{aligned} \tag{1.2}$$

Here θ^d has the obvious definition, it acts by +1 on $\mathbf{k}^d$ and by -1 on $\mathbf{p}^d$.

Because our knowledge of the structure of the Riemannian symmetric space $\mathbf{g}^d = \mathbf{k}^d \oplus \mathbf{p}^d$ and its complexification $\mathbf{g_C} = \mathbf{h_C} \oplus \mathbf{q_C}$ is extensive we can use this lemma to obtain a vast amount of structure for $\mathbf{g} = \mathbf{h} \oplus \mathbf{q}$.

Given a θ stable Cartan subspace $\mathbf{a}$ we define

$$\mathbf{a}^d = (\mathbf{a}\cap\mathbf{p})\oplus i(\mathbf{a}\cap\mathbf{k}) \subset \mathbf{p}^d. \tag{1.3}$$

Clearly $\mathbf{a}^d$ is a Cartan subspace of the (Riemannian) symmetric space $\mathbf{g}^d = \mathbf{k}^d \oplus \mathbf{p}^d$ and $\mathbf{a}^d_{\mathbf{C}} = \mathbf{a_C}$.

Let $\Delta(\mathbf{g_C}, \mathbf{a_C})$ be the set of roots of $\mathbf{a_C}$ in $\mathbf{g_C}$. Since the roots are real on $\mathbf{a}^d$ we can choose an ordering of the roots and let Δ^+ be the positive roots. For a root α let $\mathbf{g}^\alpha_{\mathbf{C}}$ be the corresponding root space.

For each root space $\mathbf{g}^\alpha_{\mathbf{C}}$ we choose a basis $\{X^i_\alpha\}$, where i runs over some index set. We define the following subspaces of $\mathbf{g_C}$,

$$\begin{aligned} \mathbf{m_C} &= z_{\mathbf{h_C}}(\mathbf{a_C}) \\ \mathbf{m} &= z_{\mathbf{h}}(\mathbf{a}) = \mathbf{m_C}\cap\mathbf{h} \end{aligned} \tag{1.4}$$

and

$$\begin{aligned} \mathbf{m_C}^\perp &= \textstyle\sum_{\alpha\in\Delta^+} \mathbf{C}(X^i_\alpha + \theta^d_{\mathbf{C}} X^i_\alpha) \\ \mathbf{m}^\perp &= \left(\textstyle\sum_{\alpha\in\Delta^+} \mathbf{C}(X^i_\alpha + \theta^d_{\mathbf{C}} X^i_\alpha)\right)\cap\mathbf{h} \end{aligned} \tag{1.5}$$

and

$$\begin{aligned} \mathbf{a_C}^\perp &= \textstyle\sum_{\alpha\in\Delta^+} \mathbf{C}(X^i_\alpha - \theta^d_{\mathbf{C}} X^i_\alpha) \\ \mathbf{a}^\perp &= \left(\textstyle\sum_{\alpha\in\Delta^+} \mathbf{C}(X^i_\alpha - \theta^d_{\mathbf{C}} X^i_\alpha)\right)\cap\mathbf{q}. \end{aligned} \tag{1.6}$$

Lemma 1.10: *We have the following direct sum decompositions;*

1. $\mathbf{g_C} = \mathbf{m_C} \oplus \mathbf{a_C} \oplus \sum_{\alpha\in\Delta} \mathbf{g}^\alpha_{\mathbf{C}}$.
2. $\mathbf{h_C} = \mathbf{m_C} \oplus \mathbf{m_C}^\perp$.
3. $\mathbf{q_C} = \mathbf{a_C} \oplus \mathbf{a_C}^\perp$.
4. $\mathbf{m}^\perp \oplus \mathbf{m} = \mathbf{h}$.
5. $\mathbf{a}^\perp \oplus \mathbf{a} = \mathbf{q}$.

Proof: 1,2 and 3 are obvious from the similar statements for the Riemannian symmetric space $\mathbf{g}^d = \mathbf{k}^d + \mathbf{p}^d$. The last two follow because the sets $\mathbf{m_C}^\perp$ and $\mathbf{a_C}^\perp$ are stable under complex conjugation with respect to $\mathbf{g_C} = \mathbf{g} + i\mathbf{g}$. ∎

Fix a Cartan subspace, $\mathbf{a}$ of $\mathbf{q}$. We let $M = Z_H(\mathbf{a})$. It is clear that $\mathbf{m} \stackrel{\text{def}}{=} z_{\mathbf{h}}(\mathbf{a}) = Lie(\mathrm{M})$.

Lemma 1.11: *The following all describe* $\mathbf{a}'$

1. $\{X \in \mathbf{a} \mid Z_H(X) = M\}$,
2. $\{X \in \mathbf{a} \mid z_{\mathbf{h}}(X) = \mathbf{m}\}$,
3. $\{X \in \mathbf{a} \mid \alpha(X) \neq 0 \text{ for every } \alpha \in \Delta(\mathbf{g_C}, \mathbf{a_C})\}$.

Proof: It is clear from the direct sum decompositions in Lemma 1.10 that 2 and 3 are equal to $\mathbf{a}'$. Denote set 1 by $\tilde{\mathbf{a}}$. We must show $\tilde{\mathbf{a}} = \mathbf{a}'$.
That is we must show

$$M = Z_H(X) \text{ if and only if } X \in \mathbf{a}'.$$

If $X \in \mathbf{a}$ but $X \notin \mathbf{a}'$ then 3 shows that $\mathbf{m}$ is strictly contained in $z_{\mathbf{h}}(X)$ thus M is strictly contained in $Z_H(X)$.

Conversely suppose $X \in \mathbf{a}'$ then we must show $M = Z_H(X)$. Once again we are going to complexify. This time things are a bit trickier, because we are at the group level we cannot assume G is contained in $Int(\mathbf{g_C})$. Therefore we proceed in two steps. First, let $G_C = Int(\mathbf{g_C})$ and assume $G \subset G_C$. From Kostant-Rallis [1971] we know $Z_{H_C}(X) = M_C$ and thus $M = M_C \cap H = Z_H(X)$. Next we will use the adjoint map to reduce to this case.

Let $G_C = Int(\mathbf{g_C})$. From what we just said we know that $Ad(M) = Z_{Ad(H)}(adX)$. But it's clear that $M = (Ad^{-1}AdM) \cap H$ and $Z_H(X) = Ad^{-1}Z_{Ad(H)}(adX) \cap H$. (We need only the fact that ad is an isomorphism and $ker(Ad) = Z$.) This proves the lemma. ∎

Corollary 1.12: $\mathbf{a}' = \mathbf{a} -$ *(a finite union of proper subspaces).*

Corollary 1.13: *For any* $X \in \mathbf{a}'$, $\mathrm{ad}X$ *is an isomorphism from* $\mathbf{m}^\perp$ *to* $\mathbf{a}^\perp$.

Corollary 1.14: *For any* $X \in \mathbf{a}'$, $\mathrm{ad}X \cdot \mathbf{h} = \mathbf{a}^\perp$.

Let M' be the normalizer of $\mathbf{a}$ in H. As usual for Weyl groups we have:

Lemma 1.15: M'/M *is finite.*

Proof: Since M' normalizes $\mathbf{a}$ it acts on $\mathbf{a_C}^*$ and it stabilizes $\Delta(\mathbf{g_C}, \mathbf{a_C})$. We know the roots span $\mathbf{a_C}^*$, so M is the subgroup of M' which acts trivially on the roots. Thus M'/M is a subgroup of the permutation group of $\Delta(\mathbf{g_C}, \mathbf{a_C})$. ∎

Lemma 1.16: *If* $X \in \mathbf{a}'$ *then there is a neighborhood* $V \subset \mathbf{a}'$ *of* X *such that*

$$\varphi : H/M \times V \to \mathbf{q} \quad (\varphi(\dot{h}, Y) = h \cdot Y)$$

is a diffeomorphism onto its image.

Proof: Since $M = Z_H(\mathbf{a})$, M'/M acts on $\mathbf{a}$. Lemma 1.11 implies that M'/M has no fixed points in $\mathbf{a}'$. Thus, by the finiteness of M'/M, we can choose a neighborhood $V \subset \mathbf{a}'$ of X such that

$w \cdot V \cap V = \emptyset$, for every $e \neq w \in M'/M$. Suppose $h, h' \in H$, $Y, Y' \in V$ and $h \cdot Y = h' \cdot Y'$. Then we have

$$Y = h^{-1}h' \cdot Y'. \tag{1.7}$$

Thus $Ad(h^{-1}h')$ carries the 0 eigenspace of adY' to that of adY. That is, it preserves $\mathbf{m} + \mathbf{a}$. Since $h^{-1}h'$ is in H this implies it normalizes $\mathbf{a}$. That is, it is in M'. But then our choice of V implies that $h^{-1}h' \in M$. This implies that $Y = Y'$. Thus φ is injective.

To complete the proof we have to check that $d\varphi$ is always nonsingular. This follows from the decompositions in Lemma 1.10 in the following way. We can equate $\mathbf{m}^{\perp}$ with $T_{eM}(H/M)$. (For the definition of $\mathbf{m}^{\perp}$ see equation 1.5.) For $T \in \mathbf{m}^{\perp}$, $Y \in \mathbf{a}$, $hM \in H/M$ and $X \in V$ we have

$$\begin{aligned} d\varphi_{(h,X)}(d\tau_h T, Y) &= \tfrac{d}{dt}|_0(h(\exp tT) \cdot (X + tY)) \\ &= h \cdot [T, X] + h \cdot Y \\ &= h \cdot ([T, X] + Y). \end{aligned} \tag{1.8}$$

Since $X \in \mathbf{a}'$ Corollary 1.13 says adX is an isomorphism from $\mathbf{m}^{\perp}$ to $\mathbf{a}^{\perp}$, also we know from Lemma 1.10 that $\mathbf{q} = \mathbf{a}^{\perp} \oplus \mathbf{a}$. Thus

$$h \cdot ([T, X] + Y) = 0 \text{ if and only if } T = 0 \text{ and } Y = 0.$$

This completes the proof of the lemma. ∎

Corollary 1.17: *Regular semisimple orbits are closed.*

It is possible to show that all semisimple orbits are closed.

The following lemma is trivial..

Lemma 1.18: *Let* V *be a real vector space with a symmetric bilinear form* B. *Let* $V_{\mathbf{C}}$ *and* $B_{\mathbf{C}}$ *be their complexifications. Then* B *is nondegenerate if and only if* $B_{\mathbf{C}}$ *is also nondegenerate.*

Lemma 1.19: *Our setup of* H, $\mathbf{q}$, $\mathbf{a}'$, *and* M *satisfies the following five conditions for every* $X \in \mathbf{a}'$.

1) $(H \cdot X) \cap \mathbf{a}'$ *is finite and of constant cardinality.*
2) $T_X\mathbf{q} = T_X(H \cdot X) \oplus T_X\mathbf{a}'$.
3) $Z_H(\mathbf{a}') = Z_H(X) = M$.
4) $H \cdot X$ *is closed and* $\langle\, , \,\rangle$ *is nondegenerate on it.*
5) $\langle\, , \,\rangle$ *restricted to* $\mathbf{a}'$ *is nondegenerate.*

Proof: To see 2 first recall for $T \in H$ the equation

$$(\frac{d}{dt} \exp tT \cdot X)_{t=0} = [T, X] = -adX(T). \tag{1.9}$$

Then couple this with Corollary 1.14 which says $adX \cdot \mathbf{h} = \mathbf{a}^{\perp}$.

Lemma 1.11 (1) is 3.

For 4, the closedness of $H \cdot X$ is asserted in Corollary 1.17 . The nondegeneracy of $\langle\ ,\ \rangle$ will follow from H invariance if we show $\langle\ ,\ \rangle$ is nondegenerate at the single point X. As we saw in the proof of 2, $T_X(H \cdot X) = \mathbf{a}^{\perp}$. But $\mathbf{a}_{\mathbf{C}}^{\perp} = (\mathbf{a}^d)_{\mathbf{C}}^{\perp}$ and since $\langle\ ,\ \rangle_{\mathbf{C}}$ is positive definite on $\mathbf{p}^d$ it is positive definite on $(\mathbf{a}^d)^{\perp}$. Lemma 1.18 then shows it is nondegenerate on $\mathbf{a}^{\perp}$.

The nondegeneracy of $\langle\ ,\ \rangle$ on $\mathbf{a}'$ follows just as in the proof of 4 by noting that $\mathbf{a}_{\mathbf{C}} = \mathbf{a}_{\mathbf{C}}{}^d$ and $\langle\ ,\ \rangle_{\mathbf{C}}$ is positive definite on $\mathbf{a}^d$.

Finally we have to prove 1. We showed in Lemma 1.15 that M'/M is finite, 1 will follow if we show $\mathrm{card}(H \cdot X \cap \mathbf{a}') = \mathrm{card}(M'/M)$. First note, if $h \in H$ and $h \cdot X \in \mathbf{a}'$ then $h \in M'$. This follows because

$$h^{-1} \cdot \mathbf{a} = h^{-1} \cdot z_{\mathbf{q}}(h \cdot X) = z_{\mathbf{q}}(X) = \mathbf{a}.$$

Thus $H \cdot X \cap \mathbf{a}' = M' \cdot X$. Now, since $Z_H(X) = M \subset M'$, we have $M' \cdot X \approx M'/M$. This finishes the proof of 1. ∎

Lemma 1.20: *The homogenous space* H/M *has an* H *invariant measure.*

Proof: From Lemma 1.11 (1) we know, $Z_H(X) = M$, for $X \in \mathbf{a}'$. But, $H \cdot X \approx H/M$. Thus, it suffices to show that the (H invariant) inner product on $\mathbf{q}$ is nondegenerate when restricted to $H \cdot X$. This is proved in Lemma 1.19 (4). ∎

This measure is necessarily unique up to a scalar multiple. We fix one and call it $d\dot{h}$.

The orbit $H \cdot X$ has an induced pseudoriemannian measure which we label $d\sigma_X$. The uniqueness of H invariant measure insures the existence of a function $\delta(X)$ on $\mathbf{a}'$ such that

$$d\sigma_X = \delta(X) d\dot{h}. \tag{1.10}$$

The following lemma occurs as part of the proof of Theorem 2.11 in Helgason [1972].

Lemma 1.21: *If* dZ *is the pseudoriemannian measure on* $\mathbf{q}$ *and* dX *that on* $\mathbf{a}$ *then, for* φ *as in Lemma 1.16* ,

$$\varphi^* dZ = \delta(X) d\dot{h} dX.$$

Definition 1.22: For f a function on $\mathbf{q}$ we put

$$M_{\mathbf{a}} f(X) = \int_{H/M} f(h \cdot X) d\dot{h} \ \text{ for } X \in \mathbf{a}', \tag{1.11}$$

whenever the integral converges.

$M_{\mathbf{a}} f(X)$ is called the *orbital integral of f at X with respect to* $\mathbf{a}$, more simply $M_{\mathbf{a}} f$ will be called the *orbital integral* of f.

The first thing we need to know about our new toy is what it does to smooth compactly supported functions.

Theorem 1.23: *If* $\mathrm{f} \in C_c^{\infty}(\mathbf{q})$ *then the orbital integral* $\mathrm{M}_{\mathbf{a}}\mathrm{f}$ *is smooth with bounded support on* $\mathbf{a}'$.

Proof: Fix $X \in \mathbf{a}'$. We want to show that in the definition

$$M_{\mathbf{a}}f(X) = \int_{H/M} f(h \cdot X) d\dot{h}$$

we can differentiate under the integral sign. For this choose a neighborhood V' of X as in Lemma 1.16. Then choose a closed neighborhood $V \subset V'$ of X. Then Lemma 1.16 implies that $H \cdot V$ is closed. Hence $\mathrm{supp} f \cap H \cdot V$ is compact. Again invoking Lemma 1.16 we see there is a compact set $U \subset H/M$ such that $\mathrm{supp} f_Y \subset U$ for all $Y \in V$. Here $f_Y(\dot{h}) = f(h \cdot Y)$ is a (smooth) function on H/M. That is

$$M_{\mathbf{a}}f(Y) = \int_U f \circ \varphi(\dot{h}, Y) d\dot{h} \tag{1.12}$$

for all $Y \in V$.

We can clearly differentiate in Equation 1.12 (with respect to Y) under the integral sign in this last expression. Therefore $M_{\mathbf{a}}f$ is smooth.

Let $B_\theta(X, X) = -B(X, \theta X)$. We know that B_θ is a positive definite inner product on $\mathbf{q}$. Let S_1 be the unit sphere in $\mathbf{q}$ with respect to this norm.

In order to show the support of $M_{\mathbf{a}}f$ is bounded we have to show that

$$\{X \in \mathbf{a} \mid H \cdot X \cap supp(f) \neq \emptyset\}$$

is bounded. Suppose this isn't true. This means there are sequences $\{X_i\} \subset S_1$, $\{h_i\} \subset H$, and $\{r_i\} \subset \mathbf{R}^+$ such that

1. $r_i \to \infty$ as $i \to \infty$.
2. $h_i \cdot (r_i X_i) \in S_1$ for all i.

This implies $\lim_{i \to \infty} h_i \cdot X_i = 0$. Thus, it is enough to show that $0 \notin \overline{H \cdot (S_1 \cap \mathbf{a})}$. (The bar indicates closure.) To prove this consider the function ψ on $\mathbf{q}$ defined by $\psi(Y) = \sum |\lambda|$, where the sum is taken over the set of eigenvalues of adX. Since $S_1 \cap \mathbf{a}$ is compact, ψ takes a nonzero minimum on it. Since ψ is H invariant and $\psi(0) = 0$ we have shown $0 \notin \overline{\psi(H \cdot (S_1 \cap \mathbf{a}))}$. This implies that $0 \notin \overline{H \cdot (S_1 \cap \mathbf{a})}$ as needed. ∎

Remark: It is not hard to show, for X in a θ stable Cartan subspace, that X minimizes B_θ on $H \cdot X$.

Statement of the Problem

We can now state our problem explicitly.

Let $\mathbf{a}_1 \cdots \mathbf{a}_r$ be representatives from each of the H conjugacy classes of Cartan subspaces in $\mathbf{q}$. (Proposition 1.6 says there are only a finite number of them.) Lemma 1.9 tells us we can assume each of these Cartan subspaces is θ stable, we therefore make this assumption.

The problem is to "invert" the orbital integrals. More explicitly, it is to compute $f(0)$ given the r functions $M_{\mathbf{a}_1}f \cdots M_{\mathbf{a}_r}f$ on $\mathbf{a}'_1 \cdots \mathbf{a}'_r$ respectively. Even more explicitly, the problem is to find differential operators D_i on $\mathbf{a}'_i$ such that

$$f(0) = \sum_{i=1}^{r} \lim_{X_i \to 0} D_i M_{\mathbf{a}_i} f(X_i). \tag{1.13}$$

The limits being taken in the appropriate $\mathbf{a}'_i$. Note that it is necessary to take limits because $M_{\mathbf{a}} f(X)$ is not defined at $X = 0 \notin \mathbf{a}'$.

The second half of the problem is to use the Exponentiation Theorem (Lemma 1.2) to lift this formula to the symmetric space.

Equation 1.13 will be called a *limit formula.* Our strategy for proving limit formulas will be to find an analytic family of distributions that contains the delta function (the left side of 1.13), a distribution related to the orbital integral and a functional equation to relate the two.

Radial Parts of Differential Operators

As a final project for this section we are going to discuss the radial parts of differential operators and relate them to orbital integrals. Radial parts of differential operators are a generalization of the term $\frac{d^2}{dr^2} + \frac{n-1}{r}\frac{d}{dr}$ occurring in the formula for the Laplacian on $\mathbf{R}^n$ in spherical coordinates. Other special cases were constructed by Harish-Chandra in [1956], [1957] and [1964], Berezin [1957], Methée [1954] and others. A general formulation involving Lie groups acting on pseudoriemannian manifolds is first described in Helgason [1965]. With a minor modification the treatment we use comes from Helgason [1972], see also Helgason [1984].

Definition 1.24: On an open set $U \subset \mathbf{q}$ a function $f \in C^\infty(U)$ is called *locally invariant* if

$$\left(\frac{d}{dt} f(\exp tX \cdot u)\right)_{t=0} = 0 \tag{1.14}$$

for all $u \in U$ and $X \in \mathbf{h}$.

Let $L_{\mathbf{q}}$ and $L_{\mathbf{a}}$ be the Laplace-Beltrami operators on $\mathbf{q}$ and $\mathbf{a}$ respectively.

Using Lemma 1.19 the following theorem is in Helgason [1972] (part 1 is Proposition 2.1 Ch. 1, part 2 is the Corollary on page 18, part 3 is Theorem 2.11). We have made a small modification of his hypotheses by allowing $(H \cdot X) \cap \mathbf{a}'$ to have cardinality different from one.

Theorem 1.25: 1) *If* D *is any differential operator on* $\mathbf{q}$ *then there exists a unique differential operator* $\Delta(\mathrm{D})$ *on* $\mathbf{a}'$ *such that* $\overline{\mathrm{Df}} = \Delta(\mathrm{D})\bar{\mathrm{f}}$ *for any locally invariant function* f *on a neighborhood of* $\mathbf{a}'$ *(bar denoting restriction to* $\mathbf{a}'$*).*

2) If D *is also* H *invariant then* $\mathrm{M_a Df(X)} = \Delta(\mathrm{D})(\mathrm{M_a f(X)})$.

3) $\Delta(\mathrm{L_q}) = \delta^{\frac{1}{2}}\mathrm{L_a} \circ \delta^{-\frac{1}{2}} - \delta^{\frac{1}{2}}\mathrm{L_a}(\delta^{-\frac{1}{2}})$.

The operator $\Delta(D)$ is called the radial part of D.

Helgason actually proves a more general theorem which we will need later. We can phrase it as follows.

Theorem 1.26: *Suppose* $\mathfrak{q}$ *is a pseudoriemannian manifold and* H *a Lie group acting on* $\mathfrak{q}$ *by isometries. If* $\mathfrak{a}'$ *is a submanifold of* $\mathfrak{q}$ *such that conditions 1-5 in Lemma 1.19 are true then 1-3 in Theorem 1.25 are still valid.*

We end this section with a very valuable integral formula which gives an explicit formula for the function δ used in equation 1.10 above.

Let dZ be the Riemannian (Lebesgue) measure on the pseudoriemannian vector space $\mathfrak{q}$. Let dX be the same thing on $\mathfrak{a}$. Fix an H invariant measure $d\dot{h}$ on H/M. As before we have $\Delta(\mathfrak{g}_{\mathbf{C}}, \mathfrak{a}_{\mathbf{C}})$ and a choice of positive roots Δ^+. Define

$$\Pi = \prod_{\alpha\in\Delta^+} \alpha^{m_\alpha}. \tag{1.15}$$

Theorem 1.27: *There is a constant* c *such that*

$$\int_{H\cdot\mathfrak{a}} f(Z)dZ = c\int_{\mathfrak{a}'} M_{\mathfrak{a}}f(X)|\Pi(X)|dX. \tag{1.16}$$

Thus, $c|\Pi| = \delta$ in the notation of the last theorem.

A few remarks are in order. First, $H\cdot\mathfrak{a}'$ is an open set in $\mathfrak{q}$ therefore the integral on the right in equation 1.16 is well defined, dZ being the ordinary Lebesgue measure on $\mathfrak{q}$ defined above. Second, if $\mathfrak{a}_1\cdots\mathfrak{a}_r$ are reresentatives from each of the H conjugacy classes of Cartan subspaces in $\mathfrak{q}$ then $\mathfrak{q}' = \bigcup_i H\cdot\mathfrak{a}_i'$ (Proposition 1.8). Therefore

$$\int_{\mathfrak{q}} f(Z)dZ = \sum_i \int_{H\cdot\mathfrak{a}_i} f(Z)dZ. \tag{1.17}$$

The proof is a standard argument (see for example Theorem 5.17 p.195 in Helgason [1984] one has to compute the 'determinant' of the map $adX : \mathfrak{m}^\perp \to \mathfrak{a}^\perp$.

2.

In this section we are going to give a complete answer to the limit formula problem described in section 1 for the case of rank one symmetric spaces. (Recall the rank of X is the dimension of any Cartan subspace .)

Our main tool will be generalized Riesz potentials. The ingredients for a generalized Riesz potential are a manifold M, a smooth function P and a measure dx on M. We define

$$I^s f = \int_{P>0} f(x)P^s(x)dx.$$

We call this a generalized Riesz potential. It would also be appropriate to call the generalized Riesz potentials Mellin transforms and write $I^s f = Z(s,f)$. This is particularly apparent in the example in the following section.

The Mellin Transform

In this subsection we consider the Mellin transform, sometimes called the Riemann-Liouiville integral, on **R**. All the statements we make about the Mellin transform are classical. It is defined by

$$I^s f = \frac{1}{\Gamma(s)} \int_0^\infty f(x) x^{s-1} dx. \tag{2.1}$$

This is well defined for $s > 0$ and $f \in \mathcal{S}(\mathbf{R})$. Note the Γ factor. The path we will follow with equation 2.1 is the same one we will follow later with other potentials. First we analytically continue using a well chosen differential operator. Then using well chosen coordinates we locate the delta function. In the present case this is not difficult.

Theorem 2.1: *The transform* $\mathrm{I}^s\mathrm{f}$ *given by equation 2.1 has an analytic continuation to the entire complex plane and* $\mathrm{I}^0\mathrm{f} = \mathrm{f}(0)$. *That is* $\mathrm{I}^0 = \delta$. *In fact* $\mathrm{I}^{-n}\mathrm{f} = (-1)^n \frac{d^n f}{dx^n}(0)$ *for all positive integers* n.

Proof: For the analytic continuation we use the equation

$$\frac{d}{dx} x^s = s x^{s-1}.$$

For $Re(s) > 0$ we have, using integration by parts,

$$-I^{s+1} f' = I^s f. \tag{2.2}$$

The left hand side of this equation is defined and analytic for $Re(s) > -2$. Thus we have an analytic continuation to this region. Iterating we analytically continue to the entire complex plane.

By equation 2.2 we have

$$I^0 f = -I^1 f' = \frac{-1}{\Gamma(1)} \int_0^\infty f'(x) dx = -f(x) \mid_0^\infty = f(0)$$

as claimed. The final statement follows easily from this and equation 2.2 ∎

The following easy modification of Theorem 2.1 will come in handy later.

Corollary 2.2: *If we let* $\mathrm{I}^s_A\mathrm{f} = \frac{1}{\Gamma(s)} \int_0^A \mathrm{f}(x) x^{s-1} dx$ *then* $\mathrm{I}^s_A\mathrm{f}$ *is entire in* s *and* $\mathrm{I}^{-n}_A\mathrm{f} = (-1)^n \frac{d^n f}{dx^n}(0)$ *for* $n = 0, 1, 2 \ldots$.

Corollary 2.3: *The transform*

$$\mathrm{I}^s_- \mathrm{f} = \frac{1}{\Gamma(s)} \int_{-\infty}^0 \mathrm{f}(x) x^{s-1} dx$$

is entire and

$$\mathrm{I}^{-n}_- \mathrm{f} = \frac{d^n f}{dx^n}(0)$$

for all nonnegative integers n

Instead of using Γ factors to cancel the poles in the integral we can combine I^s and I^s_-.

Corollary 2.4: *For* $A > 0$ *and* $B > 0$ *(possibly* ∞*), the transform*

$$I^s_{A,B}f = \int_0^A f(x)x^{s-1}dx - \cos(\pi s)\int_{-B}^0 f(x)(-x)^{s-1}dx$$

is entire in s.

We will need the above theorems with parameters. The next lemma can be found in Hormander [1966] (Lemma 2.2.11 in Chapter 2).

Lemma 2.5: *Let* u *be a complex valued function on the polydisc*

$$D = \{z \mid |z_j - z_j^0| < R, j = 1, \ldots, n\} \subset \mathbf{C}^n.$$

Assume u *is analytic in* $z' = (z_1, \ldots, z_{n-1})$ *if* z_n *is fixed. Also assume that* u *is analytic in*

$$D' = \{z \mid |z_j - z_j^0| < r, j = 1, \ldots, n-1,\ |z_n - z_n^0| < R\}$$

for some $r > 0$. *Then* u *is analytic in* D.

Theorem 2.6: *Let* x *be a real variable and* (s, s') *complex ones. Suppose* $f(x, s, s')$ *is smooth,* f *and all its derivatives in* x *are holomorphic in* (s, s')*, and* f *and all its derivatives in* (s, s') *are uniformly rapidly decreasing in* x. *Then*

$$I^{s,s'}f = \frac{1}{\Gamma(s)}\int_0^\infty f(x, s, s')x^{s-1}dx$$

extends to an entire function in (s, s') *and* $I^{0,s'}f = f(0, 0, s')$.

Proof: The proof is practically identical to that of Theorem 2.1, just drag along the extra variables, and use Lemma 2.5 . ∎

Remark: We also get the analogues of Corollaries 1.2, 1.3 and 1.4.

Returning now to $\mathbf{R}^n$, suppose $f \in C_c^\infty(\mathbf{R}^n)$ and $t \in \mathbf{R}^+$. Define $f^t(x) = f(tx)$. Then $\delta(f^t) = f(0) = \delta(f)$. Now suppose we have a homogeneous polynomial P of degree m. Let

$$I^s f = \int_{P>0} f(x)P^s(x)dx. \tag{2.3}$$

This is well defined for $Re(s) > 0$.

It follows easily that

$$I^s f^t = t^{-(ms+n)} I^s f.$$

Thus the delta function can only occur at the point $s = -n/m$. Because we don't have a factor in equation 2.3 corresponding to $1/\Gamma(s)$ in equation 2.1 we expect the potential 2.3 to have poles and the delta function to appear as a coefficient in the Laurent expansion at $-n/m$.

Heuristically we can make the following argument. Let (ρ, θ) be polar coordinates on $\mathbf{R}^n$ ($\rho \in \mathbf{R}^+$, $\theta \in S^{n-1} =$ the $n-1$ sphere). Let

$$S = S^{n-1} \cap \{x \mid P(x) > 0\}.$$

Then

$$I^s f = \int_{\rho=0}^{\infty} \int_{\theta \in S} f(\rho\theta) \rho^{ms+n-1} P^s(\theta) d\rho d\theta.$$

We then let

$$J^{s,t} f = \int_{\rho=0}^{\infty} \int_{\theta \in S} f(\rho\theta) \rho^{ms+n-1} P^t(\theta) d\rho d\theta.$$

Of course $I^s f = J^{s,s} f$. Considering just the integral in ρ our analysis of the Mellin transform equation 2.1 tells us that at $s = -n/m$ (and $Re(t) > 0$) we have a simple pole whose residue is the value of the integral in θ at $\rho = 0$. That is at $s = -n/m$

$$\operatorname{Res}(J^{s,t}, s = -n/m) = \int_{\theta \in S} f(0) P^t(\theta) d\theta.$$

Now taking into account the possible poles of this integral at $t = -n/m$ we guess that in the Laurent series of $I^s f$ around $s = -n/m$ the first nonzero coefficient is $cf(0)$ where c is a constant independent of f.

Looking at the simple example of

$$I^s f = \int_0^{\infty} \int_0^{\infty} f(x,y) y^{s-1} dx dy$$

on $\mathbf{R}^2$ we see that the above analysis can't hope to work in general. Nonetheless it will be the way we approach the analysis in our particular cases.

Riesz Potentials Associated to a Quadratic Form

We let $z_1 \cdots z_n$ be the standard coordinates on $\mathbf{R}^n$ and $\langle\ ,\ \rangle$ the standard quadratic form of signature (p, q),

$$\langle z, z \rangle = \sum_{i=1}^{p} z_i^2 - \sum_{i=p+1}^{n} z_i^2. \tag{2.4}$$

We let

$$R_+ = \{z \mid \langle z, z \rangle > 0\} \tag{2.5}$$

and

$$R_- = \{z \mid \langle z, z \rangle < 0\}. \tag{2.6}$$

On R_+ we define $r = r(z) = \langle z, z \rangle^{\frac{1}{2}}$.

For the symmetric space $G/H = SO_e(p, q+1)/SO_e(p, q)$ it is easy to see that the representation of H on $\mathbf{q}$ is just $SO_e(p, q)$ on $\mathbf{R}^n$.

The Laplace Beltrami operator on $\mathbf{R}^n$ is

$$L = \frac{\partial^2}{\partial z_1^2} + \cdots + \frac{\partial^2}{\partial z_p^2} - \frac{\partial^2}{\partial z_{p+1}^2} - \cdots - \frac{\partial^2}{\partial z_n^2}.$$

We let

$$H_+ = \{z \mid \langle z, z\rangle = 1\}.$$

We have $(H_+, \mathbf{R}^+)$ as "polar coordinates" on R_+ by the map

$$(z, r) \mapsto rz. \tag{2.7}$$

It is clear is that $H_+ = H \cdot e_1 \approx H/M$ and that (2.7) above is the map

$$\varphi : H/M \times W \to V$$

from Lemma 1.16 in section 1.

Just as in the case of the usual polar coordinates on $\mathbf{R}^n$ we have $dz = (r)^{n-1} dr d\dot{h}$. (Or use Theorem 1.27 in section 1.) Here r is the coordinate on W given in (2.7) above. Thus Theorem 1.25 section 1 gives us the next lemma.

Lemma 2.7: *The radial part of* L *(on* W*) is*

$$\Delta(\mathrm{L}) = \frac{\mathrm{d}^2}{\mathrm{dr}^2} + \frac{\mathrm{n}-1}{\mathrm{r}}\frac{\mathrm{d}}{\mathrm{dr}}.$$

Corollary 2.8: *We have* $\mathrm{L}(\mathrm{r}^{\mathrm{s}-\mathrm{n}}) = (\mathrm{s}-2)(\mathrm{s}-\mathrm{n})\mathrm{r}^{\mathrm{s}-\mathrm{n}-2}$.

For $s \in \mathbf{C}$ let

$$H_n(s) = 2^{s-1}\pi^{\frac{n-2}{2}}\Gamma(\frac{s}{2})\Gamma(\frac{s-n+2}{2}). \tag{2.8}$$

We note

$$(s-2)(s-n)H_n(s-2) = H_n(s). \tag{2.9}$$

We can now define our first Riesz potential. For $f \in C^\infty(\mathbf{R}^n)$ which is rapidly decreasing on R_+, define

$$I_+^s f = \frac{1}{H_n(s)} \int_{R_+} f(z) r^{s-n} dz. \tag{2.10}$$

It is clear that $I_+^s f$ is well defined for $Re(s) > n$. Equally clear is the fact that it is analytic in this region. Our next task is to show that it extends to an entire function on $\mathbf{C}$.

Analytic Continuation

Proposition 2.9: *The generalized Riesz potential* $\mathrm{I}_+^{\mathrm{s}}\mathrm{f}$ *defined in equation 2.10 is well defined as an integral and analytic in the region* $\mathrm{Re(s)} > \mathrm{n}$. *It can be analytically continued to the entire complex plane with functional equation given by equation 2.12.*

Proof: Since f is rapidly decreasing in R_+, for $Re(s)$ large enough, Green's formula gives

$$\int_{R_+} (Lf)r^{s-n} - fL(r^{s-n})dz \tag{2.11}$$

as an integral over the cone $\langle z, z\rangle = 0$, that is, over the boundary of R_+. On this cone r^{s-n} and its first few derivatives are 0. Thus (2.11) equals 0.

We saw that $L(r^{s-n}) = (s-2)(s-n)r^{s-n-2}$. Hence (for $Re(s)$ large enough)

$$\begin{aligned} \textstyle\int_{R_+}(Lf)r^{s-n}dz &= \textstyle\int_{R_+} f(z)L(r^{s-n})dz \\ &= \textstyle\int_{R+}(s-2)(s-n)f(z)r^{s-n-2}dz. \end{aligned}$$

Thus equation 2.9 gives the following *functional equation*

$$I_+^s Lf = I_+^{s-2} f. \tag{2.12}$$

Iterating we get

$$I_+^{s+2m}(L^m f) = I_+^s f. \tag{2.13}$$

Using this equation we can analytically continue $I_+^s f$ to the entire complex plane. ∎

Lemma 2.10: *For fixed* s *the map* f $\mapsto$ I_+^sf *is a tempered distribution.*

Proof: This is clear for $Re(s)$ large. It follows for any s by equation 2.13. ∎

With this lemma we can often prove things for rapidly decreasing functions by considering only compactly supported ones.

$I_+^0 f$

If we examine $H_n(s) = 2^{s-1}\pi^{\frac{n-2}{2}}\Gamma(\frac{s}{2})\Gamma(\frac{s-n+2}{2})$ we see that it has a pole (of order 1 or 2) at $s = 0$. Hence $\frac{1}{H_n(s)}$ has a zero there. Thus an investigation of $I_+^0 f$ entails an investigation of the pole at 0 of the integral

$$\int_{R_+} f(z)r^{s-n}dz.$$

It will turn out that $I_+^0 f = cf(0)$, where c is a constant depending only on p and q. Said slightly differently $I_+^0 = c\delta$ as distributions. Recall that this is part of our program for finding limit formulas.

Unfortunately, c is sometimes equal to 0. This necessitates defining two more types of Riesz potentials. For now we content ourselves with showing $I_+^0 f = cf(0)$. The method used to prove this is not the most straightforward but it reflects some geometry that we feel is important. We will describe this geometry at the end of this subsection.

In order to show $I_+^0 f = cf(0)$ we need coordinates on R_+. We use

$$0 < t \le 1, \quad \theta_1 \in S^{p-1}, \quad \theta_2 \in S^{q-1} \quad \text{and} \quad 0 < r < \infty,$$

where S^{p-1} is the Euclidean sphere of radius one in $\mathbf{R}^p$. We put

$$z=\left(\frac{r(t^{-1}+t)}{2}\theta_1, \frac{r(t^{-1}-t)}{2}\theta_2\right)\in \mathbf{R}^{p+q}. \tag{2.14}$$

We note, $r^2=\langle z,z\rangle$. Also, let $T=t^2$ (so $0<T\leq 1$).

Given $z\in R_+$, $a(z)=a$ is defined as the intersection of the ray $\vec{0z}$ with S^{n-1}.

The Euclidean norm of z is $|z|=r\sqrt{\frac{t^{-2}+t^2}{2}}$. Thus

$$a=\frac{1}{r(\frac{t^{-2}+t^2}{2})^{\frac{1}{2}}}\cdot z=\frac{\sqrt{2}}{2\sqrt{1+T^2}}((1+T)\theta_1,(1-T)\theta_2). \tag{2.15}$$

Note that a is smooth in T, θ_i and r and extends smoothly to all values of T. Let $r_a^2=\langle a,a\rangle$ ($\langle\ ,\ \rangle=(p,q)$ metric). Let

$$v_+=\frac{r_a}{t} \quad \text{and} \quad \sigma=\frac{r}{r_a}. \tag{2.16}$$

An easy calculation shows

$$r_a=\langle a,a\rangle^{\frac{1}{2}}=\frac{\sqrt{2}T^{\frac{1}{2}}}{(1+T^2)^{\frac{1}{2}}}, \quad \text{and} \quad v_+=\frac{\sqrt{2}}{(1+T^2)^{\frac{1}{2}}}. \tag{2.17}$$

Thus we have the following lemma.

Lemma 2.11: *The function v_+ is bounded away from zero. That is, there are positive constants* ϵ and E *such that* $0<\epsilon\leq \mathrm{v}_+\leq \mathrm{E}$.

Since $\langle z,z\rangle=r^2$ we have

$$\frac{r_a}{r}z=a \quad \text{and} \quad \frac{r}{r_a}a=\sigma a=z. \tag{2.18}$$

Another easy calculation shows,

$$r^{s-n}dz \;=\; \tfrac{1}{2^{n-1}}\sigma^{s-1}v_+^s T^{\frac{s-n}{2}}(1+T)^{p-1}(1-T)^{q-1}d\sigma dT d\theta_1 d\theta_2\ \cdot \tag{2.19}$$

Thus we have

$$I_+^s f=\tfrac{1}{2^{n-1}H_n(s)}\textstyle\int_{T=0}^1\int_{\sigma=0}^\infty\left(\int_{S^{p-1}\times S^{q-1}}f(\sigma\cdot a)d\theta_1 d\theta_2\right)\sigma^{s-1}v_+^s T^{\frac{s-n}{2}}(1+T)^{p-1}(1-T)^{q-1}d\sigma dT. \tag{2.20}$$

Let

$$G_+(\sigma,T,s)=(1+T)^{p-1}(1-T)^{q-1}v_+^s\int_{\theta_1,\theta_2}f(\sigma\cdot a)d\theta_1 d\theta_2. \tag{2.21}$$

Because of Lemma 2.11 and equation 2.15 G_+ is smooth in (σ,T,s) and G_+ and all its partial derivatives with respect to (σ,T) are holomorphic in s. Also since σ is the Euclidean norm and $a\in R_+$ is independent of σ, $G_+(\sigma,T,s)$ is rapidly decreasing in σ.

For $Re(s)>0$ and $Re(s')>0$ define

$$J^{s,s'}G_+=\frac{1}{H(s,s')}\int_{T=0}^1\int_{\sigma=0}^\infty G_+(\sigma,T,s)T^{s'}\sigma^{s-1}d\sigma dT, \tag{2.22}$$

where

$$H(s,s') = \pi^{\frac{n-2}{2}} 2^{s-1} 2^{n-1} \Gamma(s)\Gamma(s').$$

Lemma 2.12: *As a function in* (s, s'), $J^{s,s'}G_+$ *extends to an entire holomorphic function.*

Proof: First we write

$$\begin{aligned} J^{s,s'}G_+ &= \tfrac{1}{H(s,s')} \textstyle\int_{T=0}^{1} \int_{\sigma=0}^{1} G_+(\sigma,T,s) T^{s'} \sigma^{s-1} d\sigma dT \\ &+ \tfrac{1}{H(s,s')} \textstyle\int_{T=0}^{1} \int_{\sigma=1}^{\infty} G_+(\sigma,T,s) T^{s'} \sigma^{s-1} d\sigma dT. \end{aligned}$$

The integral in the second summand on the right is holomorphic in s without its Γ factor from $\frac{1}{H(s,s')}$. Thus this term becomes $\frac{1}{\Gamma(s)}\varphi(s,s')$ where φ is holomorphic in both variables.

We can write

$$\begin{aligned} G_+(\sigma,T,s) &= \textstyle\sum_{i=0}^{N} \sum_{j=0}^{M} a_{i,j}(s) \sigma^i T^j \quad + \quad T^{M+1} \sum_{i=0}^{N} h_i(T,s) \sigma^i \\ &+ \quad \sigma^{N+1} \textstyle\sum_{j=0}^{M} k_j(\sigma,s) \quad + \quad \sigma^{N+1} T^{M+1} l(\sigma,T,s), \end{aligned}$$

where $a_{i,j}$, h_i, k_j, and l are smooth in their respective variables and all their partial derivitaves with respect to (σ,T) are holomorphic in s. Thus (for $Re(s)$ and $Re(s')$ large)

$$\begin{aligned} J^{s,s'}G_+ &= \tfrac{1}{H(s,s')} \Big[\textstyle\sum_{i=0}^{N} \sum_{j=0}^{M} \frac{a_{i,j}(s)}{(s+i)(s'+j)} \quad + \quad \sum_{i=0}^{N} \int_0^1 T^{M+s'} h_i(T,s) dT \\ &+ \textstyle\sum_{j=0}^{M} \int_0^1 \sigma^{N+s} k_j(\sigma,s) d\sigma + \int_0^1 \int_0^1 \sigma^{N+s} T^{M+s'} l(\sigma,T,s) d\sigma dT \Big] \\ &+ \tfrac{1}{\Gamma(s)} \varphi(s,s'). \end{aligned} \tag{2.23}$$

The integrals above converge if $Re(s) > -N$ and $Re(s') > -M$. Looking at the definition of $H(s,s')$ we see that the poles of $\Gamma(s')$ cancel those of $\frac{1}{s'+j}$ and the poles of $\Gamma(s)$ cancel those of $\frac{1}{s+i}$. Therefore equation 2.23 gives the analytic continuation of $J^{s,s'}G_+$. ∎

Lemma 2.13: *At* s = 0, *we have* $I_+^0 f = cf(0)$ *for some constant* c.

Proof: Given the definition of $J^{s,s'}G_+$ (equation 2.22) and the definition of $H_n(s)$ (equation 2.8) we see from equation 2.20 that

$$I_+^s f = \frac{\Gamma(s)}{\Gamma(s/2)} J^{s,\frac{s+2-n}{2}} G_+, \tag{2.24}$$

and

$$I_+^0 f = \frac{1}{2} J^{0,\frac{2-n}{2}} G_+. \tag{2.25}$$

We saw in the Corollary 2.2 that

$$K^s f = \frac{1}{\Gamma(s)} \int_0^A f(x) x^{s-1} dx \tag{2.26}$$

(we use K since I is taken) is holomorphic in s and $K^0 f = f(0)$.

Thus for s' fixed with large real part we have by Theorem 2.6 and equation 2.22,

$$J^{0,s'}G_+ = \frac{2\Omega_{p-1}\Omega_{q-1}}{\pi^{\frac{n-2}{2}} 2^{n-1} \Gamma(s')} \int_0^1 (1+T)^{p-1} (1-T)^{q-1} T^{s'-1} f(0) dT. \tag{2.27}$$

Here Ω_{p-1} is the surface area of the sphere S^{p-1}.

This last equation is in the form of the potentials K^s in equation 2.26 . Thus it's analytic in s' and letting $s' = (2-n)/2$ we have

$$I_+^0 f = \frac{1}{2} J^{0,\frac{2-n}{2}} G = cf(0), \tag{2.28}$$

where

$$c = \frac{\Omega_{p-1}\Omega_{q-1}}{\pi^{\frac{n-2}{2}} 2^{n-1}\Gamma(s')} \int_0^1 (1+T)^{p-1}(1-T)^{q-1}T^{s'-1}dT|_{s'=\frac{2-n}{2}}. \tag{2.29}$$

∎

Earlier we alluded to some geometric content in the proof that $I_+^0 f = cf(0)$. A main point in that proof is the mapping of H_+ to S^{n-1} given by $z \mapsto a$. This is a special case of the map $\psi : H/M \hookrightarrow K^d/M^d$ (see equation 1.2 in section 1 for the definitions of K^d etc.) given by the following prescription. We have a map $\mathbf{q} \to \mathbf{p}^d$, given by $X_k + X_p \mapsto iX_k + X_p$ ($X_k \in \mathbf{q} \cap \mathbf{k}$, $X_p \in \mathbf{q} \cap \mathbf{p}$). Its restriction maps $\mathbf{a}$ to $\mathbf{a}^d$. Write $X \mapsto X^d$ under this map. Choose a positive Weyl chamber, $\mathbf{a}_+^d$ in $\mathbf{a}^d$. We say $\psi(\dot{h}) = \dot{k}$ if and only if $(h \cdot X)^d = k \cdot Y^d$ for some pair $X, Y \in \mathbf{a}$ such that X^d and Y^d are in $\mathbf{a}_+^d$. This is well defined since $M'/M \approx (M' \cap K)/(M \cap K) \subset W^d$, the Weyl group of G^d/K^d.

Computation of c

Lemma 2.14: *The constant* c *from the last section is* $2\sin(p\pi/2)$.

Proof: We will find c by computing $I_+^s f$ for a specific f. Using the bipolar coordinates $z = (s_1\theta_1, s_2\theta_2)$ we let $f = e^{-s_1^2}$. This function is rapidly decreasing on R_+ (but not on all of $\mathbf{R}^n$).

Using the sequence of substitutions $z = (s_1\theta_1, s_2\theta_2)$, $s_2 = s_1 x$ and $y = x^2$ we get (for this f)

$$I_+^s f = \frac{1}{2}\frac{\Omega_{p-1}\Omega_{q-1}}{H_n(s)} \int_{s_1=0}^{\infty}\int_{y=0}^{1} e^{-s_1^2} s_1^{s-1}(1-y)^{\frac{s-n}{2}} y^{\frac{q-2}{2}} dy ds_1.$$

Both integrals are well known,

$$\int_{s_1=0}^{\infty} e^{-s_1^2} s_1^{s-1} ds_1 = \frac{1}{2}\Gamma(\frac{s}{2}),$$

and

$$\int_{y=0}^{1} (1-y)^{\frac{s-n}{2}} y^{\frac{q-2}{2}} dy = \frac{\Gamma(\frac{q}{2})\Gamma(\frac{s-n+2}{2})}{\Gamma(\frac{s-n+2+q}{2})}.$$

It is also well known that

$$\Omega_m = \frac{2\pi^{\frac{m+1}{2}}}{\Gamma(\frac{m+1}{2})}.$$

Thus

$$I_+^s f = \frac{\pi}{2^{s-1}\Gamma(\frac{p}{2})\Gamma(\frac{s-p+2}{2})}. \tag{2.30}$$

Thus

$$I^0_+ f = \frac{\pi}{2^{-1}\Gamma(\frac{p}{2})\Gamma(1-\frac{p}{2})} = 2sin(\frac{p\pi}{2}) = cf(0) = c. \tag{2.31}$$

∎

Remark: If $p = 1$ then $c = 2$ and this corresponds to the case in Riesz [1949]. The reason we get 2 and he gets 1 is that he integrates over the retrograde cone while we integrate over both the forward and retrograde cones. In the cases where $p > 2$ there is only one cone (i.e. R_+ is connected), thus it makes sense for us to integrate over both parts in the special case $p = 1$.

$I^s_- f$

Define $I^s_- f$ as $I^s_+ f$ on the negative metric. More precisely, for $z \in R_- = \{z \mid \langle z,z\rangle < 0\}$ let $r^2_- = -\langle z,z\rangle$, (when necessary we will use r_+ where we previously used r). Then

$$I^s_- f = \frac{1}{H_n(s)}\int_{R_-} f(z) r_-^{s-n} dz. \tag{2.32}$$

The analysis above carries over by reversing p and q. In particular we have

$$I^0_- f = 2sin\left(\frac{q\pi}{2}\right) f(0). \tag{2.33}$$

Thus if q is odd then $I^0_- f = (-1)^{\frac{q-1}{2}} f(0)$ and if p is odd then $I^0_+ f = (-1)^{\frac{p-1}{2}} f(0)$.

$I^s_0 f$

The only case left is when both p and q are even. This case is much trickier and developing the correct potential will take some time. One of the fringe benefits of this development will be some lemmas (see for example Lemma 2.16) that will give us the relationship between the orbital integral and the Riesz potential (valid for arbitrary p and q). We continue with the assumption $pq \neq 0$, we add to this the assumption $n = p + q \geq 3$. The case $p = 1$ and $q = 1$ will be handled separately. Note, we have not yet made the assumption that p and q are even.

For $(x,y) \in \mathbf{R}^p \times \mathbf{R}^q = \mathbf{R}^n$ we have the coordinates $(r, \xi, \theta_1, \theta_2)$ on R_+ given by

$$(x,y) = (r\cosh(\xi)\theta_1, r\sinh(\xi)\theta_2). \tag{2.34}$$

Here $0 < r < \infty$, $0 \leq \xi < \infty$, $\theta_1 \in S^{p-1}$, $\theta_2 \in S^{q-1}$.

It is trivial to see that

$$dz = r^{n-1}\cosh^{p-1}(\xi)\sinh^{q-1}(\xi) dr d\xi d\theta_1 d\theta_2. \tag{2.35}$$

For $f \in C_c^\infty(\mathbf{R}^n)$ and $r > 0$ define

$$M_+ f(r) = \int_0^\infty \int_{S^{p-1}\times S^{q-1}} f(r\cosh(\xi)\theta_1, r\sinh(\xi)\theta_2)\cosh^{p-1}(\xi)\sinh^{q-1}(\xi) d\xi d\theta_1 d\theta_2. \tag{2.36}$$

Reversing the $\cosh\xi$ and $\sinh\xi$ in 2.34 we get coordinates on R_-. Therefore we define

$$M_-f(r) = \int_0^\infty \int_{S^{p-1}\times S^{q-1}} f(r\sinh(\xi)\theta_1, r\cosh(\xi)\theta_2)\sinh^{p-1}(\xi)\cosh^{q-1}(\xi)d\xi d\theta_1 d\theta_2. \tag{2.37}$$

We note that M_+f and M_-f are smooth in $r > 0$.

We have the following important lemma.

Lemma 2.15: *For* $n \geq 3$ *and* $f \in C_c^\infty(\mathbf{R}^n)$ *the limits*

$$\lim_{r\to 0+} r^{n-2}M_+f(r) \quad \textit{and} \quad \lim_{r\to 0+} r^{n-2}M_-f(r)$$

exist and are equal.

Proof: Let

$$\tilde{f}(s_1, s_2) = \int_{S^{p-1}\times S^{q-1}} f(s_1\theta_1, s_2\theta_2)d\theta_1 d\theta_2.$$

Clearly this is smooth and compactly supported.

We then have

$$M_+f(r) = \int_0^\infty \tilde{f}(r\cosh\xi, r\sinh\xi)\cosh^{p-1}(\xi)\sinh^{q-1}(\xi)d\xi, \tag{2.38}$$

and

$$M_-f(r) = \int_0^\infty \tilde{f}(r\sinh\xi, r\cosh\xi)\sinh^{p-1}(\xi)\cosh^{q-1}(\xi)d\xi. \tag{2.39}$$

Make the substitution $t = r\sinh\xi$. Since $\tilde{f}$ is compactly supported these integrals become

$$\begin{aligned} r^{n-2}M_+f(r) &= \textstyle\int_0^A \tilde{f}(\sqrt{r^2+t^2}, t)(r^2+t^2)^{\frac{p-2}{2}}t^{q-1}dt \quad \text{and} \\ r^{n-2}M_-f(r) &= \textstyle\int_0^A \tilde{f}(t, \sqrt{r^2+t^2})(r^2+t^2)^{\frac{q-2}{2}}t^{p-1}dt \end{aligned} \tag{2.40}$$

for some constant A.

Set $r = 0$ in the right hand sides of equation 2.40. They both become

$$\int_0^A \tilde{f}(t,t)t^{n-3}dt.$$

Since we have assumed $n \geq 3$ this integral converges. This proves the lemma. ∎

We can use this lemma to prove the lemma alluded to earlier.

Lemma 2.16: *For* $f \in C_c^\infty(\mathbf{R}^n)$ *we have*

$$\lim_{r\to 0+} r^{n-2}M_+f(r) = cI_+^{n-2}f = cI_-^{n-2}f = \lim_{r\to 0+} r^{n-2}M_-f(r),$$

where $c \neq 0$ *is a constant independent of* f.

Proof: By Lemma 2.15 all we have to prove is the first equality. Also by Lemma 2.15 we know the limit exists, call it a.

Fix r_0 such that $suppf \subset \{z \mid \langle z,z\rangle < r_0\}$. For $Re(s)$ large enough we have

$$\begin{array}{lcl} I_+^s f & = & \frac{1}{H_n(s)} \int_{R_+} f(z) r_+^{s-n} dz \\ \text{(by eq. 2.35)} & = & \frac{1}{H_n(s)} \int_0^{r_0} \int\int f(r\cosh\xi\theta_1, r\sinh\xi\theta_2) r^{s-1} \cosh^{p-1}(\xi)\sinh^{q-1}(\xi)\, d\theta_1\, d\theta_2\, d\xi\, dr \\ \text{(by eq. 2.36)} & = & \frac{1}{H_n(s)} \int_{r=0}^{r_0} (r^{n-2} M_+ f(r)) r^{s-n+1} dr. \end{array} \tag{2.41}$$

Since $H_n(s)$ has a simple pole at $s = n-2$ we take the residue of the integral there. That is we take

$$\lim_{s\to(n-2)^+} (s-n+2) \int_{r=0}^{r_0} (r^{n-2} M_+ f(r)) r^{s-n+1} dr. \tag{2.42}$$

In taking this residue we can assume s is real and greater than $n-2$. This is convenient since by Lemma 2.15 the integral is absolutely convergent there and we don't have to think of it as an implicitly given holomorphic extension.

Since $r^{n-2}M_+f(r)$ is continuous on $[0,\infty]$ the theory of the Melin transform implies 2.42 equals a. (Briefly, for $f \in C([0,A])$, $s\int_0^A f(r) r^{s-1} dr \leq sup|f|A^s$. So the Mellin transform converges as a measure to the delta function as s approaches 0 from above.) All that remains is a computation of c. Taking into account the definition of $H_n(s)$ we get $c = (\pi^{\frac{n-2}{2}} 2^{n-2}\Gamma(n-2))^{-1} \neq 0$. This proves the first equation of the lemma and therefore the entire lemma. ∎

Remark: We don't use it, but the equation 2.40 (with $A=\infty$ shows that for $r>0$, $f \mapsto M_+f(r)$ is actually a tempered distribution.

We note here that changing $\langle\ ,\ \rangle$ to $-\langle\ ,\ \rangle$ interchanges R_+ and R_- and changes the sign of the Laplace-Beltrami operator. Thus, in analogy to Lemma 2.7, in polar coordinates $(r_-, z) \in \mathbf{R}^+ \times H_-$ on R_- ($H_- = \{\langle z,z\rangle = -1\}$ and $(r_-,z) \mapsto r_- z$), the radial part of the Laplace-Beltrami operator is

$$\Delta(L) = -\left(\frac{d^2}{dr_-^2} + \frac{n-1}{r_-}\frac{d}{dr_-}\right). \tag{2.43}$$

Here r_- is the coordinate on $\mathbf{R}^+ \cdot e_{p+1} \subset R_-$.

Thus in analogy to equation 2.13, we have

$$I_-^s L^m f = (-1)^m I_-^{s-2m} f. \tag{2.44}$$

Combining this with Lemma 2.16 we have the following lemma.

Lemma 2.17: *The following equality holds for* m = 0,1,2 ...,

$$\mathrm{I}_+^{n-2-2m}\mathrm{f} = (-1)^m \cdot \mathrm{I}_-^{n-2-2m}\mathrm{f}.$$

We're finally ready to define our last type of Riesz potential. Define,

$$I_0^s f = \Gamma\left(\frac{s-n+2}{2}\right)\left(I_+^s f - cos\left(\frac{s-n+2}{2}\pi\right) I_-^s f\right). \tag{2.45}$$

Lemma 2.18: *The potential* $\mathrm{I}_0^s\mathrm{f}$ *an entire function in* s.

Proof: (See Corollary 2.4) The function $\Gamma(\frac{s-n+2}{2})$ has poles at $n-2-2m$ for $m = 0, 1, 2, \ldots$. By Lemma 2.17 these are cancelled by the zeros of
$I_+^s f - cos(\frac{s-n+2}{2}\pi) I_-^s f$. ∎

Now we make the assumption that p and q are both even.

Lemma 2.19: *If* p *and* q *are both even then* $\mathrm{I}_0^0\mathrm{f} = \mathrm{cf}(0)$ *where* $\mathrm{c} \neq 0$ *is a constant independent of* f.

Proof: Recall equation 2.25,

$$I_+^s f = \frac{\Gamma(s)}{\Gamma(s/2)} J_+^{s,\frac{s+2-n}{2}} G_+, \quad \text{and} \quad I_+^0 f = \frac{1}{2} J_+^{0,\frac{2-n}{2}} G_+ \tag{2.46}$$

and equation 2.27

$$J_+^{0,s'} G_+ = \frac{2\Omega_{p-1}\Omega_{q-1}}{\pi^{\frac{n-2}{2}} 2^{n-1}\Gamma(s')} \left(\int_0^1 (1+T)^{p-1}(1-T)^{q-1} T^{s'-1} dT \right) f(0). \tag{2.47}$$

Thus

$$\Gamma(s') \cdot J_+^{0,s'} G_+ = k f(0) \int_0^1 (1+T)^{p-1}(1-T)^{q-1} T^{s'-1} dT, \tag{2.48}$$

where k is the constant in equation 2.47. That is $\Gamma(s') \cdot J_+^{0,s'} G_+ = a_+ f(0)$, where the constant a_+ must still be determined.

To analyze the contribution of $I_-^0 f$ we reuse the trick of switching the sign of $\langle\ ,\ \rangle$ and reversing p and q. Thus from equations 2.46 and 2.47 we have;

$$I_-^s f = \frac{\Gamma(s)}{\Gamma(s/2)} J_-^{s,\frac{s+2-n}{2}} G_-, \quad \text{and} \quad I_-^0 f = \frac{1}{2} J_-^{0,\frac{2-n}{2}} G_- \tag{2.49}$$

and

$$\Gamma(s') \cdot J_-^{0,s'} G_- = k f(0) \int_0^1 (1+T)^{q-1}(1-T)^{p-1} T^{s'-1} dT, \tag{2.50}$$

here the constant k is the same as in 2.48.

We would like to combine equations 2.48 and 2.50 to prove the lemma. In order to do this we need to show that

$$J_0^{s,s'} G \stackrel{\text{def}}{=} \Gamma(s')(J_+^{s,s'} G_+ - \cos(\pi s') J_-^{s,s'} G_-) \tag{2.51}$$

is analytic in (s, s'). (We should remark that the function G has not been defined, but this is not important, we will never use the definition of G just $J_0^{s,s'} G$.) Assuming this for the moment, we have from equations 2.46 and 2.49

$$I_0^0 f = \frac{1}{2} J_0^{0,\frac{2-n}{2}} G.$$

But then from equations 2.48 and 2.50 there is a constant c such that

$$J_0^{0,s'} G = \Gamma(s')(J_+^{0,s'} G_+ - \cos(\pi s') J_-^{0,s'} G_-) = c f(0).$$

As before we compute c by using a particular function f. Let $\varphi \in C^\infty(\mathbf{R})$ be such that

$$\begin{aligned}\varphi(r) &= 1 \quad \text{if} \quad r > -1\\ \varphi(r) &= 0 \quad \text{if} \quad r < -2\end{aligned}$$

In bipolar coordinates $((x,y) = (s_1\theta_1, s_2\theta_2))$ let
$f(z) = \varphi(\langle z, z\rangle)e^{s_1^2}$. This function is rapidly decreasing on $\mathbf{R}^n$.

From equation 2.30, we have

$$\Gamma(\frac{s-n+2}{2})I_+^s f = \frac{\pi}{2^{s-1}\Gamma(p/2)} \tag{2.52}$$

and at $s = 0$

$$\left[\Gamma(\frac{s-n+2}{2})I_+^s f\right]_{s=0} = \frac{2\pi(-1)^{\frac{q}{2}}}{\left(\frac{n-2}{2}\right)!}.$$

It is going to be a little harder to compute $I_-^s f$. Using bipolar coordinates and making the substitutions $r^2 = s_2^2 - s_1^2$, $s_1 = s_1$ followed by $s_1 = rx$, $r = r$ and finally $y = x^2$, $r = r$ we find

$$I_-^s f = \frac{\Omega_{p-1}\Omega_{q-1}}{2H_n(s)} \int_{y>0}\int_{r>0} e^{-yr^2}\varphi(-r^2)r^{s-1}y^{\frac{p-2}{2}} \sum_{i=0}^{\frac{q-2}{2}} a_i y^i drdy,$$

where a_i is the binomial coefficient $\begin{pmatrix} (q-2)/2 \\ i \end{pmatrix}$. Since $\int_0^\infty e^{-yr^2}y^{t-1}dy = \Gamma(t)r^{-2t}$ the definition of $H_n(s)$ yields

$$\Gamma(\frac{s-n+2}{2})I_-^s f = \frac{\Omega_{p-1}\Omega_{q-1}}{2^s\pi^{\frac{n-2}{2}}\Gamma(s/2)} \sum_{i=0}^{\frac{q-2}{2}} \int_{r>0} a_i\varphi(-r^2)r^{s-1-(p+2i)}.$$

From the theory of the Mellin transform (Theorem 2.1), at $s = 0$ this is just a sum of derivatives of φ at 0. But by its construction all derivatives of φ vanish at 0. Thus $\Gamma(\frac{s-n+2}{2})I_-^s f = 0$ at $s = 0$. Thus from equation 2.52 we have

$$I_0^0 f = \frac{2\pi(-1)^{q/2}}{\left(\frac{n-2}{2}\right)!} = cf(0) = c.$$

Except for showing that $J_0^{s,s'}G$ (see equation 2.51) is analytic this proves the lemma.

We recall the definition 2.21 for $G_+(\sigma, T, s)$,

$$G_+(\sigma, T, s) = (1+T)^{p-1}(1-T)^{q-1}v_+^s \int_{S^{p-1}\times S^{q-1}} f(\sigma \cdot a)d\theta_1 d\theta_2,$$

where $v_+ = |r_a|/\sqrt{|T|}$ and σ is Euclidean length. (see equations 2.15, 2.16 and 2.17) This is well defined for $-1 \le T \le 1$.

If we take into account all the necessary sign changes, we find that $G_-(\sigma, T, s) = G_+(\sigma, -T, s)$ and

$$J_-^{s,s'}G_- = \frac{1}{H(s,s'}\int_{\sigma=0}^\infty \int_{T=-1}^0 G_+(\sigma, T, s)(-T)^{s'-1}\sigma^{s-1}dTd\sigma.$$

Thus

$$\begin{aligned} J_0^{s,s'}G &= \Gamma(s')(J_+^{s,s'}G_+ - cos(\pi s')J_-^{s,s'}G_-) \\ &= \frac{1}{\pi^{\frac{n-2}{2}}2^{s-n+2}\Gamma(s)}\int_{\sigma=0}^{\infty}\sigma^{s-1}\left[\int_{T=0}^{1}G_+(\sigma,T,s)T^{s'-1}dT\right. \\ &\quad \left. -cos(\pi s')\int_{T=-1}^{0}G_+(\sigma,T,s)(-T)^{s'-1}dT\right]d\sigma. \end{aligned}$$

Applying Corollary 2.4 (with parameters) we conclude that the term inside the brackets is holomorphic in (s,s'). Now apply Theorem 2.6 to conclude $J_0^{s,s'}G$ is holomorphic. ∎

We collect all this into one theorem.

Theorem 2.20: *Let* $\langle\, ,\,\rangle$ *be the standard inner product of signature* (p,q) *on* $\mathbf{R}^n$. *Let* $R_+ = \{\langle z,z\rangle > 0\}$ *and* $R_- = \{\langle z,z\rangle < 0\}$. *For* $f \in \mathcal{S}(\mathbf{R}^n)$ *and for* $\mathrm{Re}(s)$ *large, define the three Riesz potentials;*

$$\begin{aligned} I_+^s f &= \frac{1}{H_n(s)}\int_{R_+} f(z)|\langle Z,Z\rangle|^{s-n}dz, \\ I_-^s f &= \frac{1}{H_n(s)}\int_{R_-} f(z)|\langle Z,Z\rangle|^{s-n}dz, \\ I_0^s f &= \Gamma(\tfrac{s-n+2}{2})(I_+^s f - cos(\tfrac{s-n+2}{2}\pi)I_-^s f), \end{aligned}$$

where $H_n(s) = 2^{s-1}\pi^{\frac{n-2}{2}}\Gamma(\frac{s}{2})\Gamma(\frac{s-n+2}{2})$. *Then*

1. I_+^s, $I_-^s f$ *and* $I_0^s f$ *extend to entire functions in* s.
2. $I_+^{s+2}Lf = I_+^s f$ *and* $I_-^{s+2}Lf = -I_-^s f$.
3. $I_+^0 f = 2\sin(\frac{p\pi}{2})f(0)$ *and* $I_-^0 f = 2\sin(\frac{q\pi}{2})f(0)$.
4. *The map* $f \mapsto I_\pm^s f$ *is a tempered distribution for all* s.

Finally, for p *and* q *both even;*

5. $I_0^0 f = \frac{(-1)^{\frac{q}{2}}2\pi}{(\frac{n-2}{2})!}f(0)$.

Limit Formulas in $\mathbf{R}^n$

In this section we are going to find limit formulas for the mean value operators $M_\pm f(r)$ defined in equations 2.38 and 2.39.

Extending Lemma 2.15 we have;

Lemma 2.21: *If* q *is even then*

$$\lim_{r\to 0+}\frac{d^i}{dr^i}(r^{n-2}M_+f(r))$$

exists for all nonnegative integers i *and* $f \in C_c^\infty(\mathbf{R}^n)$. *Likewise if* p *is even*

$$\lim_{r\to 0+}\frac{d^i}{dr^i}(r^{n-2}M_-f(r))$$

exists for all nonnegative integers i *and* $f \in C_c^\infty(\mathbf{R}^n)$.

Proof: Since $\tilde{f}(s_1,s_2)$ is even in both variables separably we write it as $f^*(s_1^2,s_2^2)$. Thus equation 2.38 gives

$$r^{n-2}M_+f(r) = \int_0^{r_0} f^*(r^2+t^2,t^2)(r^2+t^2)^{\frac{p-2}{2}}t^{q-1}dt.$$

Expanding $f^*(s_1^2, s_2^2)$ in a Taylor series with remainder we get

$$f^*(s_1^2, s_2^2) = \sum_{j=0}^{N} a_j(s_2^2) s_1^{2j} + s_1^{2N+2} G(s_1^2, s_2^2),$$

where a_j and G are smooth. Thus,

$$\begin{aligned} r^{n-2} M_+ f(r) &= \\ &\int_0^{r_0} \sum_{j=0}^{N} a_j(t^2)(r^2+t^2)^{\frac{p}{2}-1+j} t^{q-1} dt + \int_0^{r_0} (r^2+t^2)^{\frac{p+2N}{2}} t^{q-1} G(r^2+t^2, t^2) dt. \end{aligned}$$

If N is large enough the second term is i times continuously differentiable in r. For the first term note $\frac{d}{dt}(r^2+t^2)^s = s(r^2+t^2)^{s-1} 2t$. Since $q-1$ is odd, repeated integration by parts shows the first term equals

$$\begin{aligned} &\sum_{\substack{l=1 \\ l \text{ odd}}}^{q-1} \sum_{k=M}^{M+\frac{q-2}{2}} \sum_{j=0}^{N} \int_0^{r_0} t^l a_{j,k,l}(t^2)(r^2+t^2)^{\frac{p}{2}+j+k} dt + \sum_{k=0}^{M+\frac{q-2}{2}} \sum_{j=0}^{N} b_{j,k}(r^2+r_0^2)^{\frac{p}{2}+j+k} \\ &\quad + \sum_{k=\frac{q-2}{2}}^{M+\frac{q-2}{2}} \sum_{j=0}^{N} c_{j,k}(r^2)^{\frac{p}{2}+j+k}, \end{aligned}$$

for some smooth functions $a_{j,k,l}$ and constants $b_{j,k}$ and $c_{j,k}$ and M arbitrarily large. Note that the sum over k in the first term starts at M. All these terms are at least M times differentiable in r independent of t. This proves the lemma. ∎

Finally we can state the limit formulas.

Theorem 2.22: *For* $f \in C_c^\infty(\mathbf{R}^n)$ *we have the following limit formulas;*

1. *If* p *and* q *are both odd and greater than 1 then*

$$\lim_{r \to 0+} r^{n-2} \Delta_+(L)^{\frac{n-2}{2}} M_+ f(r) = cf(0)$$

and

$$\lim_{r \to 0+} r^{n-2} (-\Delta_+(L))^{\frac{n-2}{2}} M_- f(r) = cf(0),$$

where $\Delta_+(L) = \frac{d^2}{dr^2} + \frac{n-1}{r}\frac{d}{dr}$.

2. *If* p *is odd and* q *is even then*

$$\lim_{r \to 0+} \left(\frac{d}{dr}\right)^{n-2} r^{n-2} M_+ f(r) = cf(0).$$

3. *If* p *is even and* q *is odd then*

$$\lim_{r \to 0+} \left(\frac{d}{dr}\right)^{n-2} r^{n-2} M_- f(r) = cf(0).$$

4. *If* p *and* q *are both even then*

$$\begin{aligned} &\lim_{r \to 0+} \left(\frac{d}{dr}\right)^{n-2} r^{n-2} M_+ f(r) + (-1)^{\frac{n}{2}} \lim_{r \to 0+} \left(\frac{d}{dr}\right)^{n-2} r^{n-2} M_- f(r) \\ &= cf(0). \end{aligned}$$

Where in each equation c *is a nonzero constant independent of* f.

Proof: 1. Using Theorem 1.25 from section 1 we have

$$\begin{aligned}
\lim_{r\to 0^+} r^{n-2}\Delta_+(L)^{\frac{n-2}{2}}M_+f(r) &= \lim_{r\to 0^+} r^{n-2}M_+f(r)L^{\frac{n-2}{2}}f \\
\text{(by Lemma 2.16)} &= \frac{1}{\pi^{\frac{n-2}{2}}2^{n-2}\Gamma(n-2)}I_+^{n-2}L^{\frac{n-2}{2}}f \\
\text{(by equation 2.13)} &= \frac{1}{\pi^{\frac{n-2}{2}}2^{n-2}\Gamma(n-2)}I_+^{0}f \\
\text{(by Theorem 2.20)} &= \frac{2(-1)^{\frac{p-1}{2}}}{\pi^{\frac{n-2}{2}}2^{n-2}\Gamma(n-2)}f(0).
\end{aligned}$$

The other equation in part 1 is similar.

2. In the polar coordinates $H_+ \times \mathbf{R}^+$, $\frac{\partial}{\partial r}$ is a vector field on R_+. We have (see the definition of $M_+f(r)$, equation 2.36) for $r > 0$

$$M_+\frac{\partial f}{\partial r} = \frac{d}{dr}M_+f(r).$$

Also,

$$\frac{d}{dr}\circ r^{n-2} = r^{n-2}(\frac{d}{dr} + \frac{n-2}{r}).$$

For $Re(s)$ large enough the following integrals all converge (use the fact that $r\frac{\partial}{\partial r}$ extends to a vector field on $\mathbf{R}^n$). Thus from equation 2.41 and Lemma 2.15 we have

$$\begin{aligned}
\int_{R_+}(\tfrac{\partial}{\partial r} + \tfrac{n-2}{r})(f(z))r^{s-n}dz &= \int_0^\infty M_+(\tfrac{\partial}{\partial r} + \tfrac{n-2}{r})(f)(r)\, r^{s-1}dr \\
&= \int_0^\infty(\tfrac{d}{dr} + \tfrac{n-2}{r})M_+f(r)\, r^{s-1}dr \\
&= \int_0^\infty \tfrac{d}{dr}(r^{n-2}M_+f(r))r^{s-n+1}dr \\
&= -(s-n+1)\int_{R_+} f(z)r^{s-n-1}dz.
\end{aligned}$$

Thus,

$$I_+^s((\frac{\partial}{\partial r} + \frac{n-2}{r})f = \frac{-H_n(s-1)}{H_n(s)}(s-n+1)I_+^{s-1}f.$$

Iterating we get

$$\begin{aligned}
&I_+^s(\tfrac{\partial}{\partial r} + \tfrac{n-2}{r})^{n-2}f = \\
&\quad(-1)^{n-2}\tfrac{H_n(s-n+2)}{H_n(s)}(s-n+1)(s-n)\cdots(s-2n+4)I_+^{s-n+2}f.
\end{aligned} \tag{2.53}$$

Since the coefficient $H_n(s-n+2)$ is meromorphic this equation is valid for all s except possibly some integers (the poles of $H_n(s-n+2)$). In particular at $s = n-2$ the pole of $H_n(s)$ cancels that of $H_n(s-n+2)$ and so this equation is valid for that value of s.

Lemma 2.21 says $\lim(\frac{d}{dr})^{n-2}r^{n-2}M_+f(r)$ exists (we're assuming q is even). Hence the proof of Lemma 2.16 (also see Lemma 2.13 Ch. 3) will give us

$$\begin{aligned}
\lim_{r\to 0^+}(\tfrac{d}{dr})^{n-2}r^{n-2}M_+f(r) &= \lim_{r\to 0^+} r^{n-2}M_+(\tfrac{\partial}{\partial r} + \tfrac{n-2}{r})^{n-2}f(r) \\
&= \frac{1}{\pi^{\frac{n-2}{2}}2^{n-2}\Gamma(n-2)}I_+^{n-2}(\tfrac{\partial}{\partial r} + \tfrac{n-2}{r})^{n-2}f \\
\textit{(by eq. 2.53)} \quad &= \frac{1}{\pi^{\frac{n-2}{2}}2^{n-2}\Gamma(n-2)}\cdot\frac{(n-2)!\Gamma(\frac{2-n}{2})}{2^{n-2}\Gamma(\frac{n-2}{2})}I_+^0f \\
&= \frac{n-2}{\pi^{\frac{n-2}{2}}4^{n-2}}\frac{\Gamma(\frac{2-n}{2})}{\Gamma(\frac{n-2}{2})}\cdot 2(-1)^{\frac{p-1}{2}}f(0) \\
&= \frac{(-1)^{\frac{q-2}{2}}}{\pi^{\frac{n-4}{2}}4^{n-3}\Gamma(\frac{n-2}{2})^2}f(0)
\end{aligned} \tag{2.54}$$

as claimed.

3. This is the same as part 2, just switching p and q.

4. Define the differential operator V on $R_+ \cup R_-$ by

$$V = \begin{cases} (\frac{\partial}{\partial r_+} + \frac{n-2}{r_+})^{n-2} \text{ on } R_+ \\ (-1)^{\frac{n-2}{2}}(\frac{\partial}{\partial r_-} + \frac{n-2}{r_-})^{n-2} \text{ on } R_-. \end{cases}$$

Here we have to distinguish the polar coordinates r_+ and r_-.

We repeat the argument of part 2 using V in place of $\frac{\partial}{\partial r} + \frac{n-2}{r}$ For $Re(s)$ large we have

$$\frac{\Gamma(\frac{s}{2})}{\Gamma(\frac{s-n+2}{2})} I_0^s V f = \frac{1}{2^{n-2}}(s-n+1)(s-n)\cdots(s-2n+4)I_0^{s-n+2}. \tag{2.55}$$

By analytic continuation this is valid for all s. The top 3 equalities in equation 2.54 are valid for q even. The analogous statements about M_- and I_- are true provided p is even. Thus the top two equations in equation 2.54 combined with equation 2.55 give

$$\begin{aligned} &\Gamma(\tfrac{n-2}{2})(\lim_{r\to 0^+}(\tfrac{d}{dr})^{n-2}r^{n-2}M_+f(r) + (-1)^{\frac{n}{2}}\lim_{r\to 0^+}(\tfrac{d}{dr})^{n-2}r^{n-2}M_-f(r)) \\ = \;& \Gamma(\tfrac{n-2}{2})(I_+^{n-2}(\tfrac{\partial}{\partial r_+} + \tfrac{n-2}{r_+})^{n-2}f - (-1)^{\frac{n-2}{2}}I_-^{n-2}(\tfrac{\partial}{\partial r_-} + \tfrac{n-2}{r_-})^{n-2}f) \\ = \;& c_1 I_0^0 f \\ = \;& cf(0) \end{aligned}$$

as claimed. ∎

The cases $p = 1$ (or $q = 1$ have some added possibilities since R_+ (R_-) is not connected. We will not take the time to deal with these cases except to call attention to the degenerate case $p = 1$ and $q = 1$. If $p = 1$ and $q = 1$ then there are 4 mean values to consider. For $r \neq 0$ define

$$\begin{aligned} \tilde{M}_+f(r) &= \textstyle\int_{\xi=-\infty}^{\infty} f(r\cosh\xi, r\sinh\xi)d\xi \\ \tilde{M}_-f(r) &= \textstyle\int_{-\infty}^{\infty} f(r\sinh\xi, r\cosh\xi)d\xi. \end{aligned}$$

Theorem 2.23: *For* $f \in C_c^\infty(\mathbf{R}^2)$ *we have*

$$\lim_{r\to 0^+} r\frac{d}{dr}\tilde{M}_\pm f(\pm r) = -2f(0).$$

Proof: See Helgason [1984] page 220.

Limit formulas for symmetric spaces

We return now to the notation of section 1. We have a semisimple Lie group G equipped with a Cartan involution θ and an involution σ commuting with θ. For the Lie algebra $\mathbf{g}$ we have

$$\mathbf{g} = \mathbf{k} \overset{\theta}{\oplus} \mathbf{p} = \mathbf{h} \overset{\sigma}{\oplus} \mathbf{q}.$$

Finally, H is the identity component of the fixed points of σ.

We make the assumption that $X = G/H$ is of rank one. This means that any Cartan subspace has dimension one. Thus if $\mathbf{a}$ is a θ stable Cartan subspace then either $\mathbf{a} \subset \mathbf{p}$ or $\mathbf{a} \subset \mathbf{k}$.

Define $\mathbf{g}_0 = \mathbf{k} \cap \mathbf{h} \oplus \mathbf{p} \cap \mathbf{q}$.

Lemma 2.24: *Any two Cartan subspaces for* $(\mathbf{g}, \sigma)$ *contained in* $\mathbf{p}$ $(\mathbf{k})$ *are conjugate under and element of* $(\mathrm{K} \cap \mathrm{H})_0$.

Proof: The involution $\sigma|_{\mathbf{g}_0}$ is a Cartan involution. Of course, any Cartan subspace of $(\mathbf{g}, \sigma)$ contained in $\mathbf{p}$ is also a Cartan subspace of $(\mathbf{g}_0, \sigma)$. Thus the lemma follows for $\mathbf{p}$ by the conjugacy of Cartan subspaces in a Riemannian symmetric space. For Cartan subspaces contained in $\mathbf{k}$ use the Riemannian symmetric space $\mathbf{k} = \mathbf{k} \cap \mathbf{h} \overset{\sigma}{\oplus} \mathbf{k} \cap \mathbf{p}$. ∎

This lemma is a special case of a more general lemma concerning "fundamental" and "maximally split" Cartan subspaces. (see Matsuki [1978] or Flensted-Jensen [1980])

Thus there are just two H conjugacy classes of Cartan subspaces. We fix θ stable representatives of these classes, $\mathbf{a}_p \subset \mathbf{p} \cap \mathbf{q}$ and $\mathbf{a}_k \subset \mathbf{k} \cap \mathbf{q}$. In the notation of section 1, $\mathbf{a}_p^d = \mathbf{a}_p$ and $\mathbf{a}_k^d = i\mathbf{a}_k$. For the root systems we write $\Delta(\mathbf{g}_\mathbf{C}, (\mathbf{a}_p)_\mathbf{C}) = \Delta_p$ and $\Delta(\mathbf{g}_\mathbf{C}, (\mathbf{a}_k)_\mathbf{C}) = \Delta_k$.

In general $\alpha \in \Delta(\mathbf{g}_\mathbf{C}, \mathbf{a}_\mathbf{C})$ is real on $\mathbf{a}^d$. Thus $\alpha \in \Delta_p$ is real on $\mathbf{a}_P$ and $\alpha \in \Delta_k$ is pure imaginary on $\mathbf{a}_k$. Also, in general, $\Delta(\mathbf{g}^d, \mathbf{a}^d)$ is the root system of a Riemannian symmetric space. Thus for $\alpha \in \Delta(\mathbf{g}^d, \mathbf{a}^d)$ the only possible multiples of α that are also roots are $\pm\alpha$, $\pm\frac{1}{2}\alpha$ and $\pm 2\alpha$. (See, for example, Helgason [1978] Ch. 7 Cor. 2.17.)

Notational Convention

In general we will call the two cases the $\mathbf{p}$ case and the $\mathbf{k}$ case. The two cases will be distinguished by using the appropriate subscript. When both cases are considered simultaneously we will drop the subscripts.

Choose a root $\alpha_p \in \Delta_p$ such that $\frac{1}{2}\alpha_p \notin \Delta_p$. Likewise choose a root $\alpha_k \in \Delta_k$. Choose $v_p \in \mathbf{a}_p$ and $v_k \in \mathbf{a}_k$ such that $\alpha_p(v_p) = 1$ and $\alpha_k(v_k) = \sqrt{-1}$.

Lemma 2.25: *If* B *is the Killing form on* $\mathbf{g}$ *then* $\mathrm{B}(\mathrm{v_p}, \mathrm{v_p}) = -\mathrm{B}(\mathrm{v_k}, \mathrm{v_k})$.

Proof: Since $\mathbf{a}_p$ and $i\mathbf{a}_k$ are both Cartan subspaces in the Riemannian symmetric space $\mathbf{g}^d = \mathbf{k}^d \oplus \mathbf{p}^d$ they are conjugate by an element in $Int(\mathbf{g}_\mathbf{C})$. By construction (since we're in rank one) this element must take iv_k to $\pm v_p$. Since B is just the restriction of the Killing form on $\mathbf{g}_\mathbf{C}$ we are done. ∎

Thus we can assume that the pseudoriemannian structure, $\langle\ ,\ \rangle$, (which is a multiple of the Killing form) on $\mathbf{q}$ is scaled so that $\langle v_p, v_p\rangle = 1$ and $\langle v_k, v_k\rangle = -1$.

Let

$$R_\mathbf{p} = \{z \in \mathbf{q} \mid \langle z, z\rangle > 0\}$$
$$R_\mathbf{k} = \{z \in \mathbf{q} \mid \langle z, z\rangle < 0\}.$$

In order to avoid unneeded complication arising from the disconnectedness of R (recall our subscript convention) we will assume that $p > 1$ and $q > 1$. In this case R is connected. The remaining cases are not difficult to handle.

Reduction to $\mathbf{R}^n$

Suppose $\dim \mathbf{q} = n$ and the signature of $\langle\ ,\ \rangle$ is (p,q). In order to apply the results in the last section we have to show that the regular semisimple orbits of H acting on $\mathbf{q}$ are just $SO_e(p,q)$ orbits in $\mathbf{R}^n$.

For $r > 0$ define

$$\begin{aligned} S_p^r &= \{z \in \mathbf{q} \mid \langle z, z\rangle = r^2\} \\ S_k^r &= \{z \in \mathbf{q} \mid \langle z, z\rangle = -r^2\}. \end{aligned}$$

Then from the definitions ($p > 1$, $q > 1$) we have $S^r = H \cdot (rv)$.

As always, let $M = Z_H(\mathrm{a})$ (remember the subscript convention), and let $d\dot{h}$ be an H invariant measure on H/M. We will assume that $d\dot{h}$ is scaled so that the constants in the following lemma are 1.

Lemma 2.26: *For* $\mathrm{p} \geq 2$ *and* $\mathrm{q} \geq 2$ *there are constants* c *such that*

$$\begin{aligned} M_{a_p} f(rv_p) &= c_p M_+ f(r) \qquad \textit{and} \\ M_{a_k} f(rv_k) &= c_k M_- f(r). \end{aligned}$$

Here $\mathrm{M}_+\mathrm{f}$ $(\mathrm{M}_-\mathrm{f})$ *is defined in equation 2.36 (2.37 and* $\mathrm{M}_{\mathbf{a}}\mathrm{f}$ *is defined in equation 1.11 sec.1.*

Proof: We do just the $\mathbf{p}$ case, the $\mathbf{k}$ case is identical. The uniqueness of the H invariant measure $d\dot{h}$ implies it is enough to show that the measure

$$cosh^{p-1}(\xi) sinh^{q-1}(\xi) d\xi d\theta_1 d\theta_2$$

on S_p^r is H invariant. This follows because

$$dz = r^{n-1} cosh^{p-1}(\xi) sinh^{q-1}(\xi) d\xi d\theta_1 d\theta_2 dr,$$

and $r^{n-1}dr$ is H invariant and S_p^r is H stable. ∎

Theorem 2.27: *For* $\mathrm{p} \geq 2$, $\mathrm{q} \geq 2$ *and* $\mathrm{d\dot{h}}$ *scaled as above, we have*

1. *If* p *and* q *are both odd then*

$$\lim_{r \to 0+} r^{n-2} \Delta_p(L)^{\frac{n-2}{2}} M_{a_p} f(rv_p) = cf(0)$$

and

$$\lim_{r \to 0+} r^{n-2} (\Delta_k(L))^{\frac{n-2}{2}} M_{a_k} f(rv_k) = cf(0),$$

where $\Delta_p(L) = \frac{d^2}{dr^2} + \frac{n-1}{r}\frac{d}{dr} = -\Delta_k(L)$.

2. If p *is odd and* q *is even then*

$$\lim_{r\to 0+}\left(\frac{d}{dr}\right)^{n-2} r^{n-2}M_{a_p}f(rv_p) = cf(0).$$

3. If p *is even and* q *is odd then*

$$\lim_{r\to 0+}\left(\frac{d}{dr}\right)^{n-2} r^{n-2}M_{a_k}f(rv_k) = cf(0).$$

4. If p *and* q *are both even then*

$$\lim_{r\to 0+}\left(\tfrac{d}{dr}\right)^{n-2} r^{n-2}M_{a_p}f(rv_p) + (-1)^{\frac{n}{2}}\lim_{r\to 0+}\left(\tfrac{d}{dr}\right)^{n-2} r^{n-2}M_{a_k}f(rv_k)$$
$$= \quad cf(0).$$

Where in each equation c *is a nonzero constant independent of* f.

Proof: This is just Theorem 2.22.

Exponentiation to X

Using the exponentiation Theorem (Lemma 1.2 sec.1) we could exponentiate these limit formulas directly to the space X. But, because they are interesting in their own right we will exponentiate the Riesz potentials and then use them to find limit formulas. This has the advantage that the differential operators produced for the limit formulas are more geometric in nature.

We recall Theorem 1.4 from section 1, $d\mathrm{Exp}_Y = d\tau_{\exp Y}\sum(T_Y)^n/(2n+1)!$. From Lemma 1.2 section 1 we choose an H invariant neighborhood W of 0 in $\mathbf{q}$ on which Exp is a diffeomorphism. Fix R_0 small enough that $\{|\langle z,z\rangle| < R_0\} \subset W$. For $r < R_0$ we define $\bar{S}_p^r = \mathrm{Exp}S_p^r$ ($\bar{S}_k^r = \mathrm{Exp}S_k^r$). As a matter of notation, we will use a bar to indicate objects on X and no bar for the corresponding object on $\mathbf{q}$.

Lemma 2.28: (Gauss' lemma) *For* $Y \in D_p$ *and* $r < R_0$ *the geodesic* $\mathrm{Exp}tY$ *intersects* $\bar{S}_p^r$ *in a right angle at their first point of intersection. Likewise for* D_k.

Definition: Let m_1 and m_2 be the multiplicities of the roots α and 2α respectively. (They will be the same for α_p and α_k.)

Lemma 2.29: *We have the 'determinant' of* $d\mathrm{Exp}_Y$ *is*

$$\det\left(\sum\frac{T_Y^n}{(2n+1)!}\right) = \left(\frac{\sinh\langle Y,Y\rangle^{\frac{1}{2}}}{\langle Y,Y\rangle^{\frac{1}{2}}}\right)^{m_1}\left(\frac{\sinh 2\langle Y,Y\rangle^{\frac{1}{2}}}{2\langle Y,Y\rangle^{\frac{1}{2}}}\right)^{m_2},$$

where $Y \in D_p$ *or* $Y \in D_k$. *(Thus, on* D_k *we get*

$$(\sin|\langle Y,Y\rangle|^{\frac{1}{2}}/|\langle Y,Y\rangle|^{\frac{1}{2}})^{m_1}(\sin|2\langle Y,Y\rangle|^{\frac{1}{2}}/2|\langle Y,Y\rangle|^{\frac{1}{2}})^{m_2}).$$

Proof: We do $Y \in D_p$, $Y \in D_k$ is identical. We can assume $Y = cv_p$ for some $c > 0$. Then T_Y has eigenvalues 0, $\alpha^2(Y)$ and $(2\alpha)^2(Y)$ with multiplicities 1, m_1 and m_2 respectively. But $\alpha^2(cv_p) = c^2 = \langle Y,Y\rangle$. ∎

Orbital Integrals

As before we define M_p as the centralizer of $\mathbf{a}_p$ in H. Thus if $0 < r < R_0$ then for $x = \mathrm{Exp}(rv_p)$, $Z_H(x) = Z_H(rv_p) = M_p$. Thus we can define

$$\bar{M}_{\mathbf{a}_p} u(x) = \int_{H/M_p} u(h \cdot x) d\dot{h}, \tag{2.56}$$

where $u \in C_c^\infty(X)$.

As always $d\dot{h}$ is an H invariant measure on H/M. Later we will scale $d\dot{h}$ so that it equals the pseudoriemannian measure on S_p^1.

We make similar definitions for $x = \mathrm{Exp} rv_k$,

$$\bar{M}_{\mathbf{a}_k} u(x) = \int_{H/M_k} u(h \cdot x) d\dot{h}, \tag{2.57}$$

for $u \in C_c^\infty(X)$.

These integrals converge since we know the orbit $H \cdot x$ is closed.

As usual when treating both cases simultaneously we will drop the subscripts $\mathbf{p}$ and $\mathbf{k}$.

Lemma 2.30: *The orbital integral* $\bar{\mathrm{M}}_{\mathbf{a}}\mathrm{u(x)}$ *is smooth on the set* $\{\mathrm{x} = \mathrm{Exp(rv)} \mid 0 < \mathrm{r} < \mathrm{R_0}\}$.

Proof: For this range of x we have $\bar{M}_{\mathbf{a}} u(\mathrm{Exp}(rv)) = M_{\mathbf{a}} u \circ \mathrm{Exp}(rv)$. Thus the lemma follows from Theorem 1.23 sec.1. ∎

Let $d\bar{\omega}^r$ be the pseudoriemannian volume element on $\bar{S}^r$. For $0 < r < R_0$ we have $\varphi_r : H/M \approx \bar{S}^r$ $(\varphi_r(hM) = h \cdot \mathrm{Exp}(rv))$. By the uniqueness of H invariant measure on H/M

$$\varphi_r^* d\bar{\omega}^r = A(r) d\dot{h}$$

for some function $A(r)$. ($A(r)$ was called δ in section 1) Thus,

$$\bar{M} u(\mathrm{Exp}(rv)) = \frac{1}{A(r)} \int_{\bar{S}^r} u(z) d\bar{\omega}^r(z). \tag{2.58}$$

We have to split the two cases.

Lemma 2.31: *We have*

$$\begin{aligned} A_p(r) &= sinh^{m_1}(r) \left(\frac{sinh2r}{2}\right)^{m_2} \\ A_k(r) &= sin^{m_1}(r) \left(\frac{sin2r}{2}\right)^{m_2} \end{aligned}$$

for the $\mathbf{p}$ *and* $\mathbf{k}$ *cases respectively.*

Proof: The map $\mathrm{Exp} : D_p \mapsto \bar{D}_p$ (or $D_k \mapsto \bar{D}_k$) preserves the lengths of geodesics through 0. Thus, for $Y \in S_p^r$ (S_k^r) the ratio of the volume elements of $\bar{S}_p^r$ and S_p^r ($\bar{S}_k^r$ and S_k^r) at z is given by $det(d\mathrm{Exp}_Y)$. By Lemma 2.29 this equals

$$\left(\frac{sinhr}{r}\right)^{m_1} \left(\frac{sinh2r}{2r}\right)^{m_2} \quad \text{for } \mathbf{p},$$
$$\left(\frac{sinr}{r}\right)^{m_1} \left(\frac{sin2r}{2r}\right)^{m_2} \quad \text{for } \mathbf{k}.$$

We will now drop the subscripts and do both cases simultaneously. But the volume element $d\omega^r$ on S^r is $r^{m_1+m_2}d\omega^1$. (Recall $n = dimX = 1 + m_1 + m_2$.) Consider the commutative diagram,

$$\begin{array}{ccc} H/M & \xrightarrow{\varphi} & \bar{S}^r \\ & \psi \searrow & \uparrow \text{Exp} \\ & & S^r \end{array}$$

where $\varphi(\dot{h}) = h \cdot \text{Exp}(rv)$ and $\psi(\dot{h}) = h \cdot rv$. We have related the various measures as follows;

1. $\varphi^* d\bar{\omega}^r = A(r)d\dot{h}$
2. $\text{Exp}^* d\bar{\omega}^r_p = \left(\frac{sinhr}{r}\right)^{m_1} \left(\frac{sinh2r}{2r}\right)^{m_2} d\omega^r_p$
3. $\text{Exp}^* d\bar{\omega}^r_k = \left(\frac{sinr}{r}\right)^{m_1} \left(\frac{sin2r}{2r}\right)^{m_2} d\omega^r_k$.

By the assumption $d\dot{h} = \psi^* d\omega^1$ we have

$$\psi^* d\omega^r = r^{n-1} d\dot{h} = r^{m_1+m_2} d\dot{h}.$$

Thus

$$A(r)d\dot{h} = \varphi^* d\bar{\omega}^r = \psi^* \circ \text{Exp}^* d\bar{\omega}^r = \begin{cases} sinh^{m_1}(r) \left(\frac{sinh2r}{2}\right)^{m_2} d\dot{h} \\ sin^{m_1}(r) \left(\frac{sin2r}{2}\right)^{m_2} d\dot{h} \end{cases}$$

in the two cases respectively. ∎

This also gives a useful formula. Let $B^{R_0}_p = \{0 < \langle Z, Z\rangle < R_0^2\}$ and $B^{R_0}_k = \{0 > \langle Z, Z\rangle > -R_0^2\}$, and (dropping subscripts) $\overline{B}^{R_0} = \text{Exp}(B^{R_0})$. Then,

$$\int_{\overline{B}^{R_0}} f(z)dz = \int_0^{R_0} \int_{H/M} f(h \cdot \text{Exp} rv)\, A(r) d\dot{h} dr. \tag{2.59}$$

Radial Part of the Laplacian

We continue using R_0 (defined just before Lemma 5.1). Define $W_p = \{\text{Exp}(rv_p) \mid 0 < r < R_0\}$ and likewise W_k. We want to use Theorem 1.26 sec.1 to compute the radial part of the Laplace-Beltrami operator, $L_X = L$, of X on W

Lemma 2.32: *The radial parts are*

$$\begin{aligned} \Delta_p(L) &= \tfrac{d^2}{dr^2} + \tfrac{A_p'(r)}{A_p(r)} \tfrac{d}{dr}, \\ \Delta_k(L) &= -\tfrac{d^2}{dr^2} - \tfrac{A_k'(r)}{A_k(r)} \tfrac{d}{dr}. \end{aligned}$$

Proof: From Theorem 1.26 sec.1 we have

$$\Delta(L) = A^{-\frac{1}{2}} L_W \circ A^{\frac{1}{2}} - A^{-\frac{1}{2}} L_W(A^{\frac{1}{2}}).$$

But $L_{W_p} = \frac{d^2}{dr^2}$ and $L_{W_k} = -\frac{d^2}{dr^2}$. A simple calculation gives the lemma. ∎

Theorem 1.26 (2) sec.1 implies the next lemma.

Lemma 2.33: *The orbital integral and radial part of the Laplacian are intertwined by*

$$\Delta(\text{L})\bar{\text{M}}_{\mathbf{a}}\text{u} = \bar{\text{M}}_{\mathbf{a}}\text{Lu}.$$

Riesz Potentials on X

Fix R_0 as defined just before Lemma 2.28. Let

$$W' = \{|\langle Z, Z\rangle| < R_0\}$$

and

$$\bar{W}' = \mathrm{Exp} W'.$$

For u supported in $\overline{W}'$ we define

$$\begin{aligned} \bar{I}_p^s u &= \tfrac{1}{H_n(s)} \int_{\bar{D}_p} u(z) sinh^{s-n} r dz \\ \bar{I}_k^s u &= \tfrac{1}{H_n(s)} \int_{\bar{D}_k} u(z) sin^{s-n} r dz. \end{aligned}$$

Here $H_n(s)$ is from equation 2.8 and dz is a G invariant measure on X. Define

$$\bar{I}_0^s u = \Gamma(\frac{s-n+2}{2})(\bar{I}_p^s - cos(\frac{s-n+2}{2}\pi)\bar{I}_k^s u).$$

These are only well defined for $Re(s)$ large enough. We will analytically continue them momentarily.

For simplicity we will assume that $u \in C_c^\infty(X)$ and $\bar{D} \cap supp(u) \subset \overline{B}^{R_0}$. As usual we have dropped the subscripts when both cases are treated at once.

Analytic Continuation

Theorem 2.34: *The two potentials* $\bar{\mathrm{I}}_{\mathrm{p}}^{\mathrm{s}}$ *and* $\bar{\mathrm{I}}_{\mathrm{k}}^{\mathrm{s}}$ *defined above can be analytically continued to the entire complex plane. At* 0 *we have*

$$\begin{aligned} \bar{I}_p^0 u &= 2sin(\tfrac{p}{2}\pi)u(0), \\ \bar{I}_k^0 u &= 2sin(\tfrac{q}{2}\pi)u(0). \end{aligned}$$

Proof: We will just do $\bar{I}_p^s$, $\bar{I}_k^s$ is identical. From Lemma 2.29 we have

$$\bar{I}_p^s u = \frac{1}{H_n(s)} \int_{B_p^{R_0}} h(Z,s) r^{s-n} dZ,$$

where

$$h(Z,s) = u \circ \mathrm{Exp}(Z) \left(\frac{sinhr}{r}\right)^{s-1} cosh^{m_2} r.$$

Recalling the correspondence of $\mathbf{q}$ with $\mathbf{R}^{p+q}$ we get

$$\bar{I}_p^s u = I_+^s h. \tag{2.60}$$

Note that $(\frac{sinhr}{r})^{s-1}$ and all its derivatives are holomorphic in s. Thus the same is true for $h(Z,s)$. Thus Theorem 2.20 finishes the proof. (We haven't proved Theorem 2.20 with parameters but the extension is as straightforward as the extension in Theorem 2.6 ∎

Lemma 2.35: *We have the following functional equations,*

$$\begin{aligned}\bar{I}_p^s L_X u &= \bar{I}_p^{s-2} u + (s-n)(s-1+m_2)\bar{I}_p^s u,\\ \bar{I}_k^s L_X u &= -\bar{I}_k^{s-2} u + (s-n)(s-1+m_2)\bar{I}_k^s u.\end{aligned}$$

Proof: We prove this for the **p** case, the **k** case is identical. For $Re(s)$ large enough Green's Theorem says that

$$\int_{\bar{B}_p^{R_0}} u(z) L_X(sinh^{s-n} r) - L_X u(z)\, sinh^{s-n} r dz \tag{2.61}$$

is a surface integral over a part of the cone $\bar{C}$ ($\bar{C} = \mathrm{Exp} C$, $C = \{\langle Z, Z\rangle = 0\}$), where $sinh^{s-n} r$ and its first few derivatives vanish, plus an integral over the surface $\bar{S}_p^{R_0}$, where u and all its derivatives vanish. Thus 2.61 vanishes. Lemma 2.32, Lemma 2.31, and the formula $sinh 2r = 2 sinhr\, coshr$ imply

$$L_X sinh^{s-n} r = (s-n)(s-2) sinh^{s-n-2} r cosh^2 r + (s-n)(1+m_2) sinh^{s-n} r. \tag{2.62}$$

Using $cosh^2 r - sinh^2 r = 1$ and $H_n(s) = (s-2)(s-n)H_n(s-2)$ we see the lemma is valid for $Re(s)$ large. Thus the lemma is true for all s by analytic continuation. ∎

We need a generalization of Lemma 2.16. Suppose D_r is a differential operator on R^+. We have the polar coordinates $H/M \times \{rv_p \mid 0 < r < R_0\}$ on X. Thus we can use D_r to define a differential operator $\hat{D}_r$ on the open set $H \cdot \{rv_p\}$. (Have D_r act on just the r variable.)

Lemma 2.36: *With this notation suppose*

$$\lim_{r \to 0+} (\mathrm{sinhr})^{n-2} D_r \bar{M}_{a_p} u(\mathrm{Exp} r v_p)$$

exists. ($\bar{M}_{a_p} u(\mathrm{Exp} r v_p)$ is a function of r) Then the limit equals $(\pi^{\frac{n-2}{2}} 2^{n-2}\Gamma(n-2))^{-1} \bar{I}_p^{n-2} \hat{D}_r u$. *An identical statement holds for the* **k** *case.*

Proof: The proof of Lemma 2.16 works in this case:

$$\begin{aligned} c_1 \bar{I}_p^{n-2}\hat{D}_r u &= (eq.2.59)\ \lim_{s\to n-2}(s-n+2)\int_0^{R_0}\int_{H/M}(\hat{D}_r u)(h\cdot \mathrm{Exp}(rv_p)) \\ &\qquad \times (sinhr)^{m_1}(\tfrac{sinh2r}{2})^{m_2}(sinhr)^{s-n} d\dot{h}dr \\ &= c_2 \int_0^{R_0} \overline{M}_{a_p}(\hat{D}_r u)(h\cdot \mathrm{Exp}(rv_p)(sinhr)^{m_1}(\tfrac{sinh2r}{2})^{m_2}(sinhr)^{s-n} d\dot{h}dr \\ &= c_3 \int_0^{R_0} D_r(\overline{M}_{a_p} u)(\mathrm{Exp}(rv_p)(coshr)^{m_2}(sinhr)^{s-1} d\dot{h}dr \\ &= c_3 \int_0^{R_0} (sinhr)^{n-2} D_r(\overline{M}_{a_p} u)(\mathrm{Exp}(rv_p)(coshr)^{m_2}(\tfrac{sinhr}{r})^{s-n+1} r^{s-n+1}, \end{aligned}$$

where c_1, c_2 and c_3 are constants independent of u. Since

$$(sinhr)^{n-2} D_r \bar{M}_{a_p} u(\mathrm{Exp} r v_p)(coshr)^{m_2}(\frac{sinhr}{r})^{s-n+1}$$

is continuous the proof of Lemma 2.16 applies. (Or see Lemma 2.13 Ch. 3.) Keeping track of the constants gives the result. ∎

Simple calculations show,

$$\begin{aligned} sinh^{n-2}r\,cosh^{m_2-1}r\Delta_p(L_X) &= \\ &[\tfrac{d^2}{dr^2}-(n-3)\tfrac{coshr}{sinhr}\tfrac{d}{dr}-(m_2-2)\tfrac{sinhr}{coshr}\tfrac{d}{dr}-2(n+m_2-1)]sinh^{n-2}r\,cosh^{m_2-1}r, \\ sin^{n-2}r\,cos^{m_2-1}r\Delta_k(L_X) &= \\ &[-\tfrac{d^2}{dr^2}+(n-3)\tfrac{cosr}{sinr}\tfrac{d}{dr}-(m_2-2)\tfrac{sinr}{cosr}\tfrac{d}{dr}-2(n+m_2-1)]sin^{n-2}r\,cos^{m_2-1}r. \end{aligned} \tag{2.63}$$

We can now state the limit formula theorem (for $p > 1$ and $q > 1$).

Theorem 2.37: *1. For* p *and* q *both odd,*

$$\lim_{r\to 0+} \sinh^{n-2}(r)Q_p\left(\Delta_p(L)\right)\bar{M}_{a_p}u(\mathrm{Exp} rv_p) = cu(o),$$

and

$$\lim_{r\to 0+} \sin^{n-2}(r)Q_k(\Delta_k(L))\bar{M}_{a_k}u(\mathrm{Exp} rv_k) = cu(o),$$

where

$$\begin{aligned} Q_p(t) &= (t+2(n-3+m_2))(t+4(n-5+m_2))\cdots(t+(n-2)(1+m_2)) \\ &= (-1)^{\frac{n-2}{2}}Q_k(-t) \end{aligned}$$

and $c = \frac{(-1)^{\frac{p-1}{2}}2}{\pi^{\frac{n-2}{2}}2^{n-2}\Gamma(n-2)}$.

2. For p odd and q even,

$$\lim_{r\to 0+}\left(\frac{d}{dr}\right)^{n-2} r^{n-2}\bar{M}_{a_p}u(\mathrm{Exp} rv_p) = cu(o),$$

where $c = (-1)^{\frac{q-2}{2}}(\pi^{\frac{n-2}{2}}4^{n-3}\Gamma(\frac{n-2}{2})^2)^{-1}$.

3. For p even and q odd,

$$\lim_{r\to 0+}\left(\frac{d}{dr}\right)^{n-2} r^{n-2}\bar{M}_{a_k}u(\mathrm{Exp} rv_k) = cu(o),$$

where $c = (-1)^{\frac{p-2}{2}}(\pi^{\frac{n-2}{2}}4^{n-3}\Gamma(\frac{n-2}{2})^2)^{-1}$.

4. For p and q both even,

$$\begin{aligned} &\lim_{r\to 0+}\left(\tfrac{d}{dr}\right)^{n-2} r^{n-2}\bar{M}_{a_p}u(\mathrm{Exp} rv_p) \\ &\quad + (-1)^{\frac{n}{2}}\lim_{r\to 0+}\left(\tfrac{d}{dr}\right)^{n-2} r^{n-2}\bar{M}_{a_k}u(\mathrm{Exp} rv_k) = cu(o), \end{aligned}$$

where

$$c = \frac{(n-4)!\pi(-1)^{\frac{q}{2}}}{2^{n-4}((\frac{n-4}{2})!)^2}.$$

Proof: For case 1, let $D_r = 1$ in Lemma 2.36. Then we have

$$\begin{aligned} \lim_{r\to 0+} sinh^{n-2}r\bar{M}_{a_p}Q_p(L_X)u(\mathrm{Exp} rv_p) &= \lim_{r\to 0+} r^{n-2}\bar{M}_{a_p}Q_p(L_X)u(\mathrm{Exp} rv_p) \\ &= \frac{1}{\pi^{\frac{n-2}{2}}2^{n-2}\Gamma(n-2)}\cdot \bar{I}_P^{n-2}Q_p(L)u \\ &= \frac{1}{\pi^{\frac{n-2}{2}}2^{n-2}\Gamma(n-2)}\cdot \bar{I}_P^{0}u \\ &= \frac{(-1)^{\frac{p-1}{2}}2}{\pi^{\frac{n-2}{2}}2^{n-2}\Gamma(n-2)}\cdot u(0). \end{aligned}$$

The second to last equality follows from Lemma 2.35. But, Lemma 2.33 gives

$$\bar{M}_{\mathbf{a}_p} Q_p(L_X) u(\mathrm{Exp} r v_p) = Q_p(\Delta_p(L)) \bar{M}_{\mathbf{a}_p} u(\mathrm{Exp} r v_p).$$

This proves the formula for this case. The **k** case is, of course, similar.

Cases 2, 3 and 4 follow directly from the equation

$$(\overline{M}_{\mathbf{a}} u)(\mathrm{Exp} r v) = M_{\mathbf{a}}(u \circ \mathrm{Exp})(r v)$$

and the corresponding case in Theorem 2.22. ∎

Remark: When $m_2 = 0$, Helgason[1959] gave the formula in case 1 in the following two situations; if $p = 1$ he solved the **p** case, if $q = 1$ he solved the **k** case.

As a final remark we note that for the symmetric spaces of complex rank one we can make a reduction to real rank one and deduce limit formulas.

Bibliography

[1970] Atiyah, M. F. Resolution of Singularities and Division of Distributions. *Communications on Pure and Applied Math.* **23** (1970), 145-150.

[1957] Berezin, F. A. Laplace Operators on Semisimple Lie Groups. *Trudy Moskov Mat. Obsc.* **6** (1957), 371-463, English transl., *Amer. Math. Soc. Trans. (2)* **21** (1962) 239-339.

[1957] Berger, M. Les Espaces Symétriques non Compacts. *Ann. Sci. École Norm. Sup.* **74** (1957), 85-177.

[1971] Bernshtein, I. N. Modules Over a Ring of Differential Operators. Study of the Fundamental Solutions of Equations With Constant Coefficients. *Func. Anal. Akad. Nauk. CCCP* **5** (1971)(1-16)

[1972] Bernshtein, I. N. The Analytic Continuation of Generalized Functions With Respect to a Parameter. *Func. Anal. Akad. Nauk. CCCP* **6** (1972)(26-40)

[1962] Borel, A. and Harish-Chandra Arithmetic Subgroups of Algebraic Groups. *Ann. of Math.* **75** (1962), 485-535.

[1979] Bjork, J. "Rings of Differential Operators." North Holland, Amsterdam, 1979.

[1955] Chevalley, C. Invariants of Finite Groups Generated by Reflections. *Amer. J. Math.* **79** (1955), 778-782.

[1980] Flensted-Jensen, M. Discrete Series for Semisimple Symmetric Spaces. *Annals of Math.* **111** (1980), 253-311.

[1955] Gelfand, I. M. and Graev, M. I. Analogue of the Plancherel Formula for the Classical Groups. *Trudy Moskov. Mat. Obsc.*(1955), 375-404. (Also *Trans. AMS* **9**).

[1964] Gelfand, I. M. and Shilov, G. E. "Generalized Functions Vol. I." (English translation) Academic Press, New York, 1964.

[1968] Gelfand, I. M. and Shilov, G. E. "Generalized Functions Vol. II." (English translation) Academic Press, New York, 1968.

[1956] Harish-Chandra The Characters of Semisimple Lie Groups. *Trans. Amer. Math So.* **83** (1956), 98-163.

[1957] Harish-Chandra Differential Operators on a Semisimple Lie Algebra. *Amer. J. Math.* **79** (1957), 87-120.

[1964] Harish-Chandra Invariant Distributions on Lie Algebras. *Amer. J. Math.* **86** (1964), 271-309.

[1957a] Harish-Chandra A Formula for Semisimple Lie Groups. *Amer. J. Math.* **79** 91957), 733-760.

[1959] Helgason, S. Differential Operators on Homogeneous Spaces. *Acta Math.* **102** (1959), 239-299.

[1965] Helgason, S. Radon-Fourier Transforms on Symmetric Spaces and Related Group Representations. *Bulletin of the American Mathematical Society* **71** (1965), 757-763.

[1978] Helgason, S. "Differential Geometry, Lie Groups and Symmetric Spaces." Academic Press, New York, 1978.

[1984] Helgason, S. "Groups and Geometric Analysis." Academic Press, Orlando, 1984.

[1972] Helgason, S. "Analysis on Lie Groups and Homogeneous Spaces." Conf. Board Math. Sci. Series, No. 14, American Mathematical Society, Providence, Rhode Island, 1972.

[1982] Hoogenboom, B. Spherical Functions and Invariant Differential Operators on Complex Grassman Manifolds. *Ark. För Mat.* **20** (1982), 69-85.

[1966] Hörmander, L. "An Introduction to Complex Analysis in Several Variables." D. Van Nostrund Co., Inc., Princeton, 1966.

[1976] Hörmander, L. "Linear Partial Differential Operators." Springer- Verlag, Berlin, 1976.

[1971] Kostant, B. and Rallis, B. Orbits and Representations Associated With Symmetric Spaces. *Amer. J. Math.* **93** (1971) 753-809.

[1983] Kosters, M. T. "Spherical Distributions on Rank One Symmetric Spaces." Thesis, Leiden, 1983.

[1976] Lepowsky, J. and McCollum, G. W. Cartan Subspaces of Symmetric Lie Algebras. *Trans. Amer. Math. Soc.* **216** (1976), 217-228.

[1978] Matsuki, T. The Orbits of Affine Symmetric Spaces Under the Action of Minimal Parabolic Subgroups. *J. Math. Soc. Japan.* **31** (1979), 331-357.

[1954] Méthée, R. D. Sur les distributions invariants dans le groupe des rotations de Lorentz. *Comment. Math. Helv.* **28** (1954), 225-269.

[1980] Oshima, T. and Matsuki, T. Orbits of Affine Symmetric Spaces Under the Action of Isotropy Subgroups. *J. Math. Soc. Japan* **32** (1980), 399-414.

[1949] Riesz, M. L'intégral de Riemann Liouville et le Problème de Cauchy. *Acta. Math.* **81** (1949), 1-223.

[1984] Sano, S. Invariant Spherical Distributions and the Fourier Inversion Formula on GL(n,C)/GL(n,R). *J. Math. Soc. Japan* **36** (1984), 191-219

[1967] Stein, E. Analysis on Matrix Spaces and Some New Representations of SL(n,C). *Ann. of Math.* **86** (1967), 461-490.

[1966] Schwartz, L. "Théorie des distributions." Hermann, Paris, 1966.

[1977] Takahashi, R. Fonctions Sphériques Zonales Sur U(n, n+k; F). In Seminaire d'Analyse Harmonique (1976-77), Fac. des Sciences, Tunis, 1977.

[1960] Tengstrand, A. Distributions Invariant Under an Orthogonal Group of Arbitrary Signature, *Math. Scand.* bf 8 (1960), 201-218.

[1972] Warner, G. "Harmonic Analysis on Semi-Simple Lie Groups I, II." Springer-Verlag, Berlin, l972.

[1957] Whitney, H. Real Algebraic Varieties. *Ann. of Math.* **66** (1957), 545-556.

RECURRENCE RELATIONS FOR PLANCHEREL FUNCTIONS

Dale PETERSON
Department of Mathematics
M I T, Cambridge MA 02139
U S A

Michèle VERGNE
C N R S , Department of Mathematics
Paris M.I.T., Cambridge, Ma 02139
France U.S.A.

Introduction

This article originated in our desire to understand the work of Rebecca Herb on discrete series constants [He]. We will prove here a recurrence relation for the coefficients of the Plancherel function, recurrence relation formally similar to Harish-Chandra matching conditions for discrete series constants ([Ha-1]). The recurrence relation for the Plancherel function of a semi-simple linear group G (Section 1) could be used to simplify the proof of the Poisson-Plancherel formula for G [Ve]. The generalized recurrence relation for the Plancherel function of a simply-connected semi-simple Lie group G (Section 4) can be used to prove the Poisson-Plancherel formula for G ([Do] for groups of type Bn). Together with results of A. Bouaziz ([Bo]) on characters of tempered representations, these will also be used in [Du-Ve] to prove the Plancherel formula for G.

The proof of the recurrence relation is based on the following simple combinatorial idea used by the first author in his unpublished 1979 study of Kostant's partition function (see [Pe-Ka]). Let Φ be a root system and let Q be the lattice spanned by Φ. Given a subset A of Φ contained in an open half space, we define a "generating function" K_A on Q by

$$\sum_{\xi\in Q} K_A(\xi)e^{-\xi} = \prod_{\beta\in A} (1-e^{-\beta})^{-1} = \prod_{\beta\in A} (1+e^{-\beta}+e^{-2\beta}+\cdots)$$

$K_A(\gamma)$ may be interpreted as counting the ways of getting from ξ to 0 using the $\beta\in A$. K_{Φ^+} is Kostant's partition function. Clearly

$$K_{A\cup B}(\gamma) = \sum_{\beta\in Q} K_A(\beta)\, K_B(\gamma-\beta) \quad \text{if} \quad A\cap B=\emptyset .$$

For a simple root α $A=\{\alpha\}$ and $B=\Phi^+-\{\alpha\}$ this gives :

$$K_{\Phi^+}(\gamma) = \sum_{n=0}^{\infty} K_{\Phi^+-\{\alpha\}}(\gamma-n\alpha)$$

Let s_α be the reflection corresponding to α, so that $s_\alpha(\Phi^+) = (\Phi^+-\{\alpha\})\cup\{-\alpha\}$. The same procedure gives

$$K_{s_\alpha(\Phi^+)}(\gamma) = \sum_{n=0}^{\infty} K_{\Phi^+-\{\alpha\}}(\gamma+n\alpha)$$

Let $\rho = \frac{1}{2} \sum_{\beta \in \Phi^+} \beta$. Combining our equations, we get

$$K_{\Phi^+}(\xi) + K_{\Phi^+}(s_\alpha(\xi+\rho) - \rho) = K_{\Phi^+}(\xi) + K_{\Phi^+}(s_\alpha(\xi+\alpha)) = K_{\Phi^+}(\xi) + K_{s_\alpha(\Phi^+)}(\xi+\alpha))$$

$$= \sum_{n \in \mathbb{Z}} K_{\Phi^+ - \{\alpha\}}(\gamma - n\alpha) \quad .$$

This final expression may be interpreted as counting the number of ways of getting from ξ to the "line" $\mathbb{Z}\alpha$ using the $\beta \in \phi^+ - \{\alpha\}$ and hence is a partition function on the lower dimensional lattice Q mod $\mathbb{Z}\alpha$ Thus $K_{\Phi^+}(\alpha) + K_{\Phi^+}(s_\alpha(\gamma+\rho) - \rho)$ is "simpler" than $K_{\Phi^+}(\gamma)$.

In this paper, K_{Φ^+} is replaced by generating functions such that

$$\prod_{\beta \in \Phi^+} \frac{1-e^{-\beta}}{1+e^{-\beta}} = \sum_{\gamma \in Q} d(\gamma) e^{-\gamma}$$

This creates two simplifications. First $f(x) = \frac{1-x}{1+x}$ satisfies $f(x^{-1}) = - f(x)$ whereas $g(x) = \frac{1}{1-x}$ only satisfies $g(x^{-1}) = - xg(x)$ so that no shift by ρ is necessary here. Second, remarkable cancellations occur when we pass, by following the idea outlined above, to the lattice Q mod $\mathbb{Z}\alpha$, so that the "simpler" function $d(\gamma) + d(s_\alpha \gamma)$ obtained is essentially of the same type as $d(\gamma)$ for the root system Φ_α of roots stable by s_α.

We would like to thank Michel Duflo and David Vogan for helpful conversations.

1. Coefficients of the Plancherel function

1.1. Let V be a real vector space, V^* the dual vector space. Let $\Phi \subset V^*$ be a root system on V. Let Φ^+ be an ordering of Φ. We define the Plancherel function P_{Φ^+} on V by :

$$P_{\Phi^+} = \prod_{\alpha\in\Phi^+} \frac{1-e^{-\alpha}}{1+e^{-\alpha}} \quad , \text{ i.e. } \quad P_{\Phi^+}(h) = \prod_{\alpha\in\Phi^+} \frac{1-e^{-(\alpha,h)}}{1+e^{-(\alpha,h)}} \quad , \text{ for } h \in V$$

$$= \prod_{\alpha\in\Phi^+} \operatorname{th}\left(\frac{(\alpha,h)}{2}\right) .$$

Consider the factor f_α corresponding to α, $f_\alpha = \dfrac{1-e^{-\alpha}}{1+e^{-\alpha}}$. If h is such that $(\alpha,h) > 0$, we have the equality :

$$f_\alpha(h) = 1 + 2 \sum_{k>0} (-1)^k e^{-k(\alpha,h)} = \sum_{k\in\mathbb{Z}} (-1)^k e^{-|k|(\alpha,h)} .$$

Let A be a subset of Φ, say $A = \{\alpha_1,\alpha_2,\ldots,\alpha_N\}$.
We define $\mathbb{Z}A = \{\Sigma k_i\alpha_i , k_i \in \mathbb{Z}\}$, $\mathbb{Z}^+A = \{\Sigma k_i\alpha_i , k_i \in \mathbb{Z} , k_i \geq 0\}$.
Write $\vec{k} = (k_1,k_2,\ldots,k_N)$ for an element of $\mathbb{Z}^N$.
Define, for $\xi \in \mathbb{Z}^+A$

$$d(\xi,A) = \sum_{\substack{\vec{k}\in\mathbb{Z}^N \\ \Sigma|k_i|\alpha_i=\xi}} (-1)^{k_1}(-1)^{k_2} \ldots (-1)^{k_N} .$$

Define $P_A = \prod_{\alpha\in A} f_\alpha$. Let $C^+ = \{h \in V , (\alpha,h) > 0 , \forall \alpha \in \Phi^+\}$. Thus, on C^+, the function P_A is given by the convergent series

$$P_A = \sum_{\xi\in\mathbb{Z}^+A} d(\xi,A)\, e^{-\xi} .$$

In particular

$$P_{\Phi^+} = \sum_{\xi\in\mathbb{Z}^+\Phi^+} d(\xi,\Phi^+)\, e^{-\xi} .$$

We extend $\xi \to d(\xi,\Phi^+)$ to a function on V^* by defining it to be zero outside $\mathbb{Z}^+\Phi^+$.

1.2. Let ψ be a function on a root system Φ, then ψ extends to a character of the lattice $\mathbb{Z}\Phi$ if and only if

1) $\psi(\alpha+\beta) = \psi(\alpha)\psi(\beta)$, whenever $\alpha, \beta, \alpha+\beta \in \Phi$
2) $\psi(\alpha)\psi(-\alpha) = 1$.

Let $\alpha \in \Phi$, $h_\alpha \in \check{\Phi}$ the co-root. Write :

$$\Phi_\alpha = \{\beta , (\beta,h_\alpha) = 0\} \quad \text{and} \quad V_\alpha = \{v \in V , (\alpha,v) = 0\} .$$

By restriction Φ_α is a root system on V_α. If $\beta \in \Phi_\alpha$, the α-string passing through β is either $\{\beta\}$ or $\{\beta-\alpha,\beta,\beta+\alpha\}$.

1.3. We define $\psi_\alpha(\beta) = 1$ if the α-string through β is $\{\beta\}$

$\psi_\alpha(\beta) = -1$ if the α-string through β is $\{\beta-\alpha,\beta,\beta+\alpha\}$.

1.4. Lemma The function ψ_α on Φ_α extends to a character on $\mathbb{Z}\Phi_\alpha$.

Proof : Let $g = g(\Phi)$ be the split Lie algebra associated to the root system Φ with Split-Cartan $a = V$. For $\alpha \in \Phi$, choose $X_\alpha \in g(\alpha)$, $X_{-\alpha} \in g(-\alpha)$ such that $[X_\alpha, X_{-\alpha}] = h_\alpha$. Consider G_a the adjoint group with Lie algebra g and φ_α: $SL(2,\mathbb{R}) \to G_a$ the homomorphism associated to this choice of X_α, $X_{-\alpha}$. Define $\sigma_\alpha = \varphi_\alpha\begin{pmatrix} 0 & 1 \\ -1 & 0\end{pmatrix}$, then σ_α leaves V stable and $\sigma_\alpha|V = s_\alpha$. If $\beta \in \Phi_\alpha$, $s_\alpha(\beta) = \beta$, then σ_α leaves stable the root space $g(\beta)$. The subspace $\sum_{k\in\mathbb{Z}} g(\beta+k\alpha)$ is an irreducible representation space of $SL(2,\mathbb{R})$, of dimension 1 or 3 according to the length of the α-string. Furthermore g_β is the 0-weight space for h_α . Consequently, by finite dimensional representation theory of $SL(2,\mathbb{R})$ the action of σ_α on $g(\beta)$ is given by the scalar $\psi_\alpha(\beta)$. Thus, if $\beta, \gamma, \beta+\gamma \in \Phi_\alpha$, we have $g(\beta+\gamma) = [g(\beta),g(\gamma)]$ and $\psi_\alpha(\beta+\gamma) = \psi_\alpha(\beta)\,\psi_\alpha(\gamma)$. Also $\psi_\alpha(\beta)\,\psi_\alpha(-\beta) = 1$.

Let Φ^+ be an order on Φ ; let $\check{\rho} = \frac{1}{2}\sum_{\beta\in\Phi^+} h_\beta \in V$. Define the character η on V^* by

$\eta(\lambda) = e^{i\pi(\check{\rho},\lambda)}$ (η depends on the choice of Φ^+) .

We have $\eta(\alpha_i) = -1$ for all simple roots α_i of Φ^+ . Let α be a simple root of Φ^+ , let $\Phi_\alpha^+ = \Phi^+ \cap \Phi_\alpha$ be the induced order, η_α the corresponding character on $(V_\alpha)^*$.

1.5. Lemma Assume α simple. On $\mathbb{Z}\Phi_\alpha$, we have

$\eta_\alpha = \eta\psi_\alpha$.

Proof : We need to prove that for $\beta \in \Phi_\alpha$,

$\eta_\alpha(\beta) = \eta(\beta)\,\psi_\alpha(\beta)$.

Fix $\beta \in \Phi_\alpha$ and let us partition the set $\Phi - \Phi_\alpha$ in $U_1 \cup U_2$, where

$U_1 = \{\gamma \in \Phi - \Phi_\alpha , \gamma \notin (\mathbb{R}\alpha + \mathbb{R}\beta)$

$U_2 = \{\gamma \in \Phi - \Phi_\alpha , \gamma \in (\mathbb{R}\alpha + \mathbb{R}\beta)\}$.

Then U_1 and U_2 are stable under the action of s_α , s_β . Remark that s_α and s_β commute. Let $\gamma \in U_1 \cap \Phi^+$, consider the set

$S^+(\gamma) = \{\pm\gamma, \pm s_\alpha\gamma , \pm s_\beta\gamma \;\pm s_\alpha s_\beta\gamma\} \cap \Phi^+$

Let $x = \frac{1}{2}(\sum_{\delta\in S^+(\gamma)} h_\delta ,\beta)$. Let us show that $x \in 2\mathbb{Z}$. There are several cases. If $(\beta,h_\gamma) = 0$ then $x = 0$. If $(\beta,h_\gamma) \neq 0$, it is easy to see that $S^+(\gamma)$ has 4 elements. As α is simple, we have either

1) $S^+(\gamma) = \{\gamma\ ,\ s_\alpha\gamma\ ,\ s_\beta\gamma\ ,\ s_\alpha s_\beta\gamma\}$

or

2) $S^+(\gamma) = \{\gamma\ ,\ s_\alpha\gamma\ ,\ -s_\beta\gamma\ ,\ -s_\alpha s_\beta\gamma\}$.

In case 1, $x = 0$, while in case 2, $x = 2(\beta,h_\gamma) \in 2\mathbb{Z}$. Let $\Phi_1 = \Phi \cap (\mathbb{R}\alpha + \mathbb{R}\beta)$ and $\Phi_1^+ = \Phi^+ \cap \Phi_1$, then α is simple for Φ_1^+ and $\{\beta,\alpha\} \subset \Phi_1$. Let η_1 , $\eta_{1,\alpha}$ be the corresponding characters on $\mathbb{Z}\Phi_1$, $\mathbb{Z}(\Phi_1)_\alpha$. The preceding argument shows that

$$(\eta\ \eta_\alpha^{-1})\ (\beta) = (\eta_1\ (\eta_{1,\alpha})^{-1})\ (\beta)\ .$$

It is easy to verify case by case, on systems of rank two, that

$$\eta_1\ (\eta_{1,\alpha})^{-1}\ (\beta) = \psi_\alpha(\beta)\ .$$

1.6. Let α be a simple root of Φ^+ . If $\xi \in V^*$, we denote by $p_\alpha(\xi)$ the restriction of ξ to V_α . Recall that we have defined the function $d(\xi;\Phi^+)$ on all of V^* by setting $d(\xi;\Phi^+) = 0$ if $\xi \notin \mathbb{Z}^+\Phi^+$.

<u>1.7. Theorem</u>. Let Φ be a root system, Φ^+ an order on Φ , $\alpha \in \Phi^+$ a simple root, then for $\xi \in \mathbb{Z}\Phi$, $\eta(\xi)(d(\xi,\Phi^+) + d(s_\alpha\xi,\Phi^+)) = 2\eta_\alpha(p_\alpha\xi)\ d(p_\alpha\xi;\Phi_\alpha^+)$.

Remark : If $p_\alpha\xi \notin \mathbb{Z}\ \Phi_\alpha$, then the right hand side is zero.

<u>Proof</u> : Consider a subset A of Φ^+ and let

$$\prod_{\beta\in A} \frac{1-e^{-\beta}}{1+e^{-\beta}} = \Sigma\ d(\xi,A)e^{-\xi}$$

Remark that if χ is a character of $\mathbb{Z}\Phi$, we have :

$$\Sigma\ \chi(\xi)d(\xi,A)e^{-\xi} = \prod_{\beta\in A} \frac{1-\chi(\beta)e^{-\beta}}{1+\chi(\beta)e^{-\beta}}$$

as an equality of functions on C^+ . Consider $A = \Phi^+ - \{\alpha\}$. As A is stable under s_α , we have $d(s_\alpha\xi,A) = d(\xi,A)$. We write :

$$\prod_{\beta\in\Phi^+} \frac{1-e^{-\beta}}{1+e^{-\beta}} = \frac{1-e^{-\alpha}}{1+e^{-\alpha}} \prod_{\beta\in\Phi^+-\{\alpha\}} \frac{1-e^{-\beta}}{1+e^{-\beta}} = (1+2\sum_{k>0}(-1)^k e^{-k\alpha})\ (\Sigma\ d(\xi,A)e^{-\xi})\ .$$

Thus

$$d(\xi,\Phi^+) = d(\xi,A) + 2\sum_{k>0}(-1)^k\ d(\xi-k\alpha,A)\ ,$$

$$d(s_\alpha\xi,\Phi^+) = d(s_\alpha\xi,A) + 2\sum_{k>0}(-1)^k\ d(s_\alpha\xi-k\alpha,A) = d(\xi,A) + 2\sum_{k>0}(-1)^k\ d(\xi+k\alpha,A)\ .$$

Thus

$$(d(\xi,\Phi^+) + d(s_\alpha\xi,\Phi^+))\ \eta(\xi) = 2\sum_{k\in\mathbb{Z}}(-1)^k\ d(\xi+k\alpha,A)\ \eta(\xi)$$

$$= 2\sum_{k\in\mathbb{Z}}\eta(\xi+k\alpha)\ d(\xi+k\alpha,A)\ .$$

Write for $\lambda \in V^*$, $\overline{\lambda}$ for the restriction of λ to V_α . Let $\xi \in \mathbb{Z}\Phi$. As the fiber of the map $\xi \mapsto \overline{\xi}$ from $\mathbb{Z}\Phi$ to $(V_\alpha)^*$ consists of sets $\{\xi + \mathbb{Z}\alpha\}$, it is clear that

$$y(\xi) = \sum_{k \in \mathbb{Z}} \eta(\xi+k\alpha)\, d(\xi+k\alpha, A)$$

depends only on the restriction $\bar{\xi}$ of ξ to V_α. We write thus

$$y(\xi) = y(\bar{\xi}) .$$

Let us restrict the function

$$\prod_{\beta \in A} \frac{1-\eta(\beta)e^{-\beta}}{1+\eta(\beta)e^{-\beta}} = \sum_{\xi \in \mathbb{Z}^+\Phi^+} d(\xi, A)\, \eta(\xi) e^{-\xi}$$

to V_α. Let $\mathbb{Z}\bar{\Phi}$ be the lattice generated by the elements $\bar{\beta}$, $\beta \in \Phi$. We then have

$$\prod_{\beta \in A} \frac{1-\eta(\beta)e^{-\bar{\beta}}}{1+\eta(\beta)e^{-\bar{\beta}}} = \sum_{\bar{\xi} \in \mathbb{Z}^+\bar{\Phi}^+} y(\bar{\xi}) e^{-\bar{\xi}} .$$

We will compute the function on the left hand side by dividing A into α-strings. Remark that if β and $\beta + \alpha$ are roots

$$\frac{1-\eta(\beta)e^{-\bar{\beta}}}{1+\eta(\beta)e^{-\bar{\beta}}} \; \frac{1-\eta(\beta+\alpha)e^{-(\overline{\beta+\alpha})}}{1+\eta(\beta+\alpha)e^{-(\overline{\beta+\alpha})}} = \frac{1-\eta(\beta)e^{-\bar{\beta}}}{1+\eta(\beta)e^{-\bar{\beta}}} \; \frac{1+\eta(\beta)e^{-\bar{\beta}}}{1-\eta(\beta)e^{-\bar{\beta}}} = 1 .$$

Thus, we can restrict ourselves to root strings of odd length. They are of the form $\{\beta\}$ or $\{\beta-\alpha, \beta, \beta+\alpha\}$ with $\beta \in \Phi_\alpha$. We identify $\beta \in \Phi_\alpha$ with its restriction $\bar{\beta}$. By Lemma 1.5, if a root string is of length 1, then

$$\frac{1-\eta(\beta)e^{-\beta}}{1+\eta(\beta)e^{-\beta}} = \frac{1-\eta_\alpha(\beta)e^{-\beta}}{1+\eta_\alpha(\beta)e^{-\beta}} , \quad \beta \in \Phi_\alpha .$$

while if a root string is of length 3, the restriction of

$$\frac{1-\eta(\beta-\alpha)e^{-(\beta-\alpha)}}{1+\eta(\beta-\alpha)e^{-(\beta-\alpha)}} \; \frac{1-\eta(\beta)e^{-\beta}}{1+\eta(\beta)e^{-\beta}} \; \frac{1-\eta(\beta+\alpha)e^{-(\beta+\alpha)}}{1+\eta(\beta+\alpha)e^{-(\beta+\alpha)}}$$ to V_α coincides with

$$\frac{1+\eta(\beta)e^{-\beta}}{1-\eta(\beta)e^{-\beta}} = \frac{1-\eta_\alpha(\beta)e^{-\beta}}{1+\eta_\alpha(\beta)e^{-\beta}} .$$ Thus we see that

$$\prod_{\beta \in A} \frac{1-\eta(\beta)e^{-\bar{\beta}}}{1+\eta(\beta)e^{-\bar{\beta}}} = \prod_{\beta \in \Phi_\alpha^+} \frac{1-\eta_\alpha(\beta)e^{-\beta}}{1+\eta_\alpha(\beta)e^{-\beta}} = \sum_{\xi \in \mathbb{Z}^+\Phi_\alpha^+} \eta_\alpha(\xi)\, d(\xi; \Phi_\alpha^+) e^{-\xi} ,$$

and we obtain Theorem 1.7.

2. The Plancherel function of a simply connected Lie group

2.1. Let g be a real semi-simple Lie algebra, G be the simply-connected Lie group with Lie algebra g, Z be the center of G .

Let h be a Cartan subalgebra of g. We denote by $\Delta(g,h)$ the set of roots of $h_{\mathbb{C}}$ in $g_{\mathbb{C}}$. If $\alpha \in \Delta(g,h)$, we denote by $g_{\mathbb{C},\alpha}$ the corresponding root space, h_α or $\check{\alpha} \in h_{\mathbb{C}}$ the coroot. We denote by

$$h_{\mathbb{R}} = \Sigma\, \mathbb{R}\, h_\alpha$$

$$a = h_{\mathbb{R}} \cap h$$

$$t = ih_{\mathbb{R}} \cap h \quad \text{so that}$$

$$h = t + a \ .$$

Let H be the Cartan subgroup associated to h , i.e. H is the centralizer of h in G . As H centralizes h , it acts by a character $\xi_\alpha(h)$ on each root space.

Let $T = \{h \in H, |\xi_\alpha(h)| = 1$ for all $\alpha \in \Delta(g_{\mathbb{C}}, h_{\mathbb{C}})\}$. Then T has Lie algebra t. Let $A = \exp a$. Then A is diffeomorphic to a . We have $H = TA$.

2.2. Let $\Phi(g,h)$ be the system of real roots of h in g . These are the roots of h vanishing on t . If $\alpha \in \Phi$, $h_\alpha \in a$, and we can choose $X_\alpha \in g \cap g_{\mathbb{C},\alpha}$, $X_{-\alpha} \in g \cap g_{\mathbb{C},-\alpha}$ such that $[X_\alpha, X_{-\alpha}] = h_\alpha$. Define

$$\gamma_{\check{\alpha}} = \exp \pi\,(X_\alpha - X_{-\alpha})$$

$$m_{\check{\alpha}} = \exp \frac{\pi}{2}\,(X_\alpha - X_{-\alpha}) \ .$$

The set $\{\gamma_{\check{\alpha}}, \gamma_{\check{\alpha}}^{-1}\}$ depends only on $\alpha \in \Phi$ and not on the choice of X_α . We have, for $a \in \check{\Phi} \subset h$

$$\mathrm{Ad}\gamma_a = \exp\,(\mathrm{ad}\; i\pi a) \quad \text{and} \quad \mathrm{Ad}m_a|h = s_{\check{a}} \ .$$

In particular $\gamma_a \in H$ and m_a normalizes H . Furthermore

$$(\gamma_a)^2 \in Z \ .$$

Let $a \in \check{\Phi}$, $\beta \in \Phi$, then

$$\gamma_a \exp t(X_\beta - X_{-\beta})\, \gamma_a^{-1} = \exp\,(e^{i\pi(a,\beta)} tX_\beta - e^{-i\pi(a,\beta)} tX_{-\beta}) \ .$$

Thus, if $\varepsilon(a,\beta) = (-1)^{(\beta,a)}$

2.3. $\gamma_a\, m_{\check{\beta}}\, \gamma_a^{-1} = (m_{\check{\beta}})^{\varepsilon(a,\beta)}$ and $\gamma_a\, \gamma_{\check{\beta}}\, \gamma_a^{-1} = (\gamma_{\check{\beta}})^{\varepsilon(a,\beta)}$.

Let s_α be the reflexion of h corresponding to $\alpha \in \Phi$. Choose an element g_α in G such that $\mathrm{Ad}g_\alpha \mid h = s_\alpha$ (for example $m_{\check{\alpha}}$). Thus we can choose

$$X_{s_\alpha(\beta)} = g_\alpha \cdot X_\beta \quad \text{and} \quad X_{-s_\alpha(\beta)} = g_\alpha \cdot X_{-\beta} .$$

For this choice of $X_{s_\alpha(\beta)}$,

2.4. $$g_\alpha \gamma_{\check{\beta}} g_\alpha^{-1} = \gamma_{s_\alpha(\check{\beta})} .$$

Let $\Gamma^{(2)} = \langle \gamma_a^2, a \in \check{\Phi} \rangle$. Then $\Gamma^{(2)}$ is a central subgroup of Γ and $\Gamma/\Gamma^{(2)}$ is commutative. Remark that the map $a \mapsto \gamma_a$ is well defined modulo $\Gamma^{(2)}$.

<u>2.5. Lemma</u>. The map $a \mapsto \gamma_a$ defines a homomorphism of $\mathbb{Z}\check{\Phi}$ into $\Gamma/\Gamma^{(2)}$.

<u>Proof.</u> Let $(a,b,c) \in \check{\Phi}$ with $c = a \pm b$. We have to prove that $\gamma_a \gamma_b \gamma_c \in \Gamma^{(2)}$. Suppose for example that a,b have the same length and $c = a+b$, then :

$$s_{\check{c}}(b) = -a \quad , \text{ i.e. } \quad (\check{c},b) = 1 .$$

Now $$\gamma_a \equiv \gamma_{s_{\check{c}}(b)} \equiv m_c \gamma_b m_c^{-1} \qquad \text{by (2.4)}$$

$$\equiv m_c \gamma_b m_c^{-1} \gamma_b^{-1} \gamma_b$$

$$\equiv m_c^2 \gamma_b \qquad \text{by (2.3)}$$

$$\equiv \gamma_c \gamma_b .$$

We list here some additional properties of the function $a \mapsto \gamma_a$ for systems of rank two, that we will use later on.

<u>2.6. Proposition</u> [Do]

Let $\Phi_o = \{\pm 2\ell_1^*, \pm 2\ell_2^*, \pm\ell_1^* \pm \ell_2^*\}$ be a subsystem of $\Phi(g,h)$ of type C_2. Then

1) all the elements γ_a, for $a \in \check{\Phi}_o$ commute,

2) we have $\gamma_{\ell_1+\ell_2}^2 = 1$ and $\gamma_{\ell_1+\ell_2} = \gamma_{\ell_1-\ell_2}$

3) we can choose γ_{ℓ_1}, γ_{ℓ_2} such that $\gamma_{\ell_1} \gamma_{\ell_2}^{-1} = \gamma_{\ell_1+\ell_2} = \gamma_{\ell_1-\ell_2}$.

<u>Proof</u> : Let g be the Split Cartan Lie algebra of type C_2. We calculate in the simply connected Lie group G of Lie algebra g. We can choose $g = sp(2, \mathbb{R})$ i.e.

$$g = \left(\begin{array}{c|c} A & B \\ \hline C & D \end{array}\right), \text{ with } B = {}^tB ,\ C = {}^tC ,\ D = -{}^tA .$$

where A,B,C, D are 2 × 2 real matrices.

Let $\mathfrak{a} = \left\{ h = \left(\begin{array}{cc|cc} h_1 & 0 & 0 & 0 \\ 0 & h_2 & 0 & 0 \\ \hline 0 & 0 & -h_1 & 0 \\ 0 & 0 & 0 & -h_2 \end{array}\right) \right\}$, with $\ell_1^*(h) = h_1$, $\ell_2^*(h) = h_2$

Let $\mathfrak{g} = \mathfrak{k} \oplus \mathfrak{p}$ be a Cartan decomposition of $\mathfrak{g}$, with

$$\mathfrak{k} = \{ \left(\begin{array}{c|c} A & B \\ \hline -{}^tB & -{}^tA \end{array}\right) , \ A + {}^tA = 0 , \ B = {}^tB\} .$$

The Lie algebra $\mathfrak{k}$ is isomorphic to $\mathfrak{u}(2)$, by $\left(\begin{array}{c|c} A & B \\ \hline -{}^tB & -{}^tA \end{array}\right) \longmapsto A + iB$.

Let G be the simply connected group of Lie algebra $\mathfrak{g}$, K the analytic subgroup of G of Lie algebra $\mathfrak{k}$. Then $K \cong SU(2) \times \mathbb{R}$. A basis X_α of the root spaces is given as follows :

2.7. $$X_{2\ell_1^*} = \left(\begin{array}{c|c} 0 & \begin{pmatrix} 1 & 0 \\ 0 & 0 \end{pmatrix} \\ \hline 0 & 0 \end{array}\right) , \quad X_{-2\ell_1^*} = \left(\begin{array}{c|c} 0 & 0 \\ \hline \begin{pmatrix} 1 & 0 \\ 0 & 0 \end{pmatrix} & 0 \end{array}\right)$$

$$X_{2\ell_2^*} = \left(\begin{array}{c|c} 0 & \begin{pmatrix} 0 & 0 \\ 0 & 1 \end{pmatrix} \\ \hline 0 & 0 \end{array}\right) , \quad X_{-2\ell_2^*} = \left(\begin{array}{c|c} 0 & 0 \\ \hline \begin{pmatrix} 0 & 0 \\ 0 & 1 \end{pmatrix} & 0 \end{array}\right)$$

$$X_{\ell_1^*+\ell_2^*} = \left(\begin{array}{c|c} 0 & \begin{pmatrix} 0 & 1 \\ 1 & 0 \end{pmatrix} \\ \hline 0 & 0 \end{array}\right) , \quad X_{-(\ell_1^*+\ell_2^*)} = \left(\begin{array}{c|c} 0 & 0 \\ \hline \begin{pmatrix} 0 & 1 \\ 1 & 0 \end{pmatrix} & 0 \end{array}\right) .$$

$$X_{\ell_1^*-\ell_2^*} = \left(\begin{array}{c|c} \begin{pmatrix} 0 & 1 \\ 0 & 0 \end{pmatrix} & 0 \\ \hline 0 & \begin{pmatrix} 0 & 0 \\ -1 & 0 \end{pmatrix} \end{array}\right) , \quad X_{-(\ell_1^*-\ell_2^*)} = \left(\begin{array}{c|c} \begin{pmatrix} 0 & 0 \\ 1 & 0 \end{pmatrix} & 0 \\ \hline 0 & \begin{pmatrix} 0 & -1 \\ 0 & 0 \end{pmatrix} \end{array}\right) .$$

Let us write $U_{\check\alpha} = X_\alpha - X_{-\alpha}$, then $U_{\check\alpha} \in \mathfrak{k}$. In the isomorphism $\mathfrak{k} \simeq \mathfrak{u}(2)$, we have :

$$U_{\ell_1} = \begin{pmatrix} i & 0 \\ 0 & 0 \end{pmatrix} , \quad U_{\ell_2} = \begin{pmatrix} 0 & 0 \\ 0 & i \end{pmatrix} , \quad U_{\ell_1+\ell_2} = \begin{pmatrix} 0 & i \\ i & 0 \end{pmatrix} , \quad U_{\ell_1-\ell_2} = \begin{pmatrix} 0 & 1 \\ -1 & 0 \end{pmatrix} .$$

Then we see that

$$\exp \pi U_{\ell_1-\ell_2} = \exp \pi U_{\ell_1+\ell_2} = \exp \pi(U_{\ell_1} - U_{\ell_2}) = \left(\begin{pmatrix} -1 & 0 \\ 0 & -1 \end{pmatrix}, 0\right) \text{ in } K = SU(2) \times \mathbb{R}$$

This proves the proposition 2.6.

As a corollary of proposition 2.6, we note :

<u>2.8. Corollary</u>

a) Let $\Phi_o \subset \Phi$ be a subsystem of type C_2. Then,

1) If $\alpha, \beta \in \Phi_o$ are short, then $\gamma_{\check\alpha} = \gamma_{\check\beta}$ and $(\gamma_{\check\alpha})^2 = 1$

2) The set $\{\gamma_{\check\nu}^2, \gamma_{\check\nu}^{-2}\}$ is independent of the long root $\nu \in \Phi_o$.

2.9. Proposition Let $\Phi_o \subset \Phi$ be a subsystem of type A_2 or G_2. Then for all $c, d \in \Phi_o$, $\gamma_c^2 = \gamma_d^2$ and $\gamma_c^4 = \text{Id}$.

Proof : Let us remark that if c, d are such that $(c,\check d)$ is odd, then by 2.3 $\gamma_c \gamma_d \gamma_c = \gamma_d$, $\gamma_d \gamma_c \gamma_d = \gamma_c$, thus $\gamma_c \gamma_d \gamma_c \gamma_d = \gamma_c^2 = \gamma_d^2$. Furthermore

$$\gamma_c \gamma_d^2 \gamma_c^{-1} = \gamma_d^{-2} = \gamma_d^2 \text{, as } \gamma_d^2 \text{ is central. Thus } \gamma_d^4 = \text{Id}.$$

This proves the proposition in the case of A_2.

In the case of G_2, we see immediately that $\gamma_c^2 = \gamma_d^2$ with $\gamma_c^4 = \text{id}$, if c and d are long. Similarly $\gamma_c^2 = \gamma_d^2$, with $\gamma_c^4 = \text{Id}$ if c and d are short. Furthermore, there exists c short, c' long with $(\check c, c')$ odd, so we obtain our proposition.

Recall that $H = TA$; thus the set $\hat H$ of unitary representations of H is equal to $\hat T \times \mathfrak{a}^*$, where $\lambda \in \mathfrak{a}^*$ corresponds to the character $\chi_\lambda(\exp X) = e^{i(\lambda,X)}$, $X \in \mathfrak{a}$ of A. We write $\tau = (\sigma,\lambda)$ for $\tau \in \hat H$, $\sigma \in \hat T$, $\lambda \in \mathfrak{a}^*$.

Remark that if $\sigma \in \hat T$ is an irreducible representation of T, the matrix $\sigma(\gamma_\alpha) + \sigma(\gamma_\alpha^{-1})$ belongs to the commutant of σ, thus is a scalar matrix. We define a function x_σ on $\check\Phi$ by

2.10. $x_\sigma(a)\ \text{Id} = \frac{1}{2}(\sigma(\gamma_a) + \sigma(\gamma_a^{-1}))$.

The function x_σ depends only on the class of the representation σ.

Let W_Φ be the Weyl group associated to Φ. If $w \in W_\Phi$ and ψ is a function on $\check\Phi$, we define $w\cdot\psi$ by $(w\cdot\psi)(a) = \psi(w^{-1}a)$. Let g be an element of G representing w, then g normalizes T. Let $(g\cdot\sigma)$ be the representation of T defined by $(g\cdot\sigma)(h) = \sigma(g^{-1}hg)$. Its class depends only on w. We denote it by $w\cdot\sigma$. It is clear that $x_{w\cdot\sigma} = w\cdot x_\sigma$.

The following function on $\hat T \times \mathfrak{a}^*$ was introduced by Harish-Chandra in connection with the Plancherel formula [Ha]. Let us choose a system Φ^+ of positive real roots. Define for $\sigma \in \hat T$, the function $P(\sigma,\Phi^+)$ on $\mathfrak{a}^*$ by :

2.11. $$P(\sigma,\Phi^+)(\lambda) = \prod_{a\in\check\Phi^+} \frac{\operatorname{Sh}(a,\lambda)}{\operatorname{Ch}(a,\lambda)+x_\sigma(a)}.$$

Let $u_\sigma(a)$ be an eigenvalue of the matrix $\sigma(\gamma_a)$. Then $x_\sigma(a) = \frac{1}{2}(u_\sigma(a)+u_\sigma(a)^{-1})$, with $|u_\sigma(a)| = 1$. Consider the factor $f_a(\sigma)$ corresponding to a in $P(\sigma,\Phi^+)$. We have :

2.12. $$f_a(\sigma)(\lambda) = \frac{Sh(a,\lambda)}{Ch(a,\lambda)+x_\sigma(a)} ,$$

$$= \frac{1}{2}\left(\frac{1-u_\sigma(a)e^{-(a,\lambda)}}{1+u_\sigma(a)e^{-(a,\lambda)}} + \frac{1-u_\sigma(a)^{-1}e^{-(a,\lambda)}}{1+u_\sigma(a)^{-1}e^{-(a,\lambda)}}\right) ,$$

$$= \frac{1}{1+u_\sigma(a)e^{-(a,\lambda)}} + \frac{1}{1+u_\sigma(a)^{-1}e^{-(a,\lambda)}} - 1 .$$

If λ is such that $(\lambda,a) > 0$, then the function $\lambda \to f_a(\sigma)(\lambda)$ is given by the convergent series

$$f_a(\sigma) = \sum_{k \in \mathbb{Z}} (-1)^k u_\sigma(a)^k e^{-|k|a} = 1 + \sum_{k>o} (-1)^k (u_\sigma(a)^k + u_\sigma(a)^{-k}) e^{-ka} .$$

Let A be a subset of $\overset{\vee}{\Phi}{}^+$, say $A = \{a_1,a_2,\dots,a_N\}$. Define $\vec{k} = \{k_1,k_2,\dots,k_N\} \in \mathbb{Z}^N$ and for $\xi \in \mathbb{Z}^+A$, let

$$d(\sigma,\xi,A) = \sum_{\substack{\vec{k} \in \mathbb{Z}^N \\ \Sigma|k_i|a_i=\xi}} (-1)^{k_1}\dots(-1)^{k_N} u_\sigma(a_1)^{k_1}\dots u_\sigma(a_N)^{k_N} .$$

Let us define $P(\sigma,A) = \prod_{a \in A} f_a(\sigma)$. Let $C^+ = \{\lambda \in \mathfrak{a}^* , (\lambda,a) > 0 , \text{ for all } a \in \overset{\vee}{\Phi}{}^+\}$. Then, for $\lambda \in C^+$, the function $\lambda \mapsto P(\sigma,A)(\lambda)$ is given by the series

2.13. $$P(\sigma,A) = \sum_{\xi \in \mathbb{Z}^+A} d(\sigma;\xi;A)e^{-\xi} .$$

In particular,

$$P(\sigma,\overset{\vee}{\Phi}{}^+) = \sum_{\xi \in \mathbb{Z}^+\overset{\vee}{\Phi}{}^+} d(\sigma;\xi,\overset{\vee}{\Phi}{}^+)e^{-\xi} .$$

Up to now, $\mathfrak{h}$ was a fixed Cartan subalgebra of $\mathfrak{g}$. We will now study the relations of the coefficients $d^{\mathfrak{h}}(\sigma,\xi,\overset{\vee}{\Phi}{}^+)$ on adjacent Cartan subalgebras.

3. Adjacent Cartan subalgebras

Let h be a Cartan subalgebra. We denote $\Phi(g,h)$ by Φ. Let $\alpha \in \Phi$, set $a = \check{\alpha}$. We define, for X_α, $X_{-\alpha}$ as in 2.2,

$$h' = \mathbb{R}(X_\alpha - X_{-\alpha}) \oplus \operatorname{Ker} \alpha .$$

Then h' is a Cartan subalgebra of g. We have

$$h' = t' \oplus a' \quad \text{with} \quad t' = t \oplus \mathbb{R}(X_\alpha - X_{-\alpha}) , \quad a' = a \cap \operatorname{Ker} \alpha .$$

We denote by $\Phi' = \Phi(g,h')$ the real roots of h' in g. We denote by c_α the linear transformation $h_{\mathbb{C}} \to h'_{\mathbb{C}}$ such that

$$c_\alpha(X) = X \quad \text{if } X \in \operatorname{Ker} \alpha , \quad c_\alpha(h_\alpha) = i(X_\alpha - X_{-\alpha}) .$$

Let $G_{\mathbb{C}}$ be the complex adjoint group of $g_{\mathbb{C}}$, then there exists an element g of $G_{\mathbb{C}}$ such that $g|h_{\mathbb{C}} = c_\alpha$. The transformation c_α is called the Cayley transform. Let $\Phi_\alpha = \{\beta \in \Phi$ such that $(\beta, h_\alpha) = 0\}$ then $c_\alpha(\Phi_\alpha) = \Phi'$. Remark that the restriction map $a^* \to (a')^*$ induces an isomorphism of Φ_α with Φ'. Thus if an order Φ^+ is chosen on Φ, it induces an order on Φ'.

Let $H = TA$, $H' = T'A'$ be the Cartan subgroups of G corresponding to h, h'. Let $\Gamma = \langle\gamma_a , a \in \check{\Phi}\rangle$ and $\Gamma' = \langle\gamma_{a'}, a' \in \check{\Phi}'\rangle$. For $\beta \in \Phi_\alpha$, let $\beta' = c_\alpha(\beta) \in \Phi'$. We will compare the elements $\gamma_{\check{\beta}}$ and $\gamma_{\check{\beta}'}$.

Consider the subsystem $\Phi(\alpha,\beta)$ generated by (α,β). If α and β are not strongly orthogonal, then $\{\pm\beta, \pm\alpha, \pm\beta \pm \alpha\}$ are roots and the system $\Phi(\alpha,\beta)$ is of type B_2. Furthermore, by (2.6) $(\gamma_a)^2 = 1$. Let ν be a long root of $\Phi(\alpha,\beta)$, $v=\check{\nu}$ then the set $\{\gamma_v^2\gamma_a, \gamma_v^{-2}\gamma_a\}$ is independent of the choice of the long root ν.

3.1. Proposition [Do]

Let $\beta \in \Phi_\alpha$, $\beta' = c_\alpha(\beta)$, $b = \check{\beta}$, $b' = \check{\beta}'$

1) Suppose β and α are strongly orthogonal, then

$$\{\gamma_{b'}, \gamma_{b'}^{-1}\} = \{\gamma_b, \gamma_b^{-1}\} .$$

2) Suppose β and α are not strongly orthogonal ; let ν be a long root of the system $\Phi(\alpha,\beta)$, $v = \check{\nu}$, then :

$$\{\gamma_{b'}, \gamma_{b'}^{-1}\} = \{\gamma_v^2\gamma_a, \gamma_v^{-2}\gamma_a\}$$

Proof : In case 1, as $[X_\alpha - X_{-\alpha}, X_\beta] = 0$, we can choose $X_{\beta'} = X_\beta$, $X_{-\beta'} = X_{-\beta}$ and then 1) is obvious. In case 2, the system $\Phi(\alpha,\beta)$ is of type C_2. Write $\Phi^+ = \{\ell_1^* \pm \ell_2^*, 2\ell_1^*, 2\ell_2^*\}$ with $\alpha = \ell_1^* - \ell_2^*$, $\beta = \ell_1^* + \ell_2^*$. Let us consider the Split Lie algebra g of type C_2 as in the proof of 2.6 with Split Cartan $h = \mathbb{R}\ell_1 \oplus \mathbb{R}\ell_2$, where

$$\ell_1 = \left(\begin{array}{c|c} \begin{pmatrix}1&0\\0&0\end{pmatrix} & 0 \\ \hline 0 & \begin{pmatrix}-1&0\\0&0\end{pmatrix}\end{array}\right) \quad , \quad \ell_2 = \left(\begin{array}{c|c} \begin{pmatrix}0&0\\0&1\end{pmatrix} & 0 \\ \hline 0 & \begin{pmatrix}0&0\\0&-1\end{pmatrix}\end{array}\right)$$

We choose the basis of root vectors as in 2.7. The adjacent Cartan h' to h with respect to α is $h' = \mathbb{R}(\ell_1 + \ell_2) \oplus \mathbb{R}(X_{\ell_1^*-\ell_2^*} - X_{-(\ell_1^*-\ell_2^*)})$. Let $c_\alpha : h_{\mathbb{C}} \to (h')_{\mathbb{C}}$ be the Cayley transform. Let $\beta = \ell_1^* + \ell_2^*$, then $\beta' = c_\alpha(\ell_1^* + \ell_2^*)$ is such that

$$\beta'(X_{\ell_1^*-\ell_2^*} - X_{-(\ell_1^*-\ell_2^*)}) = 0 \ , \quad \beta'(\ell_1+\ell_2) = 2 \ , \quad h_{\beta'} = \ell_1+\ell_2 \ .$$

We then see that we can choose

$$X_{\beta'} = \left(\begin{array}{c|c} 0 & \begin{pmatrix}1&0\\0&1\end{pmatrix} \\ \hline 0 & 0\end{array}\right) \quad , \quad X_{-\beta'} = \left(\begin{array}{c|c} 0 & 0 \\ \hline \begin{pmatrix}1&0\\0&1\end{pmatrix} & 0\end{array}\right) .$$

Let $U_{\beta'}^{\vee} = X_{\beta'} - X_{-\beta'} \in k \simeq u(2)$. In the isomorphism $k \simeq u(2)$;

$$U_{\beta'}^{\vee} = \begin{pmatrix} i & 0 \\ 0 & i \end{pmatrix} = U_{\ell_1} + U_{\ell_2} \ .$$

Thus $\gamma_{\beta'}^{\vee} = \gamma_{\ell_1} \cdot \gamma_{\ell_2} = \gamma_{\ell_2}^2 \cdot \gamma_{\ell_1} \gamma_{\ell_2}^{-1} = \gamma_{\ell_2}^2 \gamma_{\ell_1-\ell_2}$ as stated.

As a corollary of the proposition 3.1, we note :

<u>3.2. Lemma</u> $\Gamma' \subset \Gamma$.

Let $\sigma \in \hat{T}$, $\sigma' \in \hat{T}'$. We introduce the following definition :

<u>3.3. Definition</u> . The representations σ and σ' are <u>compatible</u> if σ and σ' have an irreducible component in common on $T \cap T'$.

As $\Gamma' \subset T \cap T'$, it follows that the function $x' = x_{\sigma'}$ on $\check{\Phi}'$ is entirely determined by x . We write the formulae for x' in function of x . Let us write $x_a = \frac{1}{2}(u_a + u_a^{-1})$ with $|u_a| = 1$.

<u>3.4. Proposition</u>. Let σ and σ' be two compatible representations of T , T' . Let $\beta \in \Phi_\alpha$, $\beta' = c_\alpha(\beta)$. Then

1) If $\beta \in \Phi_\alpha$ is strongly orthogonal to α , $x'(\check{\beta}') = x(\check{\beta})$.
2) If α and β are not strongly orthogonal, then $x(\check{\alpha})^2 = 1$ and

$$x'(\check{\beta}') = \frac{1}{2}\left(\left(u_{(\beta+\alpha)^\vee}^2 + u_{(\beta+\alpha)^\vee}^{-2}\right) \cdot x(\check{\alpha})\right) .$$

<u>Proof</u> : This is an obvious corollary of 3.1.

4. A recurrence relation for adjacent Plancherel functions

Let h be a Cartan subalgebra of g, $h_{\mathbb{R}} = it \oplus a$, $\Phi = \Phi(g,h)$ its system of real roots, Φ^+ an order on Φ. Let $H = TA$ the corresponding Cartan subgroup. We have defined, for $\sigma \in \hat{T}$, functions $\xi \to d^h(\sigma,\xi,\check{\Phi}^+)$ on $\xi \in \mathbb{Z}^+\check{\Phi}^+$. Let us extend $d^h(\sigma;\xi;\check{\Phi}^+)$ as a function on a by defining it to be zero outside $\mathbb{Z}^+\check{\Phi}^+$. As T^o is central in T, there exists $\lambda_o \in t^*$ such that $\sigma(\exp X) = e^{i(\lambda_o,X)}$. Let $\xi_o \in h_{\mathbb{R}} = it \oplus a$. Write $\xi = \xi_o + \xi_1$, with $\xi_o \in it$. Extend $d^h(\sigma;\xi;\check{\Phi}^+)$ as a function on $h_{\mathbb{R}}$, by

$$d^h(\sigma;\xi_o+\xi_1;\check{\Phi}^+) = e^{(\lambda_o,\xi_o)}\, d(\sigma,\xi_1,\check{\Phi}^+) \quad .$$

(Thus, by definition, $d^h(\sigma,\xi,\check{\Phi}^+)$ is zero if $\xi_1 \notin \mathbb{Z}^+\check{\Phi}^+$).

Let α be a simple root of Φ^+, h' the adjacent Cartan to h relative to α. Consider Φ_α and the induced order $\Phi_\alpha^+ = \Phi^+ \cap \Phi_\alpha$. Consider Φ' and the corresponding order Φ'^+ on Φ'. Let $c_\alpha : h_{\mathbb{C}} \to h'_{\mathbb{C}}$ be the Cayley transform. Then $c_\alpha(h_{\mathbb{R}}) = h'_{\mathbb{R}}$.

Let $\rho = \frac{1}{2} \sum_{\beta\in\Phi^+} \beta \subset h^*$

$\eta(h) = e^{i\pi(\rho\ ,h)}$ the corresponding character on $h_{\mathbb{R}}$.

$\rho_\alpha = \frac{1}{2} \sum_{\beta\in(\Phi')^+} \beta \in (h')^*$

$\eta'(h) = e^{i\pi(\rho',h)}$ the corresponding character on $h'_{\mathbb{R}}$.

The main result of this article is :

4.1. Theorem

Let σ and σ' be two compatible representations of T and T'. Then, for $\xi \in h_{\mathbb{R}}$, $\xi = \xi_o + \xi_1$, with $\xi_1 \in \mathbb{Z}\check{\Phi}$

$$\eta(\xi).(d^h(\sigma;\xi,\check{\Phi}^+) + d^h(s_\alpha\sigma,s_\alpha\xi;\check{\Phi}^+)) = \eta'(c_\alpha\xi)(d^h(\sigma',c_\alpha\xi;\check{\Phi}'^+) + d^h(\sigma',c_\alpha s_\alpha\xi;\check{\Phi}'^+)) .$$

Proof : We will follow the same method as in the proof of Theorem 1.7. Write for $a \in \check{\Phi}$, $x_\sigma(a) = \frac{1}{2}(u_\sigma(a)+u_\sigma(a)^{-1})$, with $|u_\sigma(a)| = 1$. Define, for χ a character of $\mathbb{Z}\check{\Phi}$ and $a \in \check{\Phi}^+$

4.2. $$f_a(\sigma,\chi) = \frac{1}{2}\left(\frac{1-u_\sigma(a)\chi(a)e^{-a}}{1+u_\sigma(a)\chi(a)e^{-a}} + \frac{1-u_\sigma(a)^{-1}\chi(a)e^{-a}}{1+u_\sigma(a)^{-1}\chi(a)e^{-a}}\right)$$

where $1+u_\sigma(a)\chi(a)e^{-a}$ is inverted as $\sum_{k\geq o} (-1)^k\, u_\sigma(a)^k\, \chi(a)^k\, e^{-ka}$. We have also, as $x_\sigma(a) = \frac{1}{2}(u_\sigma(a)+u_\sigma(a)^{-1})$,

4.3. $f_a(\sigma,\chi) = \dfrac{1 - \chi(a)^2 e^{-2a}}{1+2\chi(a)x_\sigma(a)e^{-a}+\chi(a)^2 e^{-2a}}$.

If A is a subset of $\check{\Phi}^+$, define $P(\sigma,\chi,A) = \prod_{a\in A} f_a(\sigma,\chi)$. If χ = identity, we omit χ in the notation. We have

4.4. $\sum_{\xi \in \mathbb{Z}^+ A} \chi(\xi)\, d(\sigma,\xi,A)e^{-\xi} = P(\sigma,\chi,A)$.

Let α be a simple root of Φ^+ . Write $a = \check{\alpha}$. Let $A = \check{\Phi}^+ - (a)$. Write

$$P(\sigma,\check{\Phi}^+) = f_a(\sigma) \prod_{\gamma\in A} f_\gamma(\sigma) .$$

Thus, as $f_a(\sigma) = 1 + \sum_{k>o} (-1)^k (u_\sigma(a)^k+u_\sigma(a)^{-k})e^{-|k|a}$, for $\xi_1 \in \mathfrak{a}$

$$d(\sigma,\xi_1,\check{\Phi}^+) = d(\sigma,\xi_1,A) + \sum_{k>o} (-1)^k (u_\sigma(a)^k+u_\sigma(a)^{-k})d(\sigma,\xi_1 - ka,A) .$$

As $x_{s_\alpha(\sigma)}(a) = x_\sigma(a)$

$$\begin{aligned} d(s_\alpha\sigma,s_\alpha\xi_1,\check{\Phi}^+) &= d(s_\alpha\sigma,s_\alpha\xi_1,A)+ \sum_{k>o} (-1)^k (u_\sigma(a)^k+u_\sigma(a)^{-k})d(s_\alpha\sigma,s_\alpha\xi_1-ka,A) \\ &= d(\sigma,\xi_1,A) + \sum_{k>o} (-1)^k (u_\sigma(a)^k+u_\sigma(a)^{-k})\, d(\sigma,\xi_1+ka,A) . \end{aligned}$$

Thus,

$$\begin{aligned} d(\sigma,\xi_1,\check{\Phi}^+) + d(s_\alpha\sigma,s_\alpha\xi_1,\check{\Phi}^+) = &\sum_{k \in \mathbb{Z}} (-1)^k\, u_\sigma(a)^k d(\sigma,\xi_1+ka,A) \\ &+ \sum_{k \in \mathbb{Z}} (-1)^k\, u_\sigma(a)^{-k} d(\sigma,\xi_1+ka,A) . \end{aligned}$$

For $h_1 \in \mathfrak{a}$, we denote by $\overline{h}_1$ the orthogonal projection of h_1 on $\mathfrak{a}' = \mathfrak{a} \cap \ker \alpha$. Let us study for $\xi_1 \in \mathbb{Z}\check{\Phi} \subset \mathfrak{a}$

$$\Sigma(-1)^k\, u_\sigma(a)^k\, d(\sigma,\xi_1+ka,A) .$$

Let δ be a character on $\mathfrak{a}$ such that $\delta(a) = u_\sigma(a)$

$\delta(h) = 1$ for $h \in \mathfrak{a} \cap \operatorname{Ker} \alpha$.

Recall that $\eta(a) = -1$. Consider for $\xi_1 \in \mathbb{Z}\check{\Phi}$

4.5. $y_\delta(\overline{\xi}_1) = \sum_{k\in \mathbb{Z}} \delta(\xi_1+ka)\, \eta(\xi_1+ka)\, d(\sigma,\xi_1+ka,A)$

$= \delta(\xi_1)\eta(\xi_1)\, (\Sigma(-1)^k u_\sigma(a)^k\, d(\sigma,\xi_1+ka,A))$, so that

4.6. $\eta(\xi_1)\, (d(\sigma,\xi_1,\check{\Phi}^+) + d(s_\alpha\sigma,s_\alpha\xi_1,\check{\Phi}^+)) = \delta(\xi_1)^{-1}\, y_\delta(\overline{\xi}_1) + \delta(\xi_1)\, y_{\delta^{-1}}(\overline{\xi}_1)$.

Introduce the function W on $(\mathfrak{a}')^* = (\mathfrak{a} \cap \ker \alpha)^*$ defined by

$$W = \prod_{b\in A} \frac{1}{2}\left(\frac{1-\delta(b)\eta(b)u_\sigma(b)e^{-\bar{b}}}{1+\delta(b)\eta(b)u_\sigma(b)e^{-\bar{b}}} + \frac{1-\delta(b)\eta(b)u_\sigma(b)^{-1}e^{-\bar{b}}}{1+\delta(b)\eta(b)u_\sigma(b)^{-1}e^{-\bar{b}}}\right)$$

As the fiber of the map $\xi_1 \mapsto \bar{\xi}_1$ from $\mathbb{Z}\check{\Phi}$ to $\mathfrak{a}'$ consists of $\{\xi_1+ka,\ k \in \mathbb{Z}\}$, we have :

4.7. $$\sum_{\bar{\xi}_1 \in \mathbb{Z}\check{\Phi}} y_\delta(\bar{\xi}_1)\, e^{-\bar{\xi}_1} = W$$

Let w_b be the individual factor of W , i.e.

4.8. $$w_b = \frac{1}{2}\left(\frac{1-\delta(b)\eta(b)u_\sigma(b)e^{-\bar{b}}}{1+\delta(b)\eta(b)u_\sigma(b)e^{-\bar{b}}} + \frac{1-\delta(b)\eta(b)u_\sigma(b)^{-1}e^{-\bar{b}}}{1+\delta(b)\eta(b)u_\sigma(b)^{-1}e^{-\bar{b}}}\right) .$$

We have also, as $\eta(b)^2 = 1$

4.9. $$w_b = \frac{1-\delta(b)^2\, e^{-2\bar{b}}}{1+2\delta(b)\eta(b)x_\sigma(b)e^{-\bar{b}}+\delta(b)^2e^{-2\bar{b}}} .$$

To compute W , we will divide the set $\check{\Phi}$ according to systems of rank two obtained by considering $(\mathbb{R}b + \mathbb{R}a) \cap \check{\Phi}$, for $b \in A$.

Consider a representation $\sigma' \in \hat{T}'$ compatible with $\sigma \in \hat{T}$ and its associated function $x_{\sigma'}$ on $\check{\Phi}'$. The function $P(\sigma',\check{\Phi}') = \prod_{b'\in\check{\Phi}'^+} f_{b'}(\sigma')$ is a function on $(\mathfrak{a}')^*$. Its individual factor $f_{b'}(\sigma')$ for $b' \in \check{\Phi}' \subset \mathfrak{a}'$ is

$$f_{b'}(\sigma') = \frac{1-e^{-2b'}}{1+2x_{\sigma'}(b')e^{-b'}+e^{-2b'}}$$

If χ' is a character of $\mathbb{Z}\check{\Phi}'$, we consider similarly

4.10. $$f_{b'}(\sigma',\chi') = \frac{1-\chi'(b')^2\, e^{-2b'}}{1+2x_{\sigma'}(b')\chi'(b')e^{-b'} + \chi'(b')^2e^{-2b'}}$$

$$= \frac{1}{2}\left(\frac{1-u_{\sigma'}(b')\chi'(b')e^{-b'}}{1+u_{\sigma'}(b')\chi'(b')e^{-b'}} + \frac{1-u_{\sigma'}(b')^{-1}\chi'(b')e^{-b'}}{1+u_{\sigma'}(b')^{-1}\chi'(b')e^{-b'}}\right) .$$

Let $b \in A$, consider $\check{\Phi}_o = (\mathbb{R}a + \mathbb{R}b) \cap \check{\Phi}$, then Φ_o is a system of rank two. The set $(\check{\Phi}_o)_\alpha = \check{\Phi}_\alpha \cap \check{\Phi}_o$ is either empty or consists of $\{d,-d\}$ for some $d \in \check{\Phi}_\alpha \cap \check{\Phi}_o$. We write

$$\prod_{d\in(\check{\Phi}_o)^+_\alpha} f_d(\sigma',\eta') = 1 \qquad \text{if } (\check{\Phi}_o)_\alpha = \emptyset$$

$$= f_d(\sigma',\eta') \qquad \text{if } (\check{\Phi}_o)^+_\alpha = \{d\} .$$

<u>4.11. Lemma</u>. Let $b \in A$, $\check{\Phi}_o = (\mathbb{R}a + \mathbb{R}b) \cap \check{\Phi}$, then

$$\prod_{c\in\check{\Phi}_o-(a)} \omega_c = \prod_{d\in(\check{\Phi}_o)^+_\alpha} f_d(\sigma';\eta')$$

Proof : Consider the possible types of systems.

1) Suppose $\check{\Phi}_o$ is of type $A_1 \times A_1$, then $\check{\Phi}_o^+ = \{a,d\}$, with a,d strongly orthogonal as well as $\check{a}$, $\check{d}$. By definition, we have $\delta(d) = 1$.
As $\alpha = \check{a}$ and $\check{d}$ are strongly orthogonal, $x_{\sigma'}(d) = x_\sigma(d)$ (3.4).
As a and d are strongly orthogonal, $\eta(d) = \eta'(d)$ (1.5). Thus $\omega_d = f_\sigma(\sigma',\eta')$.

2) Suppose that $\check{\Phi}_o$ is of type A_2 or G_2 . Let b , $b + a$ be two consecutive coroots in $\check{\Phi}_o^+ - \{a\}$. Let us see that $\omega_b\,\omega_{b+a} = 1$. Recall (2.9) that for every $c, d \in \check{\Phi}_o$, $u_\sigma(c)^2 = u_\sigma(d)^2$ and that $u_\sigma(c)^4 = 1$. Suppose first that $u_\sigma(a)^2 = -1$. Then for all $c \in \check{\Phi}_o$, $x_\sigma(c) = 0$. As $\delta(b+a)^2 = \delta(b)^2\delta(a)^2 = -\delta(b)^2$, we have, by 4.9,

$$\omega_b = \frac{1-\delta(b)^2 e^{-2\bar{b}}}{1+\delta(b)^2 e^{-2\bar{b}}} \quad , \quad \omega_{b+a} = \frac{1+\delta(b)^2 e^{-2\bar{b}}}{1-\delta(b)^2 e^{-2\bar{b}}} \quad , \text{ and } \omega_b\,\omega_{b+a} = 1 .$$

Suppose now that $u_\sigma(a)^2 = 1$, then $u_\sigma(b)^2 = 1$ for all $b \in \check{\Phi}_o$. The matrix $\sigma(\gamma_b)$ is thus the scalar matrix $\{\pm \mathrm{Id}\}$. It follows from (2.5) that the map $b \mapsto u_\sigma(b) = \pm 1$ extends to a character on $\mathbb{Z}\check{\Phi}_o$. As $u_\sigma(b) = u_\sigma(b)^{-1}$, we have, by 4.8

$$\omega_b = \frac{1-\delta(b)\eta(b)u_\sigma(b)e^{-\bar{b}}}{1+\delta(b)\eta(b)u_\sigma(b)e^{-\bar{b}}} \quad \text{and}$$

$$\delta(b+a)\eta(b+a)u_\sigma(b+a) = \delta(b)u_\sigma(a)\eta(b)(-1)u_\sigma(b)u_\sigma(a) = -\delta(b)\eta(b)u_\sigma(b) .$$

Thus $\omega_b\,\omega_{b+a} = 1$.

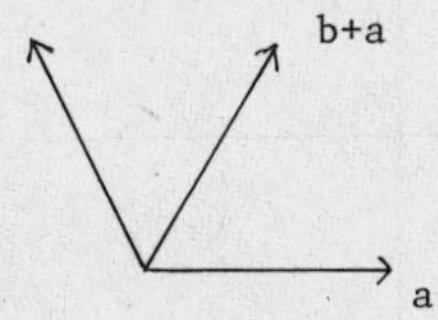

Fig.1

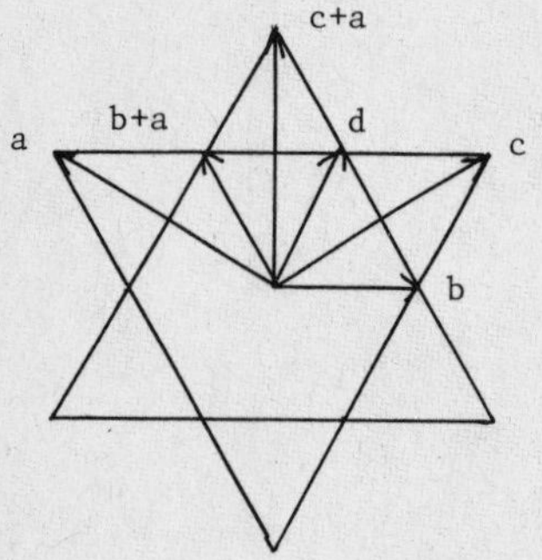

Fig.2

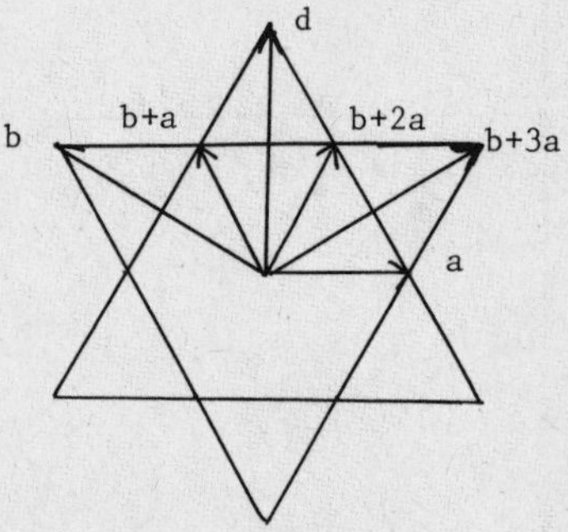

Fig.3

In the case where $\check{\Phi}_o = A_2$, $(\check{\Phi}_o)_\alpha = \emptyset$ and $\check{\Phi}_o - (a)$ is a chain consisting of two roots (b, b+a) (Fig.1). So Lemma 4.11 holds. In the case where $\check{\Phi}_o = G_2$ and a is a long root (Fig. 2), $(\check{\Phi}_o)_\alpha$ consists of one element d such that (d, a) and $(\check{d}, \check{a})$ are strongly orthogonal so that, as before, $\omega_d = f_d(\sigma', \eta')$. Furthermore $\check{\Phi}_o - (\check{\Phi}_o)_\alpha$ consists of the union of two a-chains of length two {b, b+a} and {c, c+a} . So Lemma 4.11 holds. In the case where $\Phi_o = G_2$ and a is a short root of $\check{\Phi}_o$ (Fig. 3), $(\check{\Phi}_o)_\alpha$ consists of one element $\check{d}$ such that (a, d) and $(\check{a}, \check{d})$ are strongly orthogonal, while $\check{\Phi}_o - (\check{\Phi}_o)_\alpha$ is the a-chain of length 4 {b, b+a, b+2a, b+3a} . So Lemma 4.11 holds.

3) Let us suppose finally that $\check{\Phi}_o$ is of type B_2 (Fig. 4).

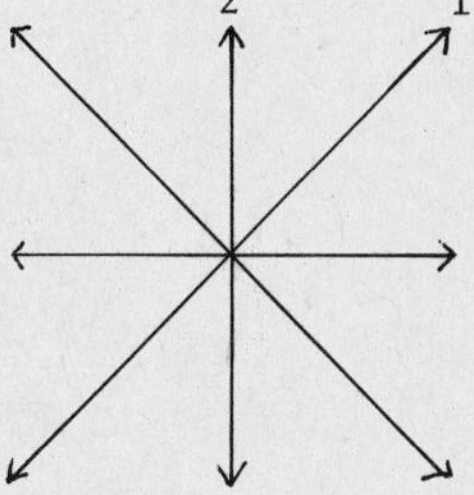

Fig.4

We label the elements of $\check{\Phi}_o$ such that $\check{\Phi}_o^+ = \{\ell_1, \ell_2, \ell_1+\ell_2, \ell_1-\ell_2\}$. From (2.6), we know that $\gamma^2_{\ell_1+\ell_2} = 1$, $\gamma_{\ell_1+\ell_2} = \gamma_{\ell_1-\ell_2}$. Thus $u_\sigma(\ell_1+\ell_2) = u_\sigma(\ell_1-\ell_2) = \pm 1$. Furthermore we can choose $u_\sigma(\ell_1), u_\sigma(\ell_2)$ such that $u_\sigma(\ell_1) = u_\sigma(\ell_1-\ell_2).u_\sigma(\ell_2)$. Consider first the case where $a = \ell_2$. We have $\check{\Phi}_o^+ - (a) = \{\ell_1, \ell_1+\ell_2, \ell_1-\ell_2\}$

$$\delta(\ell_1) = 1 , \qquad \delta(\ell_1+\ell_2) = u_\sigma(\ell_2) , \qquad \delta(\ell_1-\ell_2) = u_\sigma(\ell_2)^{-1} .$$

We compute w_{ℓ_1} . We have, by 4.8,

$$w_{\ell_1} = \frac{1}{2}(w_1 + w_2) \qquad \text{with} \qquad w_1 = \frac{1-\eta(\ell_1)u_\sigma(\ell_1)e^{-\ell_1}}{1+\eta(\ell_1)u_\sigma(\ell_1)e^{-\ell_1}}$$

$$w_2 = \frac{1-\eta(\ell_1)u_\sigma(\ell_1)^{-1}e^{-\ell_1}}{1+\eta(\ell_1)u_\sigma(\ell_1)^{-1}e^{-\ell_1}} .$$

Let us compute $f_{\ell_1}(\sigma',\eta')$. As $\check{\ell}_1$ and $\check{\ell}_2$ are strongly orthogonal $x_{\sigma'}(\ell_1) = x_\sigma(\ell_1)$, while, as (ℓ_1,ℓ_2) are not strongly orthogonal, $\eta'(\ell_1) = -\eta(\ell_1)$. Thus

4.10. $f_{\ell_1}(\sigma',\eta') = \frac{1}{2}(f^1+f^2)$

with $f_1 = \dfrac{1+u_\sigma(\ell_1)\eta(\ell_1)e^{-\ell_1}}{1-u_\sigma(\ell_1)\eta(\ell_1)e^{-\ell_1}} = \omega_1^{-1}$, $f_2 = \dfrac{1+u_\sigma(\ell_1)^{-1}\eta(\ell_1)e^{-\ell_1}}{1-u_\sigma(\ell_1)^{-1}\eta(\ell_1)e^{-\ell_1}} = \omega_2^{-1}$.

Let us compute $\omega_{\ell_1-\ell_2}$, $\omega_{\ell_1+\ell_2}$. We have

$$\delta(\ell_1-\ell_2)\eta(\ell_1-\ell_2)u_\sigma(\ell_1-\ell_2) = u_\sigma(\ell_2)^{-1}\eta(\ell_1)(-1)u_\sigma(\ell_1-\ell_2) = -\,u_\sigma(\ell_1)^{-1}\eta(\ell_1) .$$

Thus

$$\omega_{\ell_1-\ell_2} = \frac{1+u_\sigma(\ell_1)^{-1}\eta(\ell_1)e^{-\ell_1}}{1-u_\sigma(\ell_1)^{-1}\eta(\ell_1)e^{-\ell_1}} \quad \text{i.e.} \quad \omega_{\ell_1-\ell_2} = f_2 = \omega_2^{-1} .$$

Similarly

$$\delta(\ell_1+\ell_2)\eta(\ell_1+\ell_2)u_\sigma(\ell_1+\ell_2) = -\,u_\sigma(\ell_1)\eta(\ell_1)$$

$$\omega_{\ell_1+\ell_2} = \frac{1+u_\sigma(\ell_1)\eta(\ell_1)e^{-\ell_1}}{1-u_\sigma(\ell_1)\eta(\ell_1)e^{-\ell_1}} = f_1 = \omega_1^{-1} .$$

Thus we have :

$$\omega_{\ell_1-\ell_2}\,\omega_{\ell_1}\,\omega_{\ell_1+\ell_2} = \frac{1}{2}\,(\omega_2^{-1}(\omega_1+\omega_2)\omega_1^{-1}) = \frac{1}{2}\,(\omega_1^{-1}+\omega_2^{-1}) = \frac{1}{2}\,(f_1+f_2)$$
$$= f_{\ell_1}(\sigma',\eta') .$$

so the lemma holds.

It remains to consider the case where $a = \ell_1 - \ell_2$. We have $(\check{\Phi}_{o\alpha})^+ = \ell_1 + \ell_2 = d$ and $\check{\Phi}_o^+ - (a) = \{\ell_1,\ell_2,\ell_1+\ell_2\}$. We write $\theta_1 = u_\sigma(\ell_1)$, $\theta_2 = u_\sigma(\ell_2)$. We have $\theta_1\theta_2^{-1} = u_\sigma(\ell_1-\ell_2) = \varepsilon$ with $\varepsilon = \pm 1$.

Let us compute $f_d(\sigma',\eta')$. Recall (3.4) that if σ' is a representation of T' compatible with σ ,

$$u_{\sigma'}(\ell_1+\ell_2) = \theta_1^2\,\varepsilon = \theta_2^2\,\varepsilon .$$

As $\ell_1 + \ell_2$ and $\ell_1 - \ell_2$ are strongly orthogonal,

$$\eta'(\ell_1+\ell_2) = \eta(\ell_1+\ell_2) = \eta(\ell_1-\ell_2+2\ell_2) = -1 \text{ , as } a = (\ell_1-\ell_2) \text{ is simple in } \check{\Phi}+$$

Thus we have :

$$f_d(\sigma',\eta') = \frac{1}{2}\,(f_1+f_2) \qquad \text{with } f_1 = \frac{1+\theta_1^2\varepsilon e^{-d}}{1-\theta_1^2\varepsilon e^{-d}} \text{ , } f_2 = \frac{1+\theta_1^{-2}\varepsilon e^{-d}}{1-\theta_1^{-2}\varepsilon e^{-d}} .$$

Let us compute ω_{ℓ_1}, ω_{ℓ_2}, $\omega_{\ell_1+\ell_2}$. We have $\delta(\ell_1)\delta(\ell_2)^{-1} = \varepsilon = \theta_1\theta_2^{-1} = \delta(\ell_2)\delta(\ell_1)^{-1}$

$$\delta(\ell_1)\delta(\ell_2) = 1$$

so $\delta(\ell_1)^2 = \varepsilon$, $\delta(\ell_1)\theta_1 = \delta(\ell_2)\theta_2$, $\delta(\ell_1)\theta_1^{-1} = \delta(\ell_2)\theta_2^{-1}$.

We have $\ell_1 = \frac{1}{2}(\ell_1+\ell_2) + (\ell_1-\ell_2)$, $\ell_2 = \frac{1}{2}(\ell_1+\ell_2) - (\ell_1-\ell_2)$. So $\overline{\ell}_1 = \overline{\ell}_2 = \frac{d}{2}$.

As $\quad \ell_1 = \ell_2 + (\ell_1 - \ell_2)$, $\quad \eta(\ell_1) = -\eta(\ell_2)$.

So $w_{\ell_1} = \frac{1}{2}(u_1 + v_1)$, $\quad w_{\ell_2} = \frac{1}{2}(u_2 + v_2)$ with

$$u_1 = \frac{1-\eta(\ell_1)\delta(\ell_1)\theta_1 e^{-\frac{d}{2}}}{1+\eta(\ell_1)\delta(\ell_1)\theta_1 e^{-\frac{d}{2}}} , \quad v_1 = \frac{1-\eta(\ell_1)\delta(\ell_1)\theta_1^{-1} e^{-\frac{d}{2}}}{1+\eta(\ell_1)\delta(\ell_1)\theta_1^{-1} e^{-\frac{d}{2}}} , \quad u_2 = \frac{1+\eta(\ell_1)\delta(\ell_2)\theta_2 e^{-\frac{d}{2}}}{1-\eta(\ell_1)\delta(\ell_2)\theta_2 e^{-\frac{d}{2}}} = u_1^{-1}$$

$$v_2 = \frac{1+\eta(\ell_1)\delta(\ell_2)\theta_2^{-1} e^{-\frac{d}{2}}}{1-\eta(\ell_1)\delta(\ell_2)\theta_2^{-1} e^{-\frac{d}{2}}} = u_2^{-1} , \quad w_{\ell_1+\ell_2} = \frac{1+\varepsilon e^{-d}}{1-\varepsilon e^{-d}} .$$

So

$$w_{\ell_1}.w_{\ell_2}.w_{\ell_1+\ell_2} = \frac{1}{4}((u_1+v_1)(u_1^{-1}+v_1^{-1})w_{\ell_1+\ell_2}) = \frac{1}{4}(2w_{\ell_1+\ell_2}+u_1v_1^{-1}w_{\ell_1+\ell_2}+v_1u_1^{-1}w_{\ell_1+\ell_2}) .$$

Now

$$u_1v_1^{-1}w_{\ell_1+\ell_2} = \frac{1-x}{1+x}\,\frac{1+y}{1-y}\,\frac{1+xy}{1-xy} \quad \text{if} \quad x = \eta(\ell_1)\delta(\ell_1)\theta_1 e^{-\frac{d}{2}}$$

$$y = \eta(\ell_1)\delta(\ell_1)\theta_1^{-1} e^{-\frac{d}{2}}$$

$$= \frac{1-x}{1+x} + \frac{1+y}{1-y} - \frac{1+xy}{1-xy} = u_1 + v_1^{-1} - w_{\ell_1+\ell_2} .$$

Similarly :

$$v_1u_1^{-1}w_{\ell_1+\ell_2} = v_1 + u_1^{-1} - w_{\ell_1+\ell_2} , \quad \text{and} \quad w_{\ell_1}w_{\ell_2}w_{\ell_1+\ell_2} = \frac{1}{4}(u_1+u_1^{-1}+v_1+v_1^{-1}) .$$

Finally, as $\frac{1-x}{1+x} + \frac{1+x}{1-x} = 2\left(\frac{1+x^2}{1-x^2}\right)$, $\quad \frac{1}{2}(u_1+u_1^{-1}) = f_1$, $\quad \frac{1}{2}(v_1+v_1^{-1}) = f_2$,

$w_{\ell_1}.w_{\ell_2}.w_{\ell_1+\ell_2} = \frac{1}{2}(f_1+f_2) = f_\sigma(\sigma',\eta')$. So Lemma 4.11 holds.

It follows that $W = P(\sigma',\eta',\check{\Phi}'^+)$. Thus, for $\xi_1 \in \mathbb{Z}\check{\Phi}$,

$y_\delta(\overline{\xi}_1) = \eta'(\overline{\xi}_1)\, d(\sigma',\overline{\xi}_1,\check{\Phi}'^+)$. Similarly $y_{\delta^{-1}}(\overline{\xi}_1) = y_\delta(\overline{\xi}_1)$.

We thus obtain (4.6)

$$\eta(\xi_1)(d(\sigma,\xi_1,\check{\Phi}^+) + d(s_\alpha\sigma,s_\alpha\xi_1,\check{\Phi}^+)) = (\delta(\xi_1) + \delta(\xi_1)^{-1})\eta(\overline{\xi}_1)d(\sigma',\overline{\xi}_1,\check{\Phi}^+) .$$

We can restate this relation in the following theorem.

4.13 Theorem

Let $\xi_1 \in \mathbb{Z}\check{\Phi}$. Let us write $\xi_1 = ka + \overline{\xi}_1$.

a) If $\overline{\xi}_1 \notin \mathbb{Z}\check{\Phi}'$ then $d(\sigma,\xi_1,\check{\Phi}^+) + d(s_\alpha\sigma,s_\alpha\xi_1,\check{\Phi}^+) = 0$.

b) If $\overline{\xi}_1 \in \mathbb{Z}\check{\Phi}'$, then k is an integer and

$$\eta(\xi_1)(d(\sigma,\xi_1,\check{\Phi}^+) + d(s_\alpha\sigma,s_\alpha\xi_1,\check{\Phi}^+)) = \eta'(\overline{\xi}_1)(u_\sigma(a)^k + u_\sigma(a)^{-k})d(\sigma',\overline{\xi}_1;\check{\Phi}'^+) .$$

We can reformulate this theorem in the form stated as Theorem 4.1 : Let $\sigma \in \hat{T}$ and $\lambda_o \in t^*$ such that $\sigma(\exp X) = e^{i(\lambda_o,X)}\mathrm{Id}$ for $X \in t$. We have defined for $\xi = \xi_o + \xi_1$ in $h_{\mathbb{R}}$, with $\xi_o \in it$, $\xi_1 \in a$

$$d(\sigma,\xi,\check{\Phi}^+) = e^{\pi(\lambda_o,\xi_o)} d(\sigma,\xi_1,\check{\Phi}^+) \quad \text{where} \quad e^{\pi(\lambda_o,\xi_o)}\mathrm{Id} = \sigma(\exp -i\pi\xi_o) .$$

As s_α is a reflexion with respect to a real root

$$s_\alpha\xi = \xi_o + s_\alpha\xi_1 \quad \text{and}$$

$$d(\sigma,\xi,\check{\Phi}^+) + d(s_\alpha\sigma,s_\alpha\xi,\check{\Phi}^+) = e^{(\lambda_o,\xi_o)}(d(\sigma,\xi_1,\check{\Phi}^+) + d(s_\alpha\sigma,s_\alpha\xi_1,\check{\Phi}^+)) .$$

Let us write $\xi_1 = ka + \overline{\xi}_1$, with $\overline{\xi}_1 \in a'$. Then $c_\alpha\xi = ik(X_\alpha - X_{-\alpha}) + \xi_o + \overline{\xi}_1$

$$c_\alpha s_\alpha\xi = -ik(X_\alpha - X_{-\alpha}) + \xi_o + \overline{\xi}_1 .$$

If $\sigma' \in \hat{T}'$, by definition if $\overline{\xi}_1 \notin \mathbb{Z}\check{\Phi}'$

$$d(\sigma',c_\alpha\xi,\check{\Phi}'^+) = 0 .$$

If $\overline{\xi}_1 \in \mathbb{Z}\check{\Phi}'$, then k is an integer. It is clear that if σ' is a representation of T' compatible with σ , $\sigma'(\exp X) = \sigma(\exp X) = e^{i(\lambda_o,X)}\mathrm{Id}$, for $X \in t$

$$\sigma'(\exp \pi(X_\alpha - X_{-\alpha})) = u_\sigma(a)\mathrm{Id} \text{ or } u_\sigma(a)^{-1}\mathrm{Id} .$$

Thus we obtain :

$$d(\sigma',c_\alpha\xi;\check{\Phi}'^+) + d(\sigma',c_\alpha s_\alpha\xi;\check{\Phi}'^+) = (u_\sigma(a)^{-k} + u_\sigma(a)^k)e^{(\lambda_o,\xi_o)} d(\sigma',\overline{\xi}_1,\check{\Phi}'^+) .$$

So Theorem (4.1) is a reformulation of Theorem (4.13).

Bibliography

[Bo] Bouaziz, A.; Sur les caractères des groupes de Lie réductifs non connexes, to appear in Journal of Functional Analysis.

[Do] Dourmashkin, P., A Poisson-Plancherel formula for the universal covering group with Lie algebra of type Bn, Thesis M.I.T. 1984, to appear.

[Du-Ve] Duflo, M., Vergne, M., La formule de Plancherel des groupes de Lie semi-simples, Preprint 1985.

[Ha-1] Harish-Chandra, Discrete series for semi-simple Lie groups I, Acta Math. 113, 1965, 241-318.

[Ha-2] Harish-Chandra, Harmonic analysis on real reductive groups III. The Maass-Selberg relations and the Plancherel formula, Ann. of Math. 104, 1976, 117-201.

[He] Herb, R., Discrete series characters and Fourier inversion on semi-simple real Lie groups, TAMS, 277, 1983, 241-261.

[Ka-Pe] Kac,V., Peterson,D., Infinite-dimensional Lie algebras, theta functions and modular forms, Adv. in Math., Vol. 53, No.2, August 1984, 125-264.

[Ve] Vergne, M., A Poisson-Plancherel formula for semi-simple Lie groups, Ann. of Math. 115, 1982, 639-666.

A cohomological method for the determination of limit multiplicities

Jürgen Rohlfs[(1)] and Birgit Speh[(2)]

(1) Katholische Universität Eichstätt
Ostenstr. 26-28
8078 Eichstätt
Fed. Rep. of Germany

(2) Cornell University
Department of Mathematics
Ithaca, NY 14850
U.S.A.

In the first part of this paper we give a simple proof of the result of DeGeorge and Wallach on limit multiplicities of discrete series representations for cocompact lattices,[D-W]. Our proof depends on properties of the Euler-Poincaré characteristic of lattices. In the second part we indicat how our method extends to general lattices and show that the average of all the limit multiplicities of discret series representations with the same regular parameter is the corresponding formal degree. In contrast to Clozel's approach [C] we do not use the Arthur-Selberg trace formula.

(2) Sloan fellow, partially supported by NSF-Grant DMS - 8501793

1.1. We fix some notation. Let $\underline{G}/\mathbb{Q}$ be a semi-simple algebraic group defined over $\mathbb{Q}$ and abbreviate $G = \underline{G}(\mathbb{R})^o$ for the connected component of the group of real points of $\underline{G}$. We denote by K a maximal compact subgroup of G and write $\mathfrak{g}$ resp. $\mathfrak{k}$ for the Lie algebra of G resp. K. We assume throughout that $\mathrm{rank}(K) = \mathrm{rank}(G)$, i.e. that discrete series representations exist and we write $\hat{G}_d$ for the set of equivalence classes of discrete series representations. We fix a left invariant measure v on G and denote by d_ω the formal degree of $\omega \in \hat{G}_d$ with respect to the choice of v. We recall that the measure $d_\omega v$ is independent of the choice of v. As usual we write $\hat{G}$ for the set of equivalence classes of irreducible unitary representations of G. If $\Gamma \subset G$ is a lattice we write $v(\Gamma\backslash G)$ for the volume of $\Gamma\backslash G$ in the measure induced by v on $\Gamma\backslash G$. By $\{\Gamma_i\}_{i\in\mathbb{N}}$ we always denote a sequence $\Gamma_{i+1} \subset \Gamma_i$ of torsion free lattices with $\bigcap_{i=0}^{\infty} \Gamma_i = \{1\}$ such that Γ_i is normal in $\Gamma_o = \Gamma$. We call such a sequence of lattices a tower of lattices.

We will use the following result of DeGeorge and Wallach,

<u>1.2. Proposition. Let</u> $\{\Gamma_i\}_{i\in\mathbb{N}}$ <u>be a tower of cocompact lattices and denote by</u> $m(\omega,\Gamma_i)$ <u>the multiplicity of</u> $\omega \in \hat{G}$ <u>in the space</u> $L^2(\Gamma\backslash G)$ <u>of square integrable functions.</u>

<u>Then</u>
$$\limsup_{i\to\infty} \frac{m(\omega,\Gamma_i)}{v(\Gamma_i\backslash G)} \le \begin{cases} d_\omega & \text{if } \omega \in \hat{G}_d \\ 0 & \text{if } \omega \notin \hat{G}_d \end{cases}.$$

For the equality for $\omega \notin \hat{G}_d$, see [D - W : Cor. 3.3]. The inequality follows from [D - W : Cor. 3.2] and the orthogonality relation for matrix coefficients of discrete series representations.

Let V be an irreducible finite-dimensional representation of G with infinitesimal character χ_V. We denote by χ_ω the infinitesimal character of $\omega \in \hat{G}$ and abbreviate $\hat{G}(V) = \{\omega \in \hat{G} | \chi_\omega = \chi_{\tilde{V}}\}$ and $\hat{G}_d(V) = \hat{G}_d \cap \hat{G}(V)$, where $\tilde{V}$ is the contragredient representation to V.

Since $\operatorname{rank}(K) = \operatorname{rank}(G)$ the absolute value $|\chi(\Gamma_i)|$ of the Euler-Poincaré characteristic $\chi(\Gamma_i)$ of Γ_i is not zero, [S].

1.3. Proposition. Let V be a finite-dimensional irreducible representation of G. Then

$$\lim_{i\to\infty} \sum_{\omega\in\hat{G}_d(V)} \frac{m(\omega,\Gamma_i)}{|\chi(\Gamma_i)|} = \dim V .$$

Proof: We use Matsuhima's formula

$$H^\bullet(\Gamma_i, V) = \sum_{\omega\in\hat{G}(V)} H^\bullet(\mathfrak{g}, K, \omega \otimes V)^{m(\omega,\Gamma_i)} ,$$

see [B - W]. Here $H^\bullet(\Gamma_i, V)$ is the cohomology of the group Γ_i with coefficients V. It is known that

$$\chi(\Gamma_i, V) = \sum_{i=0}^{\infty} (-1)^i \dim H^i(\Gamma_i, V) = \chi(\Gamma_i) \dim V ,$$

see [S]. Moreover

$$\chi(\omega \otimes V) = \sum_{i=0}^{\infty} (-1)^i \dim H^i(\mathfrak{g}, K, \omega \otimes V) = (-1)^\ell$$

for $\omega \in \hat{G}_d(V)$, see [B - W], where $\ell = \frac{1}{2}\dim(G/K)$. Hence

$$\dim V = \frac{\chi(\Gamma_i, V)}{\chi(\Gamma_i)} = \lim_{i\to\infty} \frac{\chi(\Gamma_i, V)}{\chi(\Gamma_i)} = \lim_{i\to\infty} \sum_{\omega\in\hat{G}(V)} \frac{m(\omega,\Gamma_i)}{\chi(\Gamma_i)} \chi(\omega \otimes V) .$$

Since $\chi(\omega \otimes V) = (-1)^\ell$ for $\omega \in \hat{G}_d(V)$ and since $\chi(\Gamma_i)$ is proportional to the volume of $\Gamma_i \backslash G$ by the classical Gauß-Bonnet theorem, the result we claimed follows from 1.2. q.e.d.

Since G has a compact Cartan subgroup there exists an uniquely determined left-invariant measure $0 \neq e_\chi$ on G such that for all torsion-free cocompact lattices Γ in G we have

$$\int_{\Gamma\backslash G} e_\chi = \chi(\Gamma) \quad \text{, see [S] .}$$

<u>1.4. Proposition.</u> We have an equality of measures

$$(-1)^\ell \dim V \, e_\chi = \sum_{\omega\in\hat{G}_d(V)} d_\omega v \quad .$$

<u>Proof:</u> Let $\mathfrak{g} = \mathfrak{k} \oplus \mathfrak{p}$ be the Cartan decomposition corresponding to $K \subset G$ with Cartan involution θ and denote by $\mu = \mu_G$ the left G - and right K - invariant measure on G determined by the Riemann-metric $e(X,Y) = - B(X,\theta Y)$, $X,Y \in \mathfrak{g}$ on $\mathfrak{g}$, where B is the Killing form on $\mathfrak{g}$. Consider the real connected compact Lie group $G_u \subset G(\mathbb{C})$ corresponding to the subalgebra $\mathfrak{k} \oplus i\mathfrak{p} \subset \mathfrak{g} \otimes \mathbb{C}$. If θ_u is the Cartan involution corresponding to $G_u \subset G(\mathbb{C})$, i.e. $\theta_u(X \otimes z) = \theta(X) \otimes \overline{z}$, $z \in \mathbb{C}$, $X \in \mathfrak{g}$ and $e_{\mathbb{C}}$ the corresponding Riemann metric, then $e_{\mathbb{C}}(X,Y) = 2\, e(X,Y)$ for $X,Y \in \mathfrak{g}$. Let $\mu_{\mathbb{C}}$ be the invariant measure on G_u corresponding to $e_{\mathbb{C}}$. The argument given in [R : § 4] for $G = SO(r,s)(\mathbb{R})$ now shows without difficulty that

$$(-1)^\ell e_\chi = \frac{|W(G_u),T)|}{|W(K,T)|} \frac{2^{\dim G/2}}{\mu_{\mathbb{C}}(G_u)} \mu_G \quad ,$$

where $W(G_u,T)$ is the Weyl group of T in G_u and $W(K,T) = W(G,T)$ is the Weylgroup of T in K and $T \subset K$ is a compact Cartan subgroup.

If $G = KAN$ is an Iwasawa decomposition let λ_G be the left invariant measure given by $\lambda_G = e^{2\rho(\log a)} dk\, da\, dn$ as in [H-CH : I, § 7] where dk is an normalized invariant measure of K and $da = \mu_A$, $dn = \mu_N$ are given by the Riemann metric e. Then $\mu_G = 2^{-v/2} \mu_K(K) \lambda_G$ [H-CH: I § 37], where μ_K is again given by the Riemann metric e and $v = \dim(G/K) - \operatorname{rank}(G/K)$. Hence

$$(-1)^{\ell} e_\chi = \frac{|W(G_u,T)|}{|W(G,T)|} \frac{\mu(K)\, 2^{(\dim G - v)/2}}{\mu_{\mathbb{C}}(G_u)} \lambda_G .$$

According to [H-CH: III p. 164 Cor., I, § 27] the formal degree d_ω of $\omega \in \hat{G}_d$ with respect to measure λ_G is

$$d_\omega = C_G^{-1} |W(G,T)| \prod_{\alpha>0} (\tau,\alpha) ,$$

where $0 < \alpha \in \Delta^+(\mathfrak{g})$ are positive roots of $\mathfrak{g}$ and $\chi_V = \chi_\tau$, where τ is the Harish Chandra parameter of ω, which is a dominant weight with respect to the positive root system $\Delta^+(\mathfrak{g})$. The value of C_G is given in [H-CH I, § 37, Lemma 3] and $C_G = (2\pi)^r\, 2^{v/2}\, \mu(K)^{-1}\, \mu(T)\, |W(G,T)|$ where $r = \frac{1}{2} \dim(G/T)$.

Now all $\omega \in G_d$ with $\chi_\omega = \chi_V$ have the some formal degree and the number of such representation is $\frac{|W(G_u,T)|}{|W(G,T)|}$, see [H-Ch:III, §23, Thm. 1].

Therefore

$$\sum_{\omega \in \hat{G}_d(V)} d_\omega = \frac{|W(G_u,T)|}{|W(G,T)|} \frac{\mu(K)}{(2\pi)^r\, 2^{v/2}\, \mu(T)} \prod_{\alpha \in \Delta^+(\mathfrak{g})} (\tau,\alpha) .$$

Now we use Weyl's dimension formula $\dim V = \prod_{\alpha\in\Delta^+(g)} (\tau,\alpha)/(\rho,\alpha)$ and the formula $\mu(K)/\mu(T) = (2\pi)^{r_K} / \prod_{\alpha\in\Delta^+(k)} (\rho_K,\alpha)^{-1}$, see [H-CH, I § 37 Lemma 4] where $r_K = 1/2 \dim(K/T)$ and $\rho_K = 1/2 \sum_{\alpha\in\Delta^+(k)} \alpha$. We therefore have to prove

$$\frac{|\mu(K)|}{|\mu_{\mathbb{C}}(G_u)|} = 2^{-\dim G/2} (2\pi)^{-r+r_K} \prod_{\alpha\in\Delta^+(g)} (\tau,\alpha) \Big/ \prod_{\alpha\in\Delta^+(k)} (\rho_K,\alpha) .$$

But $\mu(K)/\mu(T) = (2\pi)^{r_K} \prod_{\alpha\in\Delta^+(k)} (\rho_K,\alpha)^{-1}$ as above and

$\mu_{\mathbb{C}}(G_u)/\mu_{\mathbb{C}}(T) = (2\pi)^r \prod_{\alpha\in\Delta^+(g)} (\tau,\alpha)_{\mathbb{C}}^{-1}$, see [H-CH : I § 37 Lemma 4] .

Now $\mu_{\mathbb{C}}(T) = 2^{\dim T/2} \mu(T)$ and $(\tau,\alpha)_{\mathbb{C}} = 2\,(\tau,\alpha)$ for all $\alpha \in \Delta^+(g)$ and the result follows. q.e.d.

If we now choose $v = (-1)^{\ell} e_{\chi}$ in Prop. 1.4 we get $\dim V = \sum_{\omega\in\hat{G}_d(V)} d_\omega$. Using Prop. 1.2 and 1.3 we see

$$\dim V = \lim_{i\to\infty} \sum_{\omega\in\hat{G}_d(V)} \frac{m(\omega,\Gamma_i)}{|\chi(\Gamma_i)|} \le \sum_{\omega\in\hat{G}_d(V)} d_\omega .$$

Hence $\lim_{i\to\infty} \frac{m(\omega,\Gamma_i)}{|\chi(\Gamma_i)|} = d_\omega$ and we obtain

<u>1.5. Corollary.</u> (DeGeorge - Wallach). <u>Let</u> $\{\Gamma_i\}_{i\in\mathbb{N}}$ <u>be a tower of cocompact lattices</u>. <u>Then</u>

$$\lim_{i\to\infty} \frac{m(\omega,\Gamma_i)}{v(\Gamma_i\backslash G)} = \begin{cases} d_\omega & \text{if } \omega \in \hat{G}_d \\ 0 & \text{if } \omega \notin \hat{G}_d \end{cases} .$$

<u>Remark:</u> If $\omega \in \hat{G}_d(V)$ is integrable Langlands shows that $m(\omega,\Gamma_i) = d_\omega \operatorname{vol}(\Gamma_i \backslash G)$, see [L] . It is known that this equality is false in general for non integrable $\omega \in \hat{G}_d(V)$.

2.1. Since we have an embedding $\underline{G}/\mathbb{Q} \longrightarrow GL_n/\mathbb{Q}$ over $\mathbb{Q}$ for some n, we can define congruence subgroups Γ_i which are the intersection of $G = \underline{G}(\mathbb{R})$ with the full congruence subgroup mod i of $GL_n(\mathbb{Z})$. Then $\{\Gamma_i\}_{i\in\mathbb{N}}$ is a tower in the sense of 1.1; it is well known that $\Gamma_i \backslash G$ has finite volume and that Γ_i is torsionfree for $i \geq 3$. We assume now that the $\mathbb{Q}$-rank $(\underline{G})$ is non zero. Then $\Gamma_i \backslash G$ is not compact.

Let $\overline{\Gamma_i \backslash G/K}$ be the Borel-Serre compactification of $\Gamma_i \backslash G/K$, see [B - S]. Then the local system $\tilde{V}$ on $\Gamma_i \backslash G/K$ given by the representation V of G extends to a local system $\tilde{V}$ on the compactification and it is well-known that naturally

$$H^\cdot(\Gamma_i,V) \xrightarrow{\sim} H^\cdot(\Gamma_i\backslash G/K,\tilde{V}) \xrightarrow{\sim} H^\cdot(\overline{\Gamma_i\backslash G/K},\tilde{V}) \xrightarrow{\sim}$$

$$\xrightarrow{\sim} H^\cdot(\mathfrak{g},K,C^\infty(\Gamma_i\backslash G) \otimes V) ,$$

where $C^\infty(\Gamma_i\backslash G)$ is the space of C^∞-functions on $\Gamma_i\backslash G$.

2.2. Denote by $L^2_{dis}(\Gamma_i\backslash G)$ resp. $L^2_{cusp}(\Gamma_i\backslash G)$ the discrete resp. cuspidal part of $L^2(\Gamma_i\backslash G)$ and by $L^2_{dis}(\Gamma_i\backslash G)^\infty$ resp. $L^2_{cusp}(\Gamma_i\backslash G)^\infty$ the intersections with $C^\infty(\Gamma_i\backslash G)$ Then we have natural maps

$$H^\cdot(\mathfrak{g},K, L^2_{cusp}(\Gamma_i\backslash G)^\infty \otimes V) \longrightarrow H^\cdot(\mathfrak{g},K,C^\infty(\Gamma_i\backslash G) \otimes V)$$

resp.

$$H^\cdot(\mathfrak{g},K,L^2_{dis}(\Gamma_i\backslash G)^\infty \otimes V) \longrightarrow H^\cdot(\mathfrak{g},K,C^\infty(\Gamma_i\backslash G) \otimes V) ,$$

whose image we denote by $H^{\cdot}_{cusp}(\Gamma_i,V)$ resp. $H^{\cdot}_{dis}(\Gamma_i,V)$.

We have an exact sequence

$$\begin{array}{c} H^{\cdot}_c(\Gamma_i\backslash G/K,\tilde{V}) \longrightarrow H^{\cdot}(\overline{\Gamma_i\backslash G/K},\tilde{V}) \longrightarrow H^{\cdot}(\partial(\overline{\Gamma_i\backslash G/K}),\tilde{V}) , \\ \uparrow\_\_\_\_\_\_\_\_\_\_\_\_\_\_\_\_\_\_\_\_\_\_\_\_\_\_\_\_\_\_\_\_\_\_\_\_\_\_\_\_\_\_\_\_\_\_\_\_| \end{array}$$

where $\partial(\overline{\Gamma_i\backslash G/K})$ denotes the boundary of the Borel-Serre compactification and $H^{\cdot}_c(\, , \,)$ stands for cohomology with compact supports. We denote the image of $H^{\cdot}_c(\Gamma_i\backslash G/K,\tilde{V})$ in $H^{\cdot}(\Gamma_i,V)$ by $H^{\cdot}_!(\Gamma_i,V)$. Therefore we have inclusions

$$H^{\cdot}_{cusp}(\Gamma_i,V) \hookrightarrow H^{\cdot}_!(\Gamma_i,V) \hookrightarrow H^{\cdot}_{dis}(\Gamma_i,V) \hookrightarrow H^{\cdot}(\Gamma_i,V)$$

and it is well known that there is an isomorphism.

$$H^{\cdot}(\mathfrak{g},K,L^2_{cusp}(\Gamma_i\backslash G)^{\infty} \otimes V) \xrightarrow{\sim} H^{\cdot}_{cusp}(\Gamma_i,V) .$$

In obvious notation we denote the Euler-Poincaré characteristic of the above subspaces of $H^{\cdot}(\Gamma,V)$ by $\chi_{cusp}(\Gamma_i,V)$ resp. $\chi_!(\Gamma_i,V)$ resp. $\chi_{dis}(\Gamma_i,V)$.

2.3. As in the first part we denote by $v(\Gamma_i\backslash G)$ the volume of $\Gamma_i\backslash G$ with respect to the left invariant measure v . In [R - S] we prove that

$$\lim_{i\to\infty} v(\Gamma_i\backslash G)^{-1} \dim H^{\cdot}(\partial(\overline{\Gamma_i\backslash G/K}),\tilde{V}) = 0$$

and that

$$\lim_{i\to\infty} v(\Gamma_i\backslash G)^{-1} (\dim H^{\cdot}_{cusp}(\Gamma_i,V) - \dim H^{\cdot}_{dis}(\Gamma_i,V)) = 0 .$$

We deduce that $\lim_{i\to\infty} v(\Gamma_i\backslash G)^{-1} (\chi(\Gamma_i,V) - \chi_{cusp}(\Gamma_i,V)) = 0$ and the same holds if χ_{cusp} is replaced by $\chi_!$.

As a consequence we have:

2.4. Theorem. *Denote by* $m(\omega,\Gamma_i)$ *the multiplicity of* $\omega \in \hat{G}$ *in* $L^2_{cusp}(\Gamma_i\backslash G)$. *Then*

$$\lim_{i\to\infty} \sum_{\omega\in\hat{G}(V)} \frac{m(\omega,\Gamma_i)}{\chi(\Gamma_i)} \chi(\omega\otimes V) = \dim V .$$

The proof uses Harder's Gauß-Bonnet theorem [H], which says that $|\chi(\Gamma_i)|$ is proportional to $v(\Gamma_i\backslash G)$ with a proportionality constant depending only on G and v .

If we assume that the highest weight of V is regular then a $\omega\in\hat{G}(V)$ with $\chi(\omega\otimes V) \neq 0$ is necessarily a discrete series representation, [V - Z]. Hence using Prop. 1.4 we have:

2.5. Corollary. *If the highest weight of* V *is regular, then*

$$\frac{1}{|\hat{G}_d(V)|} \sum_{\omega\in\hat{G}_d(V)} \lim_{i\to\infty} \frac{m(\omega,\Gamma_i)}{v(\Gamma_i\backslash G)} = d_\omega$$

Here we use that all $\omega \in G_d(V)$ have the same formal degree d_ω which of course depends on the choice of V .

Remarks

(i) If we were able to prove Prop. 1.2 in the non cocompact case our method would give the desired limit multiplicity

$$\lim_{i\to\infty} \frac{m(\omega,\Gamma_i)}{v(\Gamma_i\backslash G)} = d_\omega \quad \text{for all} \quad \omega \in \hat{G}_d \ .$$

Unfortunately Clozel's approach [C] gives the analogue of 1.2 only on a subspace of the isotypical component

$\omega \otimes \mathrm{Hom}_{(\mathfrak{g},K)}(\omega, L^2_{cusp}(\Gamma_i\backslash G))$ of $\omega \in \hat{G}_d$ in $L^2(\Gamma_i\backslash G)$.

(ii) Clozel [C] shows that $\lim\limits_{i\to\infty} \frac{m(\omega,\Gamma_i)}{v(\Gamma_i\backslash G)} \geq d_\omega\, \varepsilon$ for all $\omega \in \hat{G}_d$ and some fixed $\varepsilon > 0$. We have for V with a regular highest weight that at least for one $\omega \in \hat{G}_d(V)$

$$\lim_{i\to\infty} \frac{m(\omega,\Gamma_i)}{v(\Gamma_i\backslash G)} \geq d_\omega \ .$$

(iii) Using results of [Sp] it is possible to simplify the considerations in part two considerably in the $\mathbb{Q}$-rank 1 case.

References

[B - S] A. Borel, J.-P. Serre, Corners and arithmetic groups, Comm. Math. Helv. 48 (1973), 436-491.

[B - W] A. Borel, N. Wallach, Continuous cohomology, discrete subgroups, and representations of reductive groups, Annals of Mathematics Studies 94, Princeton University Press (1980).

[C] L. Clozel, On limit multiplicities of discrete series representations in the space of automorphic forms, preprint (1985).

[D - W] D. DeGeorge, N. Wallach, Limit formulas for multiplicities in $L^2(\Gamma \backslash G)$. Ann. of Math. 107 (1978), 133 - 150.

[H] G. Harder, A Gauss-Bonnet formula for discrete arithmetically defined groups. Ann. Sci. Ec. Norm. Sup. (4) (1971), 409 - 455.

[H-Ch] Harish-Chandra, Harmonic analysis on real reductive groups I, III, Collected Papers, Vol. IV, Springer Verlag 1984.

[L] R. Langlands, Dimensions of spaces of automorphic forms, Proc. Symp. Pure Math. IX. A.M.S. (1966), 253-257.

[R] J. Rohlfs. The Lefschetz number of an involution on the space of classes of positive definite quadratic forms. Comment. Math. Helv. 56 (1981), 272-296.

[R-S] J. Rohlfs, B. Speh, On limit multiplicities of representations with cohomology in the cuspidal spectrum, Manuscript 1985.

[S] J.-P. Serre, Cohomologie Des Groupes Discrets, In: Prospects in Mathematices, Annals of Mathematics Studies 70, Princeton University Press 1971.

[Sp] B. Speh, Automorphic representations and the Euler-Poincaré characteristic of arithmetic groups, Preprint 1984.

[V - Z] D. Vogan, G. Zuckerman, Unitary representations with non-zero cohomology, Compositio Math. 53 (1984), 51 - 90.

SPRINGER REPRESENTATIONS AND COHERENT CONTINUATION REPRESENTATIONS OF WEYL GROUPS

W. Rossmann
University of Ottawa

1. Characters as contour integrals. A character of a complex, connected, semisimple group G with infinitesimal character $\underline{\lambda}$ is of the form

$$\Theta(\underline{\lambda}) = \frac{1}{\underline{\Delta}} \sum_{\underline{w}\in\underline{W}} \underline{a}(\underline{w})e^{\underline{\lambda w}} \qquad (1)$$

As usual, this formula represents $\Theta(\underline{\lambda})$ on a fixed Cartan subgroup H of G, $\underline{\lambda}$ is an element of the complexification $\underline{h}^*$ of the dual of the Lie algebra h of H which exponentiates to a global character $e^{\underline{\lambda}}$ of H; $\underline{W}$ is the Weyl group of $\underline{h}$ in the complexification $\underline{g}$ of the Lie algebra g of G; $\underline{a}(\underline{w})$ is an integer subject to $\underline{a}(\underline{wy}) = \underline{a}(\underline{w})$ for y in the Weyl group W of h in g; $\underline{\Delta}$ is the Weyl denominator for the Cartan subgroup $\underline{H}$ in the complexification $\underline{G}$ of G:

$$\underline{\Delta} = \Pi_{\underline{\alpha} > 0}(e^{\underline{\alpha}/2} - e^{-\underline{\alpha}/2})$$

(Objects associated to the complexification $\underline{G}$ of G are generally underlined. Of course, since G is itself complex on has $\underline{G} = G\times G$ with G embedded as $g = (g,\bar{g})$, the bar denoting the conjugation in G with respect to its real form generated by the root vectors of h. Similarly $\underline{g} = g\times g$, $\underline{h} = h\times h$, $\underline{W} = W\times W$ etc.) Conversely, every $\Theta(\underline{\lambda})$ of this form represents a character, with the understanding that "character" means "virtual character", here and elsewhere. The function Θ of $\underline{\lambda} \in \underline{h}^*$ defined by formula (1) for a given $\underline{a} \in \mathbb{Z}[\underline{W}/W]$ is referred to as a *coherent family of characters*. $\Theta(\underline{\lambda})$ exists as a distribution in a neighborhood of the identity in G where exp: $g \to G$ has an inverse for *all* $\underline{\lambda} \in \underline{h}^*$. (This notion of "coherent family", which will be found convenient here, differs only inessentially form the usual one.) Note that the formula (1) may also be written as

$$\Theta(\underline{\lambda}) = \sum_{\underline{w}\in\underline{W}/W} \underline{a}(\underline{w})X_{\underline{w}}(\underline{\lambda}) \qquad (2)$$

where

$$X_{\underline{w}}(\underline{\lambda}) = \frac{1}{\underline{\Delta}} \sum_{y\in W} e^{\underline{\lambda wy}} \qquad (3)$$

$X_{\underline{w}}(\underline{\lambda}) = X_1(\underline{\lambda w})$ is the principal series character with parameter $\underline{\lambda w}$. ($X_1(\underline{\lambda}) = L(p,q)$ in [DUFLO] when $\underline{\lambda} = (p,q)$ in $\underline{h}^* = h^*\times h^*$.) Since $X_{\underline{w}}$ depends only on $\underline{w}W$, it is sufficient to consider $\underline{w}$ of the form (w,1) with $w \in W$ for which we introduce the notation

$$X_w = X_{(w,1)}.$$

Coherent families of characters admit a representation by contour integrals, as we shall now explain (see [ROSSMANN]). Let $\mathcal{B}$ = {Borel subalgebras b of g} be the flag manifold of G, $\underline{\mathcal{B}} = \mathcal{B}\times\mathcal{B}$ that of $\underline{G} = G\times G$. The cotangent bundle $\mathcal{B}^*$ of $\mathcal{B}$ consists of pairs $(\underline{b},\nu)$ with $\underline{b}$ in $\underline{\mathcal{B}}$ and $\underline{\nu} \in \underline{b}^{\perp}$ = the nilpotent radical of $\underline{b}$ when one identifies $\underline{g}^*$ with g. Introduce the subvariety S of $\underline{\mathcal{B}}^*$ consisting of pairs $(\underline{b},\nu)$ in $\underline{\mathcal{B}}^*$ with ν in $g^* \cap \underline{b}^{\perp}$. If one writes $\underline{b} = (b_1,b_2)$, this means that $\nu \in b_1 \cap \bar{b}_2{}^{\perp}$. Fix a compact form K of G and a Borel subalgebra $b_1 = h + n_1$ of g with $\bar{K} = K$ and $\bar{b}_1 = b_1$, let $\underline{K} = K\times K$ in $\underline{G}$ and $\underline{b}_1 = b_1\times b_1$ in $\underline{g}$. For each $\underline{\lambda} \in \underline{h}^*$ define a map $\underline{\pi}_{\lambda}$ from $\underline{\mathcal{B}}^*$ to the closure $\underline{\Omega}_{\lambda}$ of the regular $\underline{G}$-orbit in $\underline{g}^*$ with $\underline{\lambda} \in \underline{\Omega}_{\lambda}$ ("regular" means that $\underline{\Omega}_{\lambda}$ has maximal dimension, namely $\dim_{\mathbb{C}} \underline{\Omega}_{\lambda} = \dim_{\mathbb{C}} \underline{\mathcal{B}}^* = 4n$, $n = \dim_{\mathbb{C}} \mathcal{B}$):

$$\underline{\pi}_{\lambda}\colon \underline{\mathcal{B}}^* \to \underline{\Omega}_{\lambda},\ \underline{k}\cdot(\underline{b}_1,\underline{\nu}) \ \to\ \underline{k}\cdot(\underline{\lambda} + \underline{\nu}) \text{ for } \underline{k}\in\underline{K},\ \underline{\nu}\in\underline{b}_1{}^{\perp}.$$

When $\underline{\lambda}$ is regular in $\underline{h}^*$, then $\underline{\pi}_{\lambda}$ is bijective; when $\underline{\lambda} = 0$, then $\underline{\pi}_0\colon\underline{\mathcal{B}}^* \to \Omega_0 = \underline{N}$ is Springer's resolution of the nilpotent variety $\underline{N}$ in $\underline{g}^*$. Note that $\underline{\pi}_{\lambda}$ is not holomorphic for $\underline{\lambda} \neq 0$, and is not $\underline{G}$-equivariant, only $\underline{K}$-equivariant. If one specializes the results from [ROSSMANN] (which apply to real G) to the present situation one gets the following

THEOREM 1. *For any coherent family of characters* Θ *there is a unique* 4n-*cycle* Γ *on* S *so that for all regular* $\underline{\lambda} \in \underline{h}^*$,

$$\Theta(\underline{\lambda}) = \frac{1}{J} \int_{\underline{\pi}_{\lambda}\Gamma} e^{i\underline{\xi}}\ \underline{\mu}_{\lambda}(d\underline{\xi}) ; \qquad (4)$$

$J = \Pi_{\alpha>0} (e^{\alpha/2} - e^{-\alpha/2})/\alpha$ and $\underline{\mu}_{\lambda} = \sigma_{\lambda}{}^{2n}/(2n)!\ (2\pi)^{2n}$, σ_{λ} *the canonical (complex) symplectic form on* $\underline{\Omega}_{\lambda}$. *The map* $\Gamma \to \Theta$ *given by* (4) *defines an isomorphism of the homology group* $H_{4n}(S)$ *onto the group* CH(G) *of coherent families of characters of* G.

The integral in (4) is understood in the sense of distributions on a neighborhood of the identity in G (or, in exponential coordinates, as a Fourier transform of a distribution on g). The integral depends only on the homology class of Γ (transferred to a cycle $\underline{\pi}_{\lambda}\Gamma$ on $\underline{\Omega}_{\lambda}$ via $\underline{\pi}_{\lambda}$). The formula (4) can also be viewed as providing an isomorphism $\Gamma \to \Theta(\underline{\lambda})$ of $H_{4n}(S)$ onto the group $CH_{\underline{\lambda}}(G)$ of characters $\Theta(\underline{\lambda})$ with a fixed regular infinitesimal character $\underline{\lambda}$ (because a coherent family is evidently uniquely determined by its value at any regular $\underline{\lambda}$).

The homology group $H_{4n}(S)$ has as $\mathbb{Z}$-basis the fundamental cycles of the components of the complex variety S. As shown by STEINBERG, these are the closures S_w of the parts of S over the G-orbits $G\cdot\underline{b}_w$, $w\in\underline{W}/W$, in $\underline{\mathcal{B}}$ (here $\underline{b}_w = \underline{w}^{-1}\cdot\underline{b}_1$, $\underline{w} \in \underline{W}$; we also write S_w for $S_{(w,1)}$) and S_w for its class in $H_{4n}(S)$. Other $\mathbb{Z}$-bases for $H_{4n}(S)$ consist of the "contours" corresponding to the principal series families X_w or to the families Y_w which give the canonical irreducible subquotients of X_w for positive integral $\underline{\lambda}$. (For positive integral $\underline{\lambda}$, $Y_w(\underline{\lambda}) = V(w^{-1}p,q)$ when $\underline{\lambda} = (p,q)$ in DUFLO's notation). It is an open problem to find the formulas which express the X_w and Y_w in terms of the S_w, or the coefficients $\underline{a}(\underline{w})$ which give the S_w via formulas (1) and (4). (The relations between the X_w and the Y_w are given by the formulas of KAZHDAN and LUSZTIG [1979].)

2. Representations of Weyl groups. The Weyl group $\underline{W} = W\times W$ of $\mathfrak{h}$, $\mathfrak{g}$ acts on coherent families of characters of G in an obvious way:

$$\underline{w}\cdot\Theta(\underline{\lambda}) = \Theta(\underline{\lambda w})$$

If one identifies Θ with the corresponding $\underline{a}\in \mathbb{Z}[\underline{W}/W]$, this is just the regular representation of $\underline{W}$ on $\mathbb{Z}[\underline{W}/W]$, or equivalently the biregular representation of $W\times W$ on $\mathbb{Z}[W]$. On the other hand, there is a representation of $\underline{W}$ on $H_{4n}(S)$ defined by SPRINGER in connection with his construction of the irreducible represention of Weyl groups (1976, 1978), according to KAZHDAN and LUSZTIG [1980], who gave another construction of this representation and proved that it is isomorphic to the biregular representation of $W\times W$. The following theorem confirms what cannot be otherwise:

THEOREM 2. *The map* $H_{4n}(S) \to CH(G)$, $\Gamma \to \Theta$, *given by formula* (4) *is a* $\underline{W}$-*isomorphism.*

The proof of this theorem produces yet another construction of the representation of $\underline{W}$ on $H_{2n}(S)$ in the spirit of Kazhdan and Lusztig's. In fact, this construction shows that the representation of $\underline{W}$ on $H_{4n}(S)$ comes from a homomorphism of $\underline{W}$ into the group of proper homotopy equivalences of S, answering a question raised by Kazhdan and Lusztig for the case at hand. This implies that one actually has a representation of $\underline{W}$ on the cmplete homology $H_*(S)$ of S, a fact which Springer proved using étale cohomology. The construction is as follows.

Recall the map $\pi_\lambda: B^* \to \Omega_\lambda$, bijective for regular $\lambda \in h^*$. (We momentarily work with G rather than $\underline{G}$.) So for regular $\lambda \in h^*$ we can define $a_\lambda(w) = \pi_{w\lambda}^{-1}\pi_\lambda$: $B^* \to B^*$ (here $w\lambda = \lambda w^{-1}$). Then

$$a_\lambda(wy) = a_{y\lambda}(w)a_\lambda(y). \qquad (5)$$

If one could set $\lambda = 0$ in this equation one would get an action of W on B^*, which would furthermore leave the desingularization map π_0: $B^* \to \Omega_0 = N$ invariant. This is of course not possible. Using an idea of Kazhdan and Lusztig, one can proceed as follows: for any subet V of N denote by $B^*(V)$ the inverse image of V in B^* under π_0: $B^* \to N$. If V is a subvariety of N one can choose a neighborhood U of V

in N so that the inclusion i: $\mathcal{B}^*(V) \to \mathcal{B}^*(U)$ admits a proper homotopy inverse p: $\mathcal{B}^*(U) \to \mathcal{B}^*(V)$. One sees: for λ sufficiently close to 0 in h^* (but regular) $a_\lambda(w)\ \mathcal{B}^*(V) \subseteq \mathcal{B}^*(U)$ for all $w \in W$ from which it follows that $p \circ a_\lambda(w) \circ i$: $\mathcal{B}^*(V) \to \mathcal{B}^*(V)$ is defined for such λ. The proper homotopy class of this map is independent of λ and will be denoted $a_V(w)$. It follows from (5) that a_V is a homomorphism of W into the group $E(\mathcal{B}^*(V))$ of proper homotopy equivalences of $\mathcal{B}^*(V)$. In particular one gets an action of W on the homology $H_*(\mathcal{B}^*(V))$ with arbitray supports. When one takes $V = \{\nu\}$, a single point, then $\mathcal{B}^*(V) = \{(b,\nu)\colon \nu \in b^\perp\}$ which may be identified with the subvariety $\mathcal{B}_\nu = \{b \in \mathcal{B}\colon \nu \in b^\perp\}$ of $\mathcal{B}$, and one gets Springer's representation of W on $H_*(\mathcal{B}_\nu)$. If one replaces G by $\underline{G}$ and takes $V = N$ as subvariety of N, then $\mathcal{B}^*(V) = S$ and one gets the desired representation of $\underline{W}$ on $H_*(S)$. One even gets much more. For each G-orbit O on N, with closure $cl(O)$ and boundary $bd(O)$, set $S(cl(O)) = \underline{\mathcal{B}}^*(cl(O))$ and $S(bd(O)) = \underline{\mathcal{B}}^*(bd(O))$. $H_{4n}(S(cl(O)))$ and $H_{4n}(S(bd(O)))$ are naturally subgroups of $H_{4n}(S)$. Define $H_{4n}(S(O))$ to be the subquotient $H_{4n}(S(cl(O)))/H_{4n}(S(bd(O)))$ of $H_{4n}(S)$ (which makes some sense even as a homology group of $S(O)$). For any d = 0,1,2... denote by S^d the union of the $S(cl(O))$ with $\dim_{\mathbb{C}}(O) \leq d$. The subgroups $H_{4n}(S^d)$ of $H_{4n}(S)$ give a filtration of $H_{4n}(S)$ and the associated graded group is

$$\mathrm{gr}\ H_{4n}(S) = \textstyle\sum_O\ H_{4n}(S(O)). \tag{6}$$

$S(O)$ = the inverse image of O under $\underline{\mathcal{B}}^* \to N$ is a G-equivariant fibre bundle over O with fibre $\underline{\mathcal{B}}_\nu$ over $\nu \in O$. From this one sees that

$$H_{4n}(S(O)) = H_{4e(\nu)}(\underline{\mathcal{B}}_\nu)^{A(\nu)} \tag{7}$$

for any $\nu \in O$, with $e(\nu) = \dim_{\mathbb{C}}(\underline{\mathcal{B}}_\nu)$ and $A(\nu)$ = the fundamental group of O = the component group of the stabilizer of ν in G. The action of $\underline{A}(\nu) = A(\nu) \times A(\nu)$ on $H_{4e(\nu)}(\underline{\mathcal{B}}(\nu)) = H_{4e(\nu)}(\underline{\mathcal{B}}_\nu) \otimes H_{4e(\nu)}(\mathcal{B}_\nu)$ commutes with that of $\underline{W} = W \times W$ and one sees that the $A(\nu)$-invariants in (7) (with $A(\nu)$ embedded as diagonal in $\underline{A}(\nu)$) decompose under $W \times W$ as

$$\sum \chi_{\nu,\bar{\phi}} \otimes \chi_{\nu,\phi} \tag{8}$$

where the sum runs over the irreducible characters ϕ of $A(\nu)$ which occur in $H_{4e(\nu)}(\mathcal{B}_\nu)$ and $\chi_{\nu,\phi}$ is the character of W on the subspace which transforms according to ϕ. (A priori one may have to extend scalars from $\mathbb{Z}$ to $\mathbb{C}$ for the decomposition, but it follows from the known structure of the $A(\nu)$ that it suffices to work over $\mathbb{Q}$.) This argument (which is attributed to Springer by Kazhdan and Lusztig) shows that the $\chi_{\nu,\phi}$ are exactly the irreducible characters of W, provided one knows that the representation of $W \times W$ on $H_{4n}(S)$ is the biregular representation. In the context of the construction given above this is easily verified: one considers the cycles on $\underline{\mathcal{B}}^*$ corresponding to the cycles $K \cdot (\underline{\lambda}\ \underline{w} + b_1{}^\perp)$ on $\underline{\Omega}_{\underline{\lambda}}$ under

$\underline{\pi}_\lambda$: $\underline{\mathcal{B}}^* \to \underline{\Omega}_\lambda$ ($\underline{\lambda}$ regular). These cycles (which correspond to the principal characters $\overline{X}_w$ via formula (4)) are properly homotopic to cycles on S (also denoted X_w) and are easily seen to provide the required $\underline{W}$-isomorphism $\mathbb{Z}[\underline{W}/W] \to H_{4n}(S)$, $\underline{w}\overline{W} \to X_{\underline{w}}$.

Many problems remain open. For example, one would like to have explicit formulas for the matrices of the representation of $\underline{W}$ on $H_{4n}(S)$ in the basis of components S_w, at least for simple reflections. This is related to the problem of finding the expressions for the X_w in terms of the S_w. Kazhdan and Lusztig conjecture that for $G = GL_n$ one has $S_w = \sum_{y \leq w} (-1)^{\ell(w)-\ell(y)} P_{y,w}\; y \cdot S_1$.
Since $X_y = (-1)^n (y,1) S_1$ this would imply that

$$S_w = \sum_{y \leq w} (-1)^{\ell(w)-\ell(y)+n} P_{y,w}(1) X_y \qquad (9)$$

or equivalently

$$X_w = (-1)^n \sum y \leq w\; P_{w_o w,\, w_o w}(1) S_y. \qquad (10)$$

This would give

$$Y_w = (-1)^n (1,w_o) S_{ww_o} \qquad (11)$$

as expression for the irreducibles Y_w in terms of the S_w. The $P_{y,w}$ are the Kazhdan-Lusztig polynomials.

3. Asymtotic expansions and harmonic polynomials.

Recall that

$$\mathrm{gr}\, H_{4n}(S) = \sum_\nu H_{4e(\nu)} (\mathcal{B}_\nu)^{A(\nu)} \qquad (12)$$

where ν runs over a set of representatives for the nilpotent G-orbits in $\mathfrak{g}^*$ (see (6) and (7)). Furthermore, the inclusion $\underline{\mathcal{B}}_\nu \to \underline{\mathcal{B}}$ induces a map

$$H_{4e(\nu)} (\underline{\mathcal{B}}_\nu) \to H_{4e(\nu)} (\underline{\mathcal{B}}) \qquad (13)$$

It is not difficult to see that his map is just the $\underline{A}(\nu)$-projection onto the $\underline{A}(\nu)$-invariants $H_{4e(\nu)} (\underline{\mathcal{B}}_\nu)^{\underline{A}(\nu)}$ in $H_{4e(\nu)} (\underline{\mathcal{B}}_\nu)$, in the sense that it induces an isomorphism of $H_{4e(\nu)} (\underline{\mathcal{B}}_\nu)^{\underline{A}(\nu)}$ onto its image in $H_{4e(\nu)}(\mathcal{B})$ and annihilates the elements of $H_{4e(\nu)} (\underline{\mathcal{B}}_\nu)$ transforming according to an irreducible character $\underline{\phi}$ of $\underline{A}(\nu)$ other than the trivial character. (We may take scalars from $\mathbb{C}$ for this discussion.) The representation of $\underline{W}$ on $H_{e(\nu)} (\underline{\mathcal{B}}_\nu)^{\underline{A}(\nu)}$ is the irreducible representation associated to the trivial character $\underline{\phi} = 1$ of $\underline{A}(\nu)$ and will be denoted by $\underline{\chi}_\nu$ or $\chi_{\mathcal{O}}$; it depends only on the G-orbit $\mathcal{O}$ of ν. These representations of $\underline{W}$ will be called *orbital* representations. (We remark that for $G = GL_n$ all $A(\nu)$ are trivial so that all representations of $W = S_n$ are orbital.) The homology group $H_*(\underline{\mathcal{B}})$ of the flag manifold (with complex coefficients) may be naturally identified with the space of $\underline{W}$-harmonic polynomials on $\underline{h}^*$ (Borel's correspondence). In particular, to each element of $H_{e(\nu)}(\underline{B}_\nu)^{\underline{A}(\nu)}$ there corresponds a harmonic polynomial on $\underline{h}^*$. This gives a map from gr $H_{4n}(S)$ to harmonic polynomials on $\underline{h}^*$, which is again just the projection on the $\underline{A}(\nu) = A(\nu) \times A(\nu)$ invariants in each summand in (12). More precisely, we have defined a map

$$H_{4n}(S(\mathcal{O})) \rightarrow H_{2e(\mathcal{O})}(\underline{h}^*), \ \Gamma \rightarrow p, \qquad (14)$$

where the right side denotes the homogeneous $\underline{W}$-harmonics of degree $2e(\mathcal{O}) = 2e(\nu)$ on $\underline{h}^*$. Of course all maps, and in particular (14), are $\underline{W}$-equivariant. The map (14) has an interpretation in terms of the asymptotics of the coherent family Θ corresponding to a contour in $H_{4n}(S)$ which represents an element in the subquotient $H_{4n}(S(\mathcal{O}))$, as follows.

THEOREM 3. *Let* $\Gamma \in H_{4n}(S(\mathrm{cl}(\mathcal{O})))$, Θ *the corresponding coherent family in* CH(G) *as in* (4), p *the corresponding* $\underline{W}$-*harmonic polynomial in* $H_{2e(\mathcal{O})}(\underline{h}^*)$ *as in* (14). *Then*

$$\Theta(\underline{\lambda}) = p(\underline{\lambda}) \int_{\mathcal{O}} e^{i\xi} \mu_{\mathcal{O}}(d\xi) + o(|\underline{\lambda}|^{2e(\mathcal{O})}) \text{ as } \underline{\lambda} \rightarrow 0. \qquad (15)$$

The integral is of course again interpreted in the distribution sense, as the Fourier transform of the canonical invariant measure μ on the nilpotent G-orbit $\mathcal{O}$ in g^*. In view of what has been said in connection with equation (14), the polynomial p is non-zero if and only if Γ has a non-zero projection onto the $\underline{A}(\nu)$-invariants in $H_{4e(\nu)}(S(\mathcal{O}))$. (By definition, $\Gamma \rightarrow p$ factors through the projection of $H_{4d(\mathcal{O})}(S(\mathrm{cl}(\mathcal{O})))$ onto $H_{4d(\mathcal{O})}(S(\mathcal{O}))$.) We remark that it follows from results of BARBASCH and VOGAN [1982], that in the case when Θ represents an irreducible character for some regular integral $\underline{\lambda}$ there is a unique orbit $\mathcal{O}$ so that an equation of the type (15) holds with $p \neq 0$; at least when G is classical, in which case they compute the orbit $\mathcal{O}$ explicitly. The existance and uniqueness of $\mathcal{O}$ also follows from results of JOSEPH [1978], who showed that the formal characters of the irreducible quotients of Verma modules for g admit an analogous asymptotic formula with $p \neq 0$ belonging to an irreducible representation of W. (The irreducibility implies the uniqueness of $\mathcal{O}$. One also needs the relation between category-$\mathcal{O}$-modules and Harish-Chandra modules, see BERNSTEIN and GELFAND.)

Theorem 3 has a number of applications. For example, it can be used to prove the following property (*univalence*) of the orbital representations $\chi_{\mathcal{O}}$ of W, attached to $\mathcal{O}$ and the trivial character of $A(\nu)$.

COROLLARY 1. *The representation* $\chi_{\mathcal{O}}$ *of* W *occurs with multiplicity one in the homogeneous polynomials of degree* e $(\mathcal{O})$ *on* h^* *and does not occur in lower degree.*

This also follows from the work of BORHO and MACPHERSON, announced in [1981]. - To avoid confusion, it should be noted the $\chi_{\nu,\phi}$ here and in [BORHO, MACPHERSON] differ from Springer's by multiplication with the sign character.

A more substantial application of Theorem 3 concerns the Fourier inversion of the measures $\mu_{\mathcal{O}}$ on nilpotent orbits $\mathcal{O}$ in $\mathfrak{g}^*$, which can be stated like this:

COROLLARY 2. *Let $\mathcal{O}$ be a nilpotent orbit in $\mathfrak{g}^*$. Let $D_{\mathcal{O}}$ be the (up to scalars unique) W-invariant constant coefficient differential operators on $\underline{h}^*$ which transforms according to the orbital representation $\underline{\chi}_{\mathcal{O}} = \chi_{\mathcal{O}} \otimes \chi_{\mathcal{O}}$ of $\underline{W} = W\times W$, homogeneous of degree $e(\underline{\mathcal{O}}) = 2e(\mathcal{O})$. Then (up to as non-zero constants):*

$$\mu_{\mathcal{O}} = D_{\mathcal{O}}\, \mu_{\lambda}\Big|_{\lambda=0}, \quad D_{\mathcal{O}} = \sum_{w\in W} \chi_{\mathcal{O}}(w)\, (\partial_{\lambda}\cdot w\partial_{\lambda})^{e(\mathcal{O})} \tag{16}$$

where $\mu_{\mathcal{O}}$ is the canonical invariant measure on the G-orbit of $\lambda \in h^*$ in $\mathfrak{g}^*$, $D_{\mathcal{O}}$ operates on the variable λ, and the evaluation at $\lambda = 0$ is understood as a limit from the regular set in h^*. ∂_{λ} is the gradient in the λ-variable relative to a $\underline{W}$-invariant Hermitian inner product $(\lambda\cdot\mu)$ on h^*.

This fact has been conjectured by BARBASCH and VOGAN [1983], who had previously [1982] given a proof for classical groups and special orbits. A proof using the theory of holonomic systems has recently been given by HOTTA and KASHIWARA [1984].

Theorem 3 can be used to prove a conjecture of JOSEPH [1984, Conjecture 9.8.]. To explain this, one needs to know that for any nilpotent orbit $\mathcal{O}$ in $\mathfrak{g}^*$ the components of $\mathcal{O} \cap b_1^{\perp}$ ($b_1^{\perp}$ = the orthogonal, or radical, of the fixed Borel b_1) are varieties $V(w)$ characterized by the property that $V(w)$ intersects $b_1^{\perp} \cap w\cdot b_1^{\perp}$ densely (with $w \in W$ depending on $\mathcal{O}$). To such a variety Joseph attaches a polynomial $p_{V(w)}$ on h^*, the leading coefficient of the Hilbert-Samuel polynomial for $V(w)$. For this polynomial Joseph proposes

JOSEPH'S FORMULA. *For each $w \in W$ one has (up to a non-zero scalar)*

$$p_{V(w)} = \sum_{y\in W} A(w,y)\, y\cdot\rho^{m} \tag{17}$$

where m in the least integer ≥ 0 such that the right side is non-zero.
($m = \deg p_{V(w)} = n - \dim V(w) = e(\mathcal{O})$.)

Here $\rho = \frac{1}{2}\sum_{\alpha>0} \alpha$ (and may be replaced by any other regular element), considered as a linear function on h^*. The coefficients $A(w,y)$ are defined via the action of $\underline{W}$ on $H_{4n}(S)$ in terms of the classes S_w of the components S_w of S:

$$S_w = \sum_{y\in W} A(y,w)\, (y,1)\cdot S_1 \tag{18}$$

Note that on the right side appear the principal series contours $X_y = (-1)^n (y,1) S_1$ (except for a sign).

Joseph's formula can be proved using Theorem 3 and a result of HOTTA (1983) along these lines: Consider the projection $\underline{B}^* = B^* \times B^* \to B$ through the second factor.

It restricts to a fibration $S \to B$ with fibre $T = \{(b,\nu): b \in B,\ \nu \in b^{\perp} \cap b_1^{\perp}\}$ over the fixed base point $b_1 \in B$. The components S_w of S meet T in the components T_w of T, which are the closures of the parts of T over the Schubert cells B_w in B, i.e. the closures of the normal bundles $\{(b,\nu): b \in B_w,\ \nu \in b^{\perp} \cap b_1^{\perp}\}$ of the B_w. For any $\nu \in b_1^{\perp}$ there is a diagram

$$\begin{array}{ccccc} B_\nu & \to & T & \to & b_1^{\perp} \\ \downarrow & & \downarrow & & \downarrow \\ \underline{B}_\nu & \to & S & \to & N . \end{array}$$

S_w meets the fibre $\underline{B}_\nu$ of $S \to N$ in a single $A(\nu)$-orbit of components; T_w meets the fibre B_ν of $T \to b_1^{\perp}$ in a union of components which lie in a single $A(\nu)$-orbit and determine an element of $H_{2e(\nu)}(B_\nu)^{A(\nu)}$ (independent of ν under the identification of the $H_{2e(\nu)}(B_\nu)^{A(\nu)}$, $\nu \in O$, through the action of G). The result of Hotta referred to above implies that the map which sends $p_{V(w)}$ to this element of $H_{2e(\nu)}(B_\nu)^{A(\nu)}$ extends to a W-map on the space spanned by the $p_{V(w)}$. On the other hand, the inclusion $B_\nu \to B$ gives an injection $H_{2e(\nu)}(B_\nu)^{A(\nu)} \to H_{2e(\nu)}(B) = H_{e(\nu)}(h^*)$, as we know. Because of Corrollary 3, $p_{V(w)}$ must be the polynomial in $H_{2e(\nu)}(h^*)$ corresponding to the element in $H_{2e(\nu)}(B_\nu)^{A(\nu)}$ constructed above (up to a scalar depending only on the orbit O).

To bring in Theorem 3, note that the fibration $T \to S \to B$ leads to a $W = W\times\{1\}$-isomorphism $H_{4n}(S) \to H_{2n}(T)$ which sends S_w to T_w, the class of T_w. Theorem 3 allows one to compute the polynomial p_{S_w} on $\underline{h}^*$ associated to S_w by the map (14) from the expression (18) of S_w in terms of the X_y, using the Taylor series for the X_y in (3). The result is (up to a non-zero scalar):

$$p_{S_w}(\underline{\lambda}) = p(\lambda_1)p'(\lambda_2) \text{ if } \underline{\lambda} = (\lambda_1,\lambda_2) \text{ in } \underline{h}^* = h^*\times h^* \qquad (19)$$

where p is defined by the right side of (17). The transition $p_{S_w} \to p_{V(w)}$ from a polynomial on $\underline{h}^* = h^*\times h^*$ to one on h^* corresponds to $S_w \cap \underline{B}_\nu \to T_w \cap B_\nu$, going from a cycle on $\underline{B} = B\times B$ to one on B. One sees that this means that $p_{V(w)}$ is obtained from p_{S_w} by dropping the second factor p' in (19), which leaves $p_{V(w)} = p$, i.e. (17).

There is another conjecture of Joseph [JOSEPH (1983), Conjecture 8.13] which is related to the correspondence between characters and homology classes given by the contour integral (4). It says that the characteristic cycle (singular support in Joseph's terminology) of the $\mathcal{D}$-module corresponding to Verma module M_y is the cycle on B^* given by

$$S(M_y) = \sum_{w\in W} B(y,w)T_w \qquad (20)$$

where the coefficients B(y,w) are defined by

$$(y,1)S_1 = \sum_{w\epsilon W} B(y,w)S_w \qquad (21)$$

so that B(y,w) is the inverse matrix to A(y,w).

The right side of (21) corresponds to the right side of (20) under the $W\times\{1\}$-isomorphism $H_{4n}(S) = H_{2n}(T)$ mentioned earlier. The left side of (21) is the contour of the principal series X_y (up to a sign). So (21) turns into (20), up to a sign, if one replaces principal series X_y by Verma modules M_Y and character contours by characteristic cycles. This would mean that the category equivalence of Harish-Chandra modules and category-$\mathcal{O}$-modules (for a fixed regular integral infinitesimal character) due to Bernstein-Gelfand, Joseph, and Enright becomes the isomorphism $H_{4n}(S) \cong H_{2n}(T)$ on the level of character contours and characteristic cycles.

REFERENCES

BARBASCH D. and VOGAN D., Primitive ideals and orbital integrals in complex classical groups, Math. Ann. 259, 153-199 (1982).

BARBASCH D. and VOGAN D., Primitive ideals and orbital integrals in complex exceptional groups, J. of Alg. 80, 350-383 (1983).

BERNSTEIN J.N. and GELFAND S.I., Tensor products of finite and infinite dimensional representations of semisimple Lie algebras, Compositio Math. 41, 245-285 (1980).

BORHO W. and MACPHERSON R., Représentations des groupes de Weyl et homologie d'intersection pour les variétés nilpotentes, C.R. Acad. Sc. Paris 292, 707-710 (1981).

DUFLO M., Représentations irréductibles des groupes semi-simples complexes, LNM 497, 26-88, Springer (1975).

HOTTA A., On Joseph's construction of Weyl group representations, Tôhoku Math. J. 36, 49-74 (1984).

HOTTA A. and KASHIWARA M., The invariant holonomic system on a semisimple Lie algebra, Inventiones Math. 75, 327-358 (1984).

KAZHDAN D. and LUSZTIG G., Representations of Coxeter groups and Hecke algebras, Inventiones Math. 53, 165-184 (1979).

KAZHDAN D. and LUSZTIG G., A topological approach to Springer's representations, Advances in Math. 38, 222-228 (1980).

JOSEPH A., Goldie rank in the enveloping algebra of a semisimple Lie algebra II, J. of Alg. 65, 284-306 (1980).

JOSEPH A., On the classification of primitive ideals in the enveloping algebra of a semisimple Lie algebra, LNM 1024, 30-76, Springer (1983).

JOSEPH A., On the variety of a highest weight module, J. of Alg. 88, 238-278 (1984).

SPRINGER T.A., Trigonometric sums, Greens functions of finite groups, and representations of Weyl groups, Inventiones Math. 36, 173-207 (1976).

SPRINGER T.A., A constructions of representations of Weyl groups, Inventiones Math. 44, 279-293 (1978).

STEINBERG R., On the desingularization of the unipotent variety, Inventiones Math. 36, 209-224 (1976).

ROSSMANN W., Characters as contour integrals, LNM 1077, 375-388, Springer (1984).

Distributions sphériques invariantes sur l'espace symétrique semi-simple et son c-dual

par Shigeru SANO

Université Louis Pasteur de Strasbourg

Université de Shokugyokunren (I.V.T)

Introduction

Nous allons étudier les distributions sphériques invariantes (DSI) sur certains espaces symétriques non Riemanniens. Dans le cas particulier du groupe lui-même, il s'agit de distributions propres invariantes (DPI), et, dans ce cas, Harish-Chandra a démontré qu'il n'existe pas de DPI à support singulier et qu'une DPI est une fonction localement intégrable. Mais en général dans un espace symétrique semi-simple $\mathfrak{X}$ ce résultat est faux. Par contre il est toujours vrai que la restriction d'une DSI à $\mathfrak{X}'$ (l'ensemble des éléments réguliers de $\mathfrak{X}$) est une fonction analytique invariante. Une DSI qui n'a pas de support singulier est alors déterminé par sa restriction à $\mathfrak{X}'$. Et on peut prévoir qu'une DSI qui contribue à la formule de Plancherel est de ce type (voir par exemple [27] , [3] , [19]).

Nous déterminons une dualité entre les séries discrètes et les séries continues de DSI. Le groupe de Lie, semi-simple réel G et l'espace semi-simple G_c/G sont en relation de c-dualité. Nous étudions (§3 et §4) les DSI sur $\mathfrak{X} = G \times G/G$, qui désigne un espace symétrique soit du type G, soit du type G_c/G. Soit Φ une fonction analytique G-invariante sur $\mathfrak{X}'$. Le théorème 5.1 donne une condition nécessaire et suffisante pour que Φ définisse une DSI sur $\mathfrak{X}$.

Nous nous contentons d'étudier les distributions sphériques invariantes car le théorème de Kashiwara [13] explique que une hyperfonction sphérique invariante est en fait une distribution. Par conséquent on peut définir la transformation de Fourier d'une fonction test comme dans le cas d'un groupe.

§1. Algèbre de Lie semi-simple symétrique et sous espace de Cartan

Soit G une groupe de Lie connexe, σ un automorphisme involutif de G et H un sous groupe fermé de G satisfaisant $(G_\sigma)_0 \subset H \subset G_\sigma$ où $G_\sigma = \{x \in G : \sigma(x) = x\}$ et $(G_\sigma)_0$ est la composante connexe de l'identité de G_σ. Alors le triplet (G , H , σ) est appelé un espace symétrique affine.

Soit $\mathfrak{g}$ une algèbre de Lie, σ un automorphisme involutif et soit $\mathfrak{g} = \mathfrak{h} \oplus \mathfrak{q}$ la décomposition en sous-espaces propres pour σ, c'est-à-dire

$$\mathfrak{h} = \{ X \in \mathfrak{g} \, ; \, \sigma X = X \}, \quad \mathfrak{q} = \{ X \in \mathfrak{g} \, ; \, \sigma X = -X \}.$$

Alors le triplet ($\mathfrak{g}$, $\mathfrak{h}$, σ) est appelé algèbre de Lie symétrique. A chaque espace symétrique affine (G , H , σ) correspond un triplet ($\mathfrak{g}$, $\mathfrak{h}$, σ) où $\mathfrak{g}$ et $\mathfrak{h}$ sont les algèbres de Lie de G et de H respectivement et, σ l'automorphisme de $\mathfrak{g}$ déterminé par l'automorphisme σ de G . Pour simplifier, nous écrivons souvent ($\mathfrak{g}$, $\mathfrak{h}$) au lieu de ($\mathfrak{g}$, $\mathfrak{h}$, σ). On suppose $\mathfrak{g}$ semi-simple dès maintenant.

Définition 1.1. Soit ($\mathfrak{g}$, $\mathfrak{h}$, σ) un triplet symétrique. Un sous-espace $\mathfrak{a}$ est appelé un sous-espace de Cartan si $\mathfrak{a}$ est un sous-espace abelien maximal de $\mathfrak{q}$ et si pour chaque $X \in \mathfrak{a}$, ad(X) est un endomorphisme semi-simple de $\mathfrak{g}$.

Soit (G , H , σ) un espace symétrique affine et ($\mathfrak{g}$, $\mathfrak{h}$, σ) l'algèbre de Lie symétrique correspondante. Definissons une application φ de G dans G par

$$\varphi(g) = g\sigma(g)^{-1} \quad (g \in G).$$

L'application φ donne une injection de G/G_σ dans G. Posons $X = \varphi(G)$. Pour chaque élément $a \in G$, definissons une application différentiable l_a de G dans G par $l_a(b) = ab$ et une application différentiable B_a de X dans X par $B_a(x) = ax\sigma(a)^{-1}$. On a $\varphi(l_a(b)) = B_a(\varphi(b))$. Alors G/G_σ et X sont isomorphes, comme G-espaces symétriques.

Soit θ un automorphisme involutif de Cartan de $\mathfrak{g}$ satisfaisant $\sigma\theta = \theta\sigma$ (cf. [14]) et $\mathfrak{g} = \mathfrak{k} \oplus \mathfrak{p}$ la décomposition de Cartan correspondante. Posons

$$\mathfrak{h}_+ = \mathfrak{h} \cap \mathfrak{k} , \quad \mathfrak{h}_- = \mathfrak{h} \cap \mathfrak{p} ,$$

$$\mathfrak{q}_+ = \mathfrak{q} \cap \mathfrak{k} , \quad \mathfrak{q}_- = \mathfrak{q} \cap \mathfrak{p} .$$

Pour un sous-espace vectoriel $\mathfrak{a}$ de $\mathfrak{g}$, $\mathfrak{a}_c$ denote la complexification de $\mathfrak{a}$. Si $\mathfrak{a}$ est une sous-algèbre, $U(\mathfrak{a}_c)$ denote l'algèbre enveloppante universelle de $\mathfrak{a}_c$. Soit $\mathfrak{j}$ une sous-algèbre de Cartan de $\mathfrak{g}$, θ -invariante et σ -invariante. Pour $\mathfrak{a}$ un sous-espace θ -invariant de $\mathfrak{j}$, $\mathfrak{a}^*$ denote l'espace dual et $\mathfrak{a}_c^*$ la complexification de $\mathfrak{a}^*$. Alors pour $\lambda \in \mathfrak{a}_c^*$ on pose $\mathfrak{g}_c(\mathfrak{a} : \lambda) = \{ X \in \mathfrak{g}_c ; ad(Y)X = \lambda(Y)X$ pour tout $Y \in \mathfrak{a}_c \}$, $\mathfrak{g}(\mathfrak{a} : \lambda) = \mathfrak{g}_c(\mathfrak{a} : \lambda) \cap \mathfrak{g}$ et on écrit $\Sigma(\mathfrak{a}) = \{ \lambda \in \mathfrak{a}_c^* \setminus \{0\} ; \mathfrak{g}_c(\mathfrak{a} : \lambda) \neq \{0\} \}$. Par la forme de Killing B(,) de l'algèbre complexe $\mathfrak{g}_c$, on identifie $\mathfrak{a}_c^*$ et $\mathfrak{a}_c$, et alors $\mathfrak{a}_c^*$ est identifié avec un sous-espace de $\mathfrak{j}_c^*$. On prolonge

σ et θ sur $\mathfrak{g}_c$ en une involution $\mathbb{C}$-linéaire. Soit K le sous-groupe analytique correspondant à $\mathfrak{k}$. Soit $\mathfrak{a}$ un sous-espace de Cartan, θ-invariant. Pour $\alpha \in \Sigma(\mathfrak{a})$, il existe un unique élément $H_\alpha \in \mathfrak{a}^\alpha = \sqrt{-1}(\mathfrak{a} \cap \mathfrak{k}) \oplus (\mathfrak{a} \cap \mathfrak{p})$ tel que $B(H_\alpha . H) = \alpha(H)$ pour tout $H \in \mathfrak{a}$. Pour $\alpha \in \Sigma(\mathfrak{a})$ on pose,

$$\mathfrak{l}_c = \mathbb{C}H_\alpha \oplus \mathfrak{g}_c(\mathfrak{a}; \alpha) \oplus \mathfrak{g}_c(\mathfrak{a}; -\alpha),$$

$$\mathfrak{l} = \mathfrak{l}_c \cap \mathfrak{q}, \ \mathfrak{l}_+ = \mathfrak{l} \cap \mathfrak{k}, \ \mathfrak{l}_- = \mathfrak{l} \cap \mathfrak{p}.$$

Soit $\mathfrak{a}$ un sous-espace de Cartan θ-invariant de $\mathfrak{q}$.

Définition 1.2.

(i) Une racine α de $\Sigma(\mathfrak{a})$ est dite racine réelle si α prend des valeurs réelles sur $\mathfrak{a}$.

(ii) Une racine α de $\Sigma(\mathfrak{a})$ est dite racine imaginaire si prend des valeurs imaginaires pures sur $\mathfrak{a}$.

(iii) Une racine α de $\Sigma(\mathfrak{a})$ est dite racine complexe si α n'est ni réelle, ni imaginaire.

Définition 1.3.

(i) Une racine réelle $\alpha \in \Sigma(\mathfrak{a})$ est dite singulière si $\mathfrak{l}_+ \neq \{0\}$

(ii) Une racine réelle $\alpha \in \Sigma(\mathfrak{a})$ est dite vectorielle si $\mathfrak{l}_+ = \{0\}$ c'est à dire si $\mathfrak{l}_- = \mathfrak{l}$.

(iii) Une racine imaginaire $\alpha \in \Sigma(\mathfrak{a})$ est dite singulière si $\mathfrak{l}_- \neq \{0\}$

(iv) Une racine imaginaire $\alpha \in \Sigma(\mathfrak{a})$ est dite compacte si $\mathfrak{l}_- = \{0\}$ c'est à dire si $\mathfrak{l}_+ = \mathfrak{l}$.

Remarque. Dans le cas des espaces ($\mathfrak{g} \oplus \mathfrak{g}$, diagonal), il n'y a pas de racine réelle vectorielle. Et dans des espaces($\mathfrak{g}_c$, $\mathfrak{g}$), il n'y a pas de racine compacte (cf. Lemme 2.1).

Lemme 1.1.

(i) Soit $\alpha \in \Sigma(\mathfrak{a})$ une racine réelle singulière. Il existe un élément $X_\alpha \in \mathfrak{g}_c(\mathfrak{a}:\alpha)$ tel que $X_\alpha \in \mathfrak{h}_- \oplus \mathfrak{q}_+$. Soit $\mathfrak{l}'_c = \mathbb{C}H_\alpha \oplus \mathbb{C}X_\alpha \oplus \mathbb{C}(\sigma X_\alpha)$ $\mathfrak{l}' = \mathfrak{l}'_c \cap \mathfrak{g}$ et $\mathfrak{l}'_{\mathfrak{h}} = \mathfrak{l}' \cap \mathfrak{h}$. Alors $\sigma X_\alpha \in \mathfrak{g}_c(\mathfrak{a}:-\alpha)$ et $(\mathfrak{l}', \mathfrak{l}'_{\mathfrak{h}})$ est isomorphe à $(\mathfrak{sl}(2,\mathbb{R}), \mathfrak{so}(1,1))$.

(ii) Soit $\alpha \in \Sigma(\mathfrak{a})$ une racine imaginaire singulière. On a $X_\alpha \in \mathfrak{g}_c(\mathfrak{a};\alpha)$ tel que $X_\alpha \in \mathfrak{h}_- \oplus \sqrt{-1}\,\mathfrak{q}_-$. Soit $\mathfrak{l}'_c = \mathbb{C}H_\alpha \oplus \mathbb{C}X_\alpha \oplus \mathbb{C}(\sigma X_\alpha)$, $\mathfrak{l}' = \mathfrak{l}'_c \cap \mathfrak{g}$ et $\mathfrak{l}'_{\mathfrak{h}} = \mathfrak{l}' \cap \mathfrak{h}$. Alors $\sigma X_\alpha \in \mathfrak{g}_c(\mathfrak{a};-\alpha)$ et $(\mathfrak{l}', \mathfrak{l}'_{\mathfrak{h}})$ est isomorphe à $(\mathfrak{sl}(2,\mathbb{R}), \mathfrak{so}(1,1))$.

Démonstration. (i) Soit τ la conjugaison complexe de $\mathfrak{g}_c$ relativement à $\mathfrak{g}$, $\mathfrak{g}_c(\mathfrak{a};\alpha)$ est τ invariant et $-\sigma\theta$ invariant. Ensuite on peut choisir X_α dans $\mathfrak{h}_- \oplus \mathfrak{q}_+$ car α est singulière. D'après $H_\alpha \in \mathbb{R}[X_\alpha, \sigma X_\alpha]$, on a $\mathfrak{l}' = \mathbb{R}H_\alpha \oplus \mathbb{R}X_\alpha \oplus \mathbb{R}(\sigma X_\alpha)$ et $\mathfrak{l}'_{\mathfrak{h}} = \mathbb{R}(X_\alpha + \sigma X_\alpha)$. Alors $(\mathfrak{l}', \mathfrak{l}'_{\mathfrak{h}}) \simeq (\mathfrak{sl}(2,\mathbb{R}), \mathfrak{so}(1,1))$. (ii) On procède de manière analogue à (i). C.Q.F.D.

Soit $\mathfrak{a}$ un sous-espace de Cartan, θ-invariant de $\mathfrak{q}$ et $\alpha \in \Sigma(\mathfrak{a})$ une racine réelle singulière.

Définition 1.4. On prend X_α dans $\mathfrak{g}$ en le normalisant par $\alpha([X_\alpha, \sigma X_\alpha]) = 2$. Definissons un automorphisme ν de $\mathfrak{g}_c$ par :

$$\nu = \exp\left\{-\frac{\sqrt{-1}\,\pi}{4}\,\mathrm{ad}\,(X_\alpha + \sigma X_\alpha)\right\}.$$

Alors $\mathfrak{b} = \nu(\mathfrak{a}) \cap \mathfrak{q}$ est un sous-espace de Cartan θ-invariant qui n'est conjugé à $\mathfrak{a}$ par aucun élément de H. Et on a : $\dim(\mathfrak{b} \cap \mathfrak{k}) = \dim(\mathfrak{a} \cap \mathfrak{k}) + 1$. Pour chaque racine $\gamma \in \Sigma(\mathfrak{a})$, posons $(\nu\gamma)(X) = \gamma(\nu^{-1}(X))$ $(X \in \mathfrak{b}_c)$. Alors $\nu\cdot\gamma$ est une racine de $\mathfrak{b}$ et $\beta = \nu\cdot\alpha$ est une racine imaginaire singulière. Pour la racine réelle singulière α, on pose $\mathfrak{a}_\alpha = \{ X \in \mathfrak{a} ; \alpha(X) = 0 \}$ et $\mathfrak{z} = \mathfrak{z}_{\mathfrak{g}}(\mathfrak{a}_\alpha)$ (le centralisateur de $\mathfrak{a}_\alpha$ dans $\mathfrak{g}$). On a la décomposition en sous-espaces propres $\mathfrak{z} = \mathfrak{z}_{\mathfrak{h}} \oplus \mathfrak{z}_{\mathfrak{q}}$ où $\mathfrak{z}_{\mathfrak{h}} = \mathfrak{z} \cap \mathfrak{h}$, $\mathfrak{z}_{\mathfrak{q}} = \mathfrak{z} \cap \mathfrak{q}$. Alors ($\mathfrak{z}$, $\mathfrak{z}_{\mathfrak{h}}$, σ) est une algèbre de Lie semi-simple symétrique de rang déployé égal à 1 ([18]). En adaptant une démonstration de Harish-Chandra [9] , on peut montrer.

<u>Lemme 1.2.</u> Soit $\mathfrak{t}$ est un sous-espace de Cartan de $\mathfrak{q}$ qui est contenu dans $\mathfrak{z}_{\mathfrak{q}}$. Alors $\mathfrak{t}$ est conjugé à $\mathfrak{a}$ ou $\mathfrak{b}$ par un élément du sous-groupe analytique Z_H de H correspondant à la sous algèbre $\mathfrak{z}_{\mathfrak{h}}$.

<u>Remarque.</u> On peut définir l'automorphisme inverse. Soit $\mathfrak{b} = \theta(\mathfrak{b})$ un sous-espace de Cartan de $\mathfrak{q}$, $\beta \in \Sigma(\mathfrak{b})$ une racine imaginaire singulière et X_β un élément de $\mathfrak{g}_c(\mathfrak{b}, \beta)$ choisi dans $\mathfrak{h}_- \oplus \sqrt{-1}\,\mathfrak{q}_-$ (cf. Lemme 1.1) tel que $\beta([\sigma X_\beta, X_\beta]) = 2$. On pose :

$$\tilde{\nu} = \exp \left\{ \frac{\sqrt{-1}\,\pi}{4} \operatorname{ad}(X_\beta + \sigma X_\beta) \right\}.$$

On a alors :

1) $\tilde{\nu}(\mathfrak{b}_c) \cap \mathfrak{q}$ est un sous-espace de Cartan θ-invariant $\mathfrak{a}$ de $\mathfrak{q}$.

2) $\tilde{\nu} \cdot \beta$ est une racine réelle singulière α appartenant à $\Sigma(\mathfrak{a})$.

3) $\tilde{\nu}(X_\beta)$ appartient à $\mathfrak{g}(\mathfrak{a}; \alpha)$ et vérifie $\alpha([\tilde{\nu}(X_\beta), \sigma \tilde{\nu}(X_\beta)]) = 2$.

Si on pose $X_\alpha = \tilde{\nu}(X_\beta)$ et si on définit ν comme dans la Définition 1.4, on a $\tilde{\nu} \cdot \nu = \nu \cdot \tilde{\nu} = \mathrm{id}$ car $X_\beta + \sigma X_\beta = X_\alpha + \sigma X_\alpha \in \mathfrak{h}_-$.

<u>Définition 1.5.</u> Dans la situation de la Définition 1.4, on rappelle que $\mathfrak{b}$ est défini à l'aide de $\mathfrak{a}$, d'une racine réelle singulière α et de l'automorphisme ν . Dans ce cas on dit que

$\mathfrak{a}$ et $\mathfrak{b}$ sont adjacents et on note cette relation entre $\mathfrak{a}$ et $\mathfrak{b}$: $\mathfrak{a} \to \mathfrak{b}$. Soit Car($\mathfrak{q}$) l'ensemble des classes de conjugaison par des sous-espaces de Cartan de $\mathfrak{q}$. La classe de conjugaison de $\mathfrak{a}$ est notée $[\mathfrak{a}]$ et nous allons aussi adopter la notation $[\mathfrak{a}] \to [\mathfrak{b}]$. On obtient aussi un diagramme dans Car($\mathfrak{q}$) qu'on note $\Pi(\mathfrak{g}, \mathfrak{h})$. On définit un ordre $<$ dans Car($\mathfrak{q}$), de sorte que $[\mathfrak{c}] < [\mathfrak{c}']$ si et seulement s'il existe une chaîne $\mathfrak{c}_1, \mathfrak{c}_2, \cdots, \mathfrak{c}_n$ de sous-espaces de Cartan telle que $[\mathfrak{c}_1] = [\mathfrak{c}]$, $[\mathfrak{c}_n] = [\mathfrak{c}']$ et $[\mathfrak{c}_1] \to [\mathfrak{c}_2] \to \cdots \to [\mathfrak{c}_n]$.

Exemple. $\Pi(\mathfrak{sp}(n,\mathbb{R}), \mathfrak{s}(\mathfrak{gl}(n,\mathbb{R}) \oplus \mathbb{R}))$ et $\Pi(\mathfrak{sp}(n,\mathbb{C}), \mathfrak{sp}(n,\mathbb{R}))$ sont les mêmes diagrammes suivants :

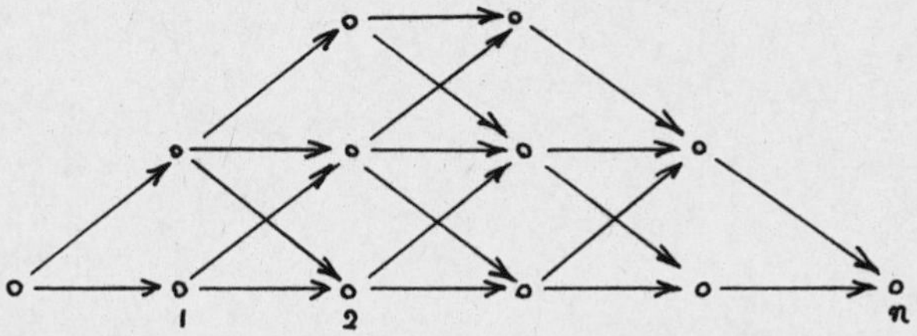

§2. c-dualité

Nous voulons étudier les distributions sphériques invariantes (DSI) et l'analyse harmonique sur l'espace semi-simple symétrique G/H. Pour cela, nous allons montrer qu'il y a dualité entre les séries discrétes de DSI et les séries continues de DSI qui contribuent à la formule de Plancherel. Donnons l'idée de la c-dualité d'algèbre de Lie semi-simple symétrique qui donne cette dualité.

Définition 2.1. Soit $(\mathfrak{g}, \mathfrak{h}, \sigma)$ un triplet symétrique. Posons $\bar{\mathfrak{g}} = \mathfrak{h} \oplus \bar{\mathfrak{q}}$ ou $\bar{\mathfrak{q}} = \sqrt{-1}\,\mathfrak{q}$. Le triplet $(\bar{\mathfrak{g}}, \mathfrak{h}, \sigma)$ est appelé son dual à cause de son caractère compact (c-dualité) de $(\mathfrak{g}, \mathfrak{h}, \sigma)$. Soit (G_c, H_c) un espace semi-simple correspond à $(\mathfrak{g}_c, \mathfrak{h}_c)$ on dit que deux espace semi-simple $(G, H), (\bar{G}, H)$ sont en c-dualité s'ils correspondent respectivement à la paire $(\mathfrak{g}, \mathfrak{h})$ et $(\bar{\mathfrak{g}}, \mathfrak{h})$ et si G/H et $\bar{G}/H$ sont formes réelles de G_c/H_c.

<u>Remarque.</u> M.Berger a défini le dual et l'associé de l'algèbre de Lie semi-simple symétrique. La relation est la suivante: L'associé $(\mathfrak{g}^a, \mathfrak{h}^a, \sigma\theta)(=(\mathfrak{g}, \mathfrak{h})^a)$ du triplet $(\mathfrak{g}, \mathfrak{h}, \sigma)$ est défini par

$$\mathfrak{g}^a = \mathfrak{g} \quad , \quad \mathfrak{h}^a = (\mathfrak{k} \cap \mathfrak{h}) \oplus (\mathfrak{p} \cap \mathfrak{q}).$$

Le dual $(\mathfrak{g}^d, \mathfrak{h}^d, \theta)(=(\mathfrak{g}, \mathfrak{h})^d)$ du triplet $(\mathfrak{g}, \mathfrak{h}, \sigma)$ est défini par

$$\mathfrak{g}^d = (\mathfrak{k} \cap \mathfrak{h}) \oplus \sqrt{-1}(\mathfrak{k} \cap \mathfrak{q}) \oplus \sqrt{-1}(\mathfrak{p} \cap \mathfrak{h}) \oplus \mathfrak{p} \cap \mathfrak{q} \, , \quad \mathfrak{h}^d = (\mathfrak{k} \cap \mathfrak{h}) \oplus \sqrt{-1}(\mathfrak{k} \cap \mathfrak{q}).$$

Alors on a $(\bar{\mathfrak{g}}, \mathfrak{h}) = (\mathfrak{g}, \mathfrak{h})^{dad} = (\mathfrak{g}, \mathfrak{h})^{ada}$. Par suite on a le diagramme suivant :

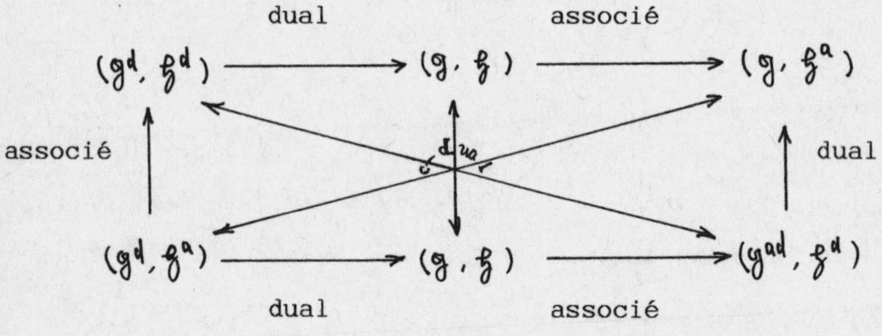

<u>Exemple.</u>

(i) Soit $(\mathfrak{g}, \mathfrak{k}, \theta)$ un triplet Riemannien de type non compact. Son c-dual est donné par le triplet Riemannien compact $(\mathfrak{u}, \mathfrak{k}, \theta)$ où $\mathfrak{u} = \mathfrak{k} \oplus \sqrt{-1}\mathfrak{p}$. Par exemple $(\mathfrak{sl}(n, \mathbb{R}), \mathfrak{so}(n))$ et $(\mathfrak{su}(n), \mathfrak{so}(n))$ sont c-duaux.

(ii) Pour la paire $(\mathfrak{g} \oplus \mathfrak{g}$, diagonal), son c-dual est donné par la paire $(\mathfrak{g}_c, \mathfrak{g})$.

(iii) Pour un paire $(\mathfrak{g}, \mathfrak{h})$ de type K_ε (cf. [17]), le c-dual est donné par $(\mathfrak{g}^{ad}, \mathfrak{h})$. Ainsi $(\mathfrak{sl}(n,\mathbb{R}), \mathfrak{so}(p,q))$ $(n = p+q)$ et $(\mathfrak{su}(p,q), \mathfrak{so}(p,q))$ sont en c-dualité et $(\mathfrak{sl}(n,\mathbb{R}), \mathfrak{s}(\mathfrak{gl}(n-r,\mathbb{R}) \oplus \mathfrak{gl}(r,\mathbb{R})))$ est c-autodual.

Soient deux paires $(\mathfrak{g}, \mathfrak{h})$ et $(\bar{\mathfrak{g}}, \mathfrak{h})$ c-duales. Soit $\mathfrak{a}$ un sous-espace de Cartan θ-invariant de $\mathfrak{q}$. Alors $\bar{\mathfrak{a}} = \sqrt{-1}\,\mathfrak{a}$ est un sous-espace de Cartan de $\bar{\mathfrak{q}}$. Soit γ l'isomorphisme linéaire de $\mathfrak{g}$ sur $\bar{\mathfrak{g}}$ défini par $\gamma(X+Y) = X + \sqrt{-1}\,Y$ $(X \in \mathfrak{h}, Y \in \mathfrak{q})$. Pour $\beta \in \mathfrak{a}^*$, on définit $\gamma.\beta \in (\bar{\mathfrak{a}})^*$ par $\gamma.\beta(\bar{H}) = \beta(\gamma^{-1}(\bar{H}))$ $(\bar{H} \in \bar{\mathfrak{a}})$. On a le lemme suivant :

Lemme 2.1.

(i) Soit $\alpha \in \Sigma(\mathfrak{a})$ une racine réelle singulière. Alors $\gamma.\alpha \in \Sigma(\bar{\mathfrak{a}})$ est une racine imaginaire singulière.

(ii) Soit $\alpha \in \Sigma(\mathfrak{a})$ une racine réelle vectorielle. Alors $\gamma.\alpha \in \Sigma(\bar{\mathfrak{a}})$ est une racine compacte.

Démonstration. (i) Soit $\bar{\theta}$ l'involution de Cartan de $\bar{\mathfrak{g}}$ donné par $\bar{\theta}(X+\sqrt{-1}\,Y) = \theta(X) - \sqrt{-1}\,\theta(Y)$ $(X \in \mathfrak{h}, Y \in \mathfrak{q})$. Soit $\bar{\mathfrak{g}} = \bar{\mathfrak{k}} \oplus \bar{\mathfrak{p}}$ la décomposition de Cartan correspondante et $\bar{\mathfrak{g}} = \bar{\mathfrak{h}} \oplus \bar{\mathfrak{q}}$ $(\bar{\mathfrak{h}} = \mathfrak{h})$ la décomposition en sous-espaces propres de σ. Alors $\bar{\mathfrak{h}}_- = \mathfrak{h} \cap \bar{\mathfrak{p}} = \mathfrak{h} \cap \mathfrak{p} = \mathfrak{h}_-$ et $\sqrt{-1}\,\bar{\mathfrak{q}}_- = \sqrt{-1}(\bar{\mathfrak{q}} \cap \bar{\mathfrak{p}}) = \mathfrak{q} \cap \mathfrak{k} = \mathfrak{q}_+$. D'après le lemme 1.1, il existe non nul élément $X_\alpha \in \mathfrak{g}_c(\mathfrak{a}; \alpha)$ tel que $X_\alpha \in \mathfrak{h}_- \oplus \mathfrak{q}_+$. Alors $X_\alpha \in \bar{\mathfrak{g}}_c(\bar{\mathfrak{a}}; \gamma.\alpha)$ et $X_\alpha \in \bar{\mathfrak{h}}_- \oplus \sqrt{-1}\,\bar{\mathfrak{q}}_-$. $\gamma.\alpha$ est une racine imaginaire singulière de $\Sigma(\bar{\mathfrak{a}})$. (ii) On procède de manière analogue à (i) C.Q.F.D.

Soit (G, H, σ) un espace symétrique semi-simple et $(\mathfrak{g}, \mathfrak{h}, \sigma)$ le triplet correspondant. Supposons $G_\sigma = H$ et posons X comme dans §1.

<u>Définition 2.2.</u> Soit $\mathfrak{a}$ un sous-espace de Cartan de $\mathfrak{q}$. Le centralisateur $A = Z_X(\mathfrak{a})$ de $\mathfrak{a}$ dans X est appelé sous-espace global de Cartan de X.

<u>Remarque.</u> Soit $\mathfrak{a}_1, \cdots, \mathfrak{a}_n$ un ensemble maximal de sous-espaces de Cartan de $\mathfrak{q}$ deux à deux non conjugués par H et $A_j = Z_X(\mathfrak{a}_j)$ $(j=1,\cdots,n)$ les sous-espaces globaux de Cartan correspondants. Alors on a $X' = \bigcup_{j=1}^{n} \bigcup_{h \in H} h A_j' h^{-1}$ (où X' est l'ensemble de tous les éléments réguliers de X et $A_j' = A_j \cap X'$ (cf. [15])). Une distribution sphérique invariante est analytique sur X'. Elle est déterminée, excepté sa partie singulière, par les valeurs qu'elle prend sur A_j' $(j=1,\cdots,n)$. Nous pensons alors la c-dualité donne la dualité entre la série discrète et la série continue. Nous allons étudier les cas ($\mathfrak{g} \oplus \mathfrak{g}$, diagonal) et ($\mathfrak{g}_c$, $\mathfrak{g}$) dans le chapitre suivant.

§3. <u>Un exemple d'espaces en c-dualité</u>

Soit $\mathfrak{g}$ une algèbre de Lie semi-simple, $\mathfrak{g}_c$ sa complexifiée, G_c un groupe de Lie connexe d'algèbre de Lie $\mathfrak{g}_c$ et G son sous-groupe analytique d'algèbre de Lie $\mathfrak{g}$.

Soit σ_1 l'involution définie sur $G \times G$ (resp. $\mathfrak{g} \oplus \mathfrak{g}$) par $\sigma_1(x, y) = (y, x)$ (resp. $\sigma_1(X, Y) = (Y, X)$). On appelle diag le sous-groupe (resp. la sous-algèbre) des points fixes de σ_1. On dit que l'espace symétrique ($G \times G$, diag, σ_1) est un espace symétrique $G \times G / G$ de cas I. On peut réaliser cet espace symétrique comme sous-variété de G_c : l'application φ_1 de $G \times G \to G_c$ qui à (x , y) associé xy^{-1} induit un G-diffeomorphisme de $G \times G$ /diag sur $G \subset G_c$. On pose $X_1 = G$ et G opére sur cet espace symétrique par conjugaison.

Soit σ_2 l'involution définie sur G_c (resp. $\mathfrak{g}_c$) par $\sigma_2(x) = \mathrm{conj}\, x$ (resp. $\sigma_2(X) = \mathrm{conj}\, X$). On dit que l'espace symétrique (G_c, G , σ_2) est un espace symétrique $G \times G / G$ de cas II. On peut réaliser cet espace symétrique comme sous-variété fermée de G_c : l'application φ_2 de $G_c \to G_c$ qui à x associe $x\sigma(x)^{-1}$ induit un G-diffeomorphisme de G_c/G sur son image qu'on note X_2. De plus G_c opére sur cet espace symétrique par $x \longrightarrow gx\sigma(g)^{-1}$ $(g \in G_c)$.

Les espaces symétriques X_1 et X_2 sont en c-dualité. Soit ψ l'application de $\mathfrak{g}_c$ sur $\mathfrak{g}\oplus\mathfrak{g}$ qui à $X+\sqrt{-1}Y$ $(X,Y\in\mathfrak{g})$associe (X,X) $+J(Y,Y)$ où $J(X,Y)=(-Y,X)$. C'est un isomorphisme d'algèbre de Lie, commutant à l'action adjointe de G et tel que $\sigma_1\cdot\psi=\psi\cdot\sigma_2$. Remarquons que J induit une structure complexe sur $\mathfrak{g}\oplus\mathfrak{g}$.

Nous étudierons simultanément les deux espaces symétriques associés à G_c. Dans le cas de cas I, on réalisera l'espace symétrique en posant $X_1=G$ et on étudiera le triplet associé $(X_1, \mathfrak{q}_1 : G_c)$ où $\mathfrak{q}_1=\mathfrak{g}$. On désigne par $\mathbb{D}(X_1)$ l'algèbre des opérateurs différentiels complexes bi-invariants sur X_1. Dans le cas de cas II, on réalisera l'espace symétrique en possant X_2 $=\{g\sigma_2(g)^{-1}; g\in G_c\}$ et on étudiera le triplet associé $(X_2, \mathfrak{q}_2 : G_c)$. L'application φ_2 de $G_c/G \to X_2$ qui à gG associe $g\sigma(g)^{-1}$ induit un G_c-diffeomorphisme de G_c/G sur X_2. On désigne par $\mathbb{D}(X_2)$ l'algèbre des opérateur différentiels complexes G_c-invariants sur X_2. Quand il n'est pas nécessaire de distinguer un espace de cas I et de cas II, nous enlevons l'index.

§4. Distribution sphérique invariante sur $G\times G/G$

Nous étudions des DSI dans la situation de §3.

Définition 4.1. Pour $x\in G_c$, on définit

$$\det\left[t-(Ad(x)-id)\right]=\sum_{i=0}^{n} D_i(x)\, t^i \qquad (n=\dim_{\mathbb{R}} G_c).$$

Si $D_0,\cdots,D_{l-1}\equiv 0$ et $D_l\not\equiv 0$, on pose $l=$ rang de G_c et $\Delta(x)=|D_l(x)|^{1/4}$.

Un élément x de G_c est appelé régulier si $D_l(x) \neq 0$. Et on pose $G_c' = \{x \in G_c ;$ régulier $\}$ et $X' = X \cap G_c'$. Ensuite un élément de est appelé régulier dans . On a le lemme suivant d'après ([15]).

Lemme 4.1. (decomposition de X' par l'action de G). Soit $\mathfrak{a}_1, \mathfrak{a}_2, \cdots, \mathfrak{a}_n$ un ensemble maximal de sous-espaces de Cartan qui ne soient pas G-conjugués. Posons $A_j = Z_X(\mathfrak{a}_j)$ $(j=1,2,\cdots,n)$. Alors on a

$$X' = \bigcup_{j=1}^{n} \bigcup_{g \in G} g A_j' g^{-1} \qquad (A_j' = A_j \cap X').$$

Soit $U(\mathfrak{g}_c)$ l'algèbre universelle enveloppant de $\mathfrak{g}_c$ et $\mathfrak{z}(\mathfrak{g}_c)$ son centre. Soit $U(\mathfrak{g}_c)^G$ la sous-algèbre des éléments de $U(\mathfrak{g}_c)$ invariants par $Ad(g)$ pour $g \in G$. Soit $\mathfrak{j}$ un sous-algèbre de Cartan de $\mathfrak{g}_c$. Pour un système de racine $\Sigma(\mathfrak{j})$, on choisit un ensemble $\Sigma^+(\mathfrak{j})$ des racines positives. Posons $\mathcal{N}^{\pm} = \sum_{\pm\alpha \in \Sigma^+(\mathfrak{j})} \mathfrak{g}_c(\mathfrak{j} : \alpha)$. Si D appartient à $\mathfrak{z}(\mathfrak{g}_c)$, on démontre qu'il existe un unique élément D_j' appartenant à $U(\mathfrak{j})$ tel que $D - D_j'$ appartient à $U(\mathfrak{g}_c)\mathcal{N}^+ + \mathcal{N}^- U(\mathfrak{g}_c)$ On identifie $U(\mathfrak{j})$ avec l'algèbre des fonctions polynômes sur $\mathfrak{j}$ et on pose pour $D \in U(\mathfrak{g}_c)$

$$[\gamma_c(D)](\lambda) = D_j'(\lambda - \rho) \qquad \text{où} \quad \rho = \frac{1}{2} \sum_{\alpha \in \Sigma^+(\mathfrak{j})} \alpha .$$

On peut alors montrer que γ_c est un isomorphisme (appelé isomorphisme de Harish-Chandra) de $\mathfrak{z}(\mathfrak{g}_c)$ sur $I(\mathfrak{j})$ (la sous-algèbre des élémemnts de l'algèbre symétrique $S(\mathfrak{j})$ invariants par le groupe de Weyl $W(\mathfrak{j})$ de la paire $(\mathfrak{g}_c, \mathfrak{j})$).

Dans le cas $X \simeq G_c/G$, pour un élément $D = \sum a X_1^{n_1} \cdots X_k^{n_k} \in U(\mathfrak{g}_c)^G$ on définit un opérateur différentiel sur X par

$$D_x f = \sum a \frac{\partial^n}{\partial t_1^{n_1} \cdots \partial t_k^{n_k}} \Big|_{t=0} f \circ \varphi_2 \left(g \exp \frac{t_1 X_1 + \cdots + t_k X_k}{2}\right)$$

$$\left(x = g\sigma(g)^{-1} \in \mathfrak{X},\ t = (t_1, \cdots, t_k),\ n = \sum_{j=1}^{k} n_j \right)$$

pour $f \in C^\infty(\mathfrak{X})$. Par l'application, on identifie $U(\mathfrak{g}_2)^G = U(\mathfrak{g}_c)^G/(U(\mathfrak{g}_c)^G \cap U(\mathfrak{g}_c)\mathfrak{g})$ avec $\mathbb{D}(\mathfrak{X})$. Soit $\mathfrak{a}$ un sous-espace de Cartan de $\mathfrak{g}_2$. Alors $\mathfrak{j} = \mathfrak{a}_c$ est un sous-espace de Cartan de $\mathfrak{g}_c$. Soit $\pi_\mathfrak{g}$ (resp. $\pi_\mathfrak{a}$) la projection naturel de $\mathfrak{Z}(\mathfrak{g}_c)$(resp. $I(\mathfrak{a}_c)$) dans $U(\mathfrak{g}_2)^G$(resp. $I(\mathfrak{a}_c)/(I(\mathfrak{a}_c) \cap I(\mathfrak{a}_c)\mathfrak{a})$). On définit un isomorphisme $\gamma^\mathfrak{a}$ de $U(\mathfrak{g}_2)^G$ sur $I(\mathfrak{a}_c)/(I(\mathfrak{a}_c) \cap I(\mathfrak{a}_c)\mathfrak{a})$ par $\gamma^\mathfrak{a}(\pi_\mathfrak{g}(D)) = \pi_\mathfrak{a}(\gamma_c(D))$ ($D \in \mathfrak{Z}(\mathfrak{g}_c)$).

Dans le cas $\mathfrak{X} = G$, pour un élément $D = \sum a X_1^{n_1} \cdots X_k^{n_k} \in \mathfrak{Z}(\mathfrak{g}_c)$ on définit un opérateur différentiel sur $\mathfrak{X}$ par

$$D_g f = \sum a \frac{\partial^n}{\partial t_1^{n_1} \cdots \partial t_k^{n_k}} \Big|_{t=0} f(g \exp(t_1 X_1 + \cdots + t_k X_k))$$

Pour $f \in C^\infty(\mathfrak{X})$. Par l'application, on identifie $U(\mathfrak{g}_1)^G = \mathfrak{Z}(\mathfrak{g}_c)$ avec $\mathbb{D}(\mathfrak{X})$. Soit $\mathfrak{a}$ un sous-algèbre de Cartan de $\mathfrak{g}_1 = \mathfrak{g}$. Un isomorphisme $\gamma^\mathfrak{a}$ de $U(\mathfrak{g}_1)^G$ sur $I(\mathfrak{a}_c)$ est donné $\gamma^\mathfrak{a}(D) = \gamma_c(D)$ ($D \in \mathfrak{Z}(\mathfrak{g}_c)$).

<u>Proposition 4.2.</u> Soit $\mathfrak{a}$ un sous-espace de Cartan de $\mathfrak{g}$ et $A = Z_\mathfrak{X}(\mathfrak{a})$. Soit $f \in C^\infty(\mathfrak{X})$ telle que $f(gxg^{-1}) = f(x)$ ($x \in \mathfrak{X}$, $g \in G$) Alors on a pour tout $D \in \mathbb{D}(\mathfrak{X})$,

$$(Df)\big|_{A'} = \Delta^{-1} \gamma^\mathfrak{a}(D)(\Delta f\big|_{A'})$$

Démonstration. Soit (π, V) une représentation holomorphe irreductible de dimension finie de G_c sur V de poids dominant et soit χ son caractère. Soit J un sous-groupe de Cartan de G associé à $\mathfrak{j} = \mathfrak{a}_c$. Si D appartient à $\mathfrak{Z}(\mathfrak{g}_c)$, on a

$$\Delta(a)\,[D\chi](a) = \gamma_c(D)\{\Delta\chi|_{J'}\}(a) \qquad (a \in J').$$

D'après le théorème de Peter-Weyl et du " unitary trich ", on a pour toute fonction holomorphe sur G_c et centrale

$$[Df](a) = \frac{1}{\Delta(a)}\left[\gamma^{\mathfrak{a}}(D)\{\Delta f|_{J'}\}\right](a) \qquad (a \in J').$$

On a donc déterminé la partie radiale de l'opérateur différentiel sur X par les restrictions à A' des fonctions holomorphes centraux sur G_c. C.Q.F.D.

Définition 4.2. Une distribution Θ sur X appelée distribution sphérique invariante si elle satisfait les conditions suivantes.

(i) $\Theta(gxg^{-1}) = \Theta(x)$ pour tout $g \in G$ et $x \in X$.

(ii) Il existe un homomorphisme λ de $U(\mathfrak{g})^G$ dans $\mathbb{C}$ tel que pour tout $D \in U(\mathfrak{g})^G$ on dit : $D\Theta = \lambda(D)\Theta$.

Posons $\mathfrak{D}_\lambda(X)$ l'ensemble de tout la distribution sphérique invariante avec un homomorphisme λ.

On peut généraliser la méthode que Atiyah a utilisée pour démontrer l'intégrabilité locale des distributions propres invariantes sur le groupe de Lie semi-simple [1] . On a :

Théorème 4.3. Soit Θ une distribution sphérique invariante sur X. On a alors : (i) La restriction de Θ à X' est une fonction analytique invariante. (ii) Si Θ est nulle sur X', Θ est identiquement nulle sur tout l'espace $\mathcal{X}$. (iii) Θ est localement intégrable sur $\mathcal{X}$.

Remarque. Le théorème 4.3 a été démomntré en collaboration avec N.Bopp [21] . On faisait ce travail à l'Institut de Recherche Mathématique Avancée. Je remercie vivement Monsieur le professeur J.Faraut de la discussion précieuse avec moi.

§5. Caractérisation des DSI sur $G\times G/G$

On utilise les notations dans §3. Soit $\mathfrak{a}$ un sous-espace de Cartan de $\mathfrak{q}$ et $A_{\mathfrak{a}} = Z_{\mathcal{X}}(\mathfrak{a})$. Si $\mathfrak{a}_c$ est la sous-algèbre de Cartan de $\mathfrak{g}_c$ qui contient $\mathfrak{a}$ et $(A_{\mathfrak{a}})_c$ le sous-groupe de Cartan de correspondant à $\mathfrak{a}_c$, alors $(A_{\mathfrak{a}})_c$ contient $A_{\mathfrak{a}}$. Soit $\Sigma^+(\mathfrak{a})$ l'ensemble des racines positives de $(\mathfrak{g}_c, \mathfrak{a}_c)$, $S_R^{\mathfrak{a}}$ l'ensemble des racines réelles singulières positives et $S_I^{\mathfrak{a}}$ l'ensemble des racines imaginaires singulières positives. Et posons $S^{\mathfrak{a}} = S_R^{\mathfrak{a}} \cup S_I^{\mathfrak{a}}$. Maintenant supposons que G est acceptable. Alors pour $\rho = 1/2 \sum_{\alpha\in\Sigma^+(\mathfrak{a})} \alpha$ on peut définir un homomorphisme ξ_ρ de $(A_{\mathfrak{a}})_c$ dans $\mathbb{C}^*\setminus\{0\}$ par $\xi_\rho(\exp X) = e^{\rho(X)}$ $(X\in\mathfrak{a}_c)$. On peut vérifier qu'alors la forme linéaire $\rho_I = 1/2 \sum_{\alpha\in S_I^{\mathfrak{a}}} \alpha$ se remonte aussi en un homomorphisme ξ_{ρ_I} de $(A_{\mathfrak{a}})_c$ dans $\mathbb{C}^*$ (on utilise la remarque qui suit la définition 4.1 du Chapitre I). On pose pour $a \in A_{\mathfrak{a}}$

$$'\Delta^{\mathfrak{a}}(a) = \prod_{\alpha\in\Sigma^+(\mathfrak{a})} (1-\xi_\alpha(a)^{-1}), \qquad \Delta^{\mathfrak{a}}(a) = \xi_\rho(a)\, '\Delta^{\mathfrak{a}}(a),$$

$$'\Delta_R^{\mathfrak{a}}(a) = \prod_{\alpha\in S_R^{\mathfrak{a}}} (1-\xi_\alpha(a)^{-1}), \qquad \varepsilon_R^{\mathfrak{a}}(a) = \operatorname{sgn}('\Delta_R^{\mathfrak{a}}(a)),$$

$$'\Delta_I^{\mathfrak{a}}(a) = \prod_{\alpha\in S_I^{\mathfrak{a}}} (1-\xi_\alpha(a)^{-1}), \qquad \varepsilon_I^{\mathfrak{a}}(a) = \operatorname{sgn}\left((\sqrt{-1})^{-n(I)}\, \xi_{\rho_I}(a)\, '\Delta_I^{\mathfrak{a}}(a)\right).$$

$$(n(I) = \# S_I^{\mathfrak{a}})$$

On remplace F par $\mathbb{R}$ (resp. I) si X est du cas I(resp. du cas II), et on a donc les notations $S_F^{\mathfrak{a}}$, $'\Delta_F^{\mathfrak{a}}$ et $\varepsilon_F^{\mathfrak{a}}$. Pour une racine $\alpha \in \Sigma(\mathfrak{a})$, on choisit $H_\alpha \in \mathfrak{a}^d = F_1(\mathfrak{a} \cap k) \oplus (\mathfrak{a} \cap q)$ tel que $B(H_\alpha, H) = \alpha(H)$ pour tout $H \in \mathfrak{a}$ on définit H'_α par $H'_\alpha = \frac{2}{|\alpha|^2} H_\alpha$. Soit $A'_{\mathfrak{a}}(F) = \{a \in A_{\mathfrak{a}} ; '\Delta_F(a) \neq 0\}$ et $W_G(A_{\mathfrak{a}}) = N_G(A_{\mathfrak{a}})/Z_G(A_{\mathfrak{a}})$. Définissons la fonction localement constante $\varepsilon^F(w : a)$ sur $A'_{\mathfrak{a}}$ par $(\varepsilon_F^{\mathfrak{a}} \Delta^{\mathfrak{a}})(wa) = \varepsilon^F(w : a)(\varepsilon_F^{\mathfrak{a}} \Delta^{\mathfrak{a}})(a)$ $(w \in W_G(A_{\mathfrak{a}}), a \in A'_{\mathfrak{a}})$.

Soit $\mathfrak{a}_1, \ldots, \mathfrak{a}_n$ un ensemble maximal de sous-espaces de Cartan θ-invariant non G-conjugués de q, fixons un ordre pour les racines de $(\mathfrak{g}_c, (\mathfrak{a}_j)_c)$ et posons $A_j = Z_X(\mathfrak{a}_j)$ où on écrit j au lieu de $\mathfrak{a}_j$.

Si Θ est une distributions sphérique invariante sur X, alors la restriction de Θ à X' (notée $\widetilde{\Theta}$) est une fonction analytique. On lui associe la famille des fonctions $(\varkappa_j)_{j=1,\ldots,n}$ définies sur par

$$\varkappa_j(a) = (\varepsilon_F^j \Delta^j)(a)\widetilde{\Theta}(a) \qquad \text{pour} \quad a \in A'_j. \tag{1}$$

Ces fonctions sont ε^F-symétriques, c'est-à-dire,

$$\varkappa_j(wa) = \varepsilon^F(w : a)\varkappa_j(a) \qquad \text{pour} \quad w \in W_G(A_j),\ a \in A'_j. \tag{2}$$

Réciproquement, si on se donne une famille de fonctions analytiques sur A'_j et ε^F-symétriques, on peut leur associer une fonction $\widetilde{\Theta}$, G-invariante et analytique sur X' par

$$\widetilde{\Theta}(gag^{-1}) = [(\varepsilon_F^j \Delta^j)(a)]^{-1} \varkappa_j(a) \qquad a \in A'_j,\ g \in G. \tag{3}$$

Le théorème ci-dessous caractérise les fonctions $\varkappa_j$ pour lesquelles l'expression

$$(\Theta, f) = \int_{X'} \widetilde{\Theta}(x) f(x)\, dx \qquad \text{où} \quad f \in C_c^\infty(X) \tag{4}$$

définit une distribution sphérique invariante sur $\mathbb{X}$.

Théorème 5.1. La fonction $\mathbf{G}$-invariante $\tilde{\Theta}$ sur $\mathbb{X}'$ associée par (3) à la famille $(\kappa_j)_{j=1,\dots,n}$ définit une distribution sphérique Θ sur $\mathbb{X}'$ par (4) si eet seulement si les fonctions vérifient les conditions suivantes :

(a-1) Il existe un homomorphisme χ de $\mathbb{D}(\mathbb{X})$ dans $\mathbb{C}$ tel que :

$$D\kappa_j = \lambda_j(D)\kappa_j \quad \text{pour } D\in I(\mathfrak{a}_j), \text{ où } \lambda_j = \chi\cdot(\gamma^j)^{-1}$$

(a-2) Chaque κ_j peut être prolongée analytiquement de A_j' à $A_j'(F)$.

(a-3) Pour tout $j\in\{1,\dots,n\}$ et $\alpha\in S_R^j$, posons $\mathfrak{a}=\mathfrak{a}_j$.
Soit $\mathfrak{b}$ le sous-espace de Cartan de β une racine imaginaire singulière de $\Sigma(\mathfrak{b})$ obtenu à l'aide de $\mathfrak{a}$, d'une racine réelle singulière α dans $\mathfrak{q}$ (cf. Définition 1.4). Prenons l'ordre des racines de pour qui satisfait à $\Sigma^+(\mathfrak{b}) = \nu\cdot\Sigma^+(\mathfrak{a})$. Définissons $\kappa_{\mathfrak{b}}$ à partir de Θ tel que

$$\kappa_{\mathfrak{b}}(a) = (\varepsilon_F^{\mathfrak{b}}\Delta^{\mathfrak{b}})(a)\,\tilde{\Theta}(a) \qquad (a\in A'_{\mathfrak{b}}).$$

Alors

$$H'_\alpha(\varepsilon_F^{\mathfrak{a}}\kappa_{\mathfrak{a}})(a_0) = H'_\beta(\varepsilon_F\kappa_{\mathfrak{b}})(a_0) \qquad \left(\begin{array}{c} a_0\in A_{\mathfrak{a}}\cap A_{\mathfrak{b}} \\ \prod_{\gamma\in\Sigma^+(\mathfrak{a})\setminus\{\alpha\}}(1-\xi_\gamma(a_0^{-1}))\neq 0\end{array}\right)$$

où chaque côté dénote la valeur limite à a_0 qui existe sous les conditions (4), (a-1) et (a-2).

Esquisse de la démonstration. Soit $\tilde{\Theta}$ la restriction de Θ à $\mathbb{X}'$.

d'après le Théorème 4.3, $\widehat{\Theta}$ est une fonction analytique localement intégrable. On pose $\chi_j(a) = (\varepsilon_F^j \Delta^j)\,\widetilde{\Theta}(a)$ $(a \in A'_j)$. Pour un élément f de $C_c^\infty(X)$, on définit une intégrale orbitale K_f^j sur A'_j par

$$K_f^j(a) = \varepsilon_F^j(a)\, \overline{\Delta^j(a)} \int_{G/Z_G(A_j)} f(gag^{-1})\, d_j \bar{g} .$$

D'après la formule de Weyl, on a

$$(\Theta, f) = \int_{X'} f(x)\, \widetilde{\Theta}(x)\, dx$$

$$= \sum_{j=1}^{n} c_j \int_{A'_j} K_f^j(a)\, \kappa_j(a)\, da$$

où les c_j sont des nombres positifs. Dans le cas où $X = G$, Hirai a démontré le théorème (cf. [12]). Supposons donc que $X \simeq G_c/G$. Les chagements de signe de $\varepsilon_I \Delta$ ont lieu sur un réseau, alors que les changements de signe de $\varepsilon_R \Delta$ ont lieu sur les murs des chambres de weyl. La situation étant ici, pour cela, différente de celle du groupe, il faut modifier la démonstration. On définit une intégrale orbitale $\psi_h^{\mathfrak{a}}$ ($h \in C_c^\infty(\mathfrak{g})$) sur $\mathfrak{a}$ par

$$\psi_h^{\mathfrak{a}}(X) = \delta_I^{\mathfrak{a}}(X)\, \pi^{\mathfrak{a}}(X) \int_{G/Z_G(\mathfrak{a})} h(Ad(g)X)\, d_{\mathfrak{a}} \bar{g}$$

où $\pi^{\mathfrak{a}}(X) = \prod_{\alpha \in \Sigma^+(\mathfrak{a})} \alpha(X)$ et $\delta_I^{\mathfrak{a}}(X) = sgn\left((\sqrt{-1})^{-n(I)} \prod_{\alpha \in S_I^{\mathfrak{a}}} \alpha(X)\right)$

On peut obtenir une relation entre les limites de $K_f^{\mathfrak{a}}$ et $\psi_{h_f}^{\mathfrak{a}}$ en un point semi-régulier (même raisonnement que le lemme 4.3 de [19]). La fonction ψ_h a essentiellement le même comportement au voisinage d'un point singulier que l'intégrale orbitale définie par Harish-Chandra pour une algèbre de Lie semi-simple. On en déduit que $\psi_h^{\mathfrak{a}}$ et $\psi_h^{\mathfrak{b}}$ vérifient le théorème 30, p51 [26] . Par consequent, $K_f^{\mathfrak{a}}$ et $K_f^{\mathfrak{b}}$ vérifient les relations de sauts données par le théorème

11, p396 [26] . A l'aide de ces relations de saut, on peut démontrer la même formule que celle du Lemme 6.4, [12] . Le théorème se démontre alors par un raisonnement analogue à celui du §9, [12].

C.Q.F.D.

§6. Le cas de $G_2 \times G_2 / G_2$

Soit $G = SL(2,\mathbb{R})$ et $\mathfrak{g} = \mathfrak{sl}(2,\mathbb{R})$. Nous allons appliquer le Théorème 5.1 aux espaces symétriques (c-duaux) $SL(2,\mathbb{R})$ et $SL(2,\mathbb{C})/SL(2,\mathbb{R})$. Précisons les notations du §3: On pose $\mathfrak{q}_1 = \mathfrak{g}$ car $\{X \in \mathfrak{g}\oplus\mathfrak{g} : \sigma_1(X) = -X\}$ s'identifie à $\mathfrak{g}$, $\mathfrak{q}_2 = i\mathfrak{g}$ (où $i = \sqrt{-1}$) car $i\mathfrak{g} = \{X \in \mathfrak{g}_c : \sigma_2(X) = -X\}$, $\mathfrak{X}_2 = \{x = \begin{pmatrix} z & ia \\ ib & \bar{z} \end{pmatrix} : a, b \in \mathbb{R}, z \in \mathbb{C}, \det x = 1\}$ en utilisant l'identification par φ_2 de $\mathfrak{X}_2$ et de $SL(2,\mathbb{C})/SL(2,\mathbb{R})$. On considére l'isomorphisme $\mathbb{R}$ -linéaire γ de $\mathfrak{g}_c$ dans $\mathfrak{g}_c$ défini par $\gamma(X) = iX$ pour $X \in \mathfrak{g}_c$. On définit les éléments X_0, Y_0, H_0, U_0, V_0 et W_0 de $\mathfrak{g}_c$ de la façon suivante:

$$X_0 = \begin{pmatrix} 0 & 1 \\ 0 & 0 \end{pmatrix}, \quad Y_0 = \begin{pmatrix} 0 & 0 \\ 1 & 0 \end{pmatrix}, \quad H_0 = \begin{pmatrix} 1 & 0 \\ 0 & -1 \end{pmatrix};$$

$$U_0 = \frac{1}{2}\begin{pmatrix} 1 & -i \\ -i & -1 \end{pmatrix}, \quad V_0 = \frac{1}{2}\begin{pmatrix} 1 & i \\ i & -1 \end{pmatrix}, \quad W_0 = \begin{pmatrix} 0 & i \\ -i & 0 \end{pmatrix}.$$

On a alors $[H_0, X_0] = 2X_0$, $[H_0, Y_0] = -2Y_0$, $[X_0, Y_0] = H_0$; $[W_0, U_0] = 2U_0$, $[W_0, V_0] = -2V_0$, $[U_0, V_0] = W_0$. Si on pose $\bar{X} = \gamma(X)$ pour $X \in \mathfrak{g}_c$, on a alors:

$$\bar{X}_0 = \begin{pmatrix} 0 & i \\ 0 & 0 \end{pmatrix}, \quad \bar{Y}_0 = \begin{pmatrix} 0 & 0 \\ i & 0 \end{pmatrix}, \quad \bar{H}_0 = \begin{pmatrix} i & 0 \\ 0 & -i \end{pmatrix};$$

$$\bar{U}_0 = \frac{1}{2}\begin{pmatrix} i & 1 \\ 1 & -i \end{pmatrix}, \quad \bar{V}_0 = \frac{1}{2}\begin{pmatrix} i & -1 \\ -1 & -i \end{pmatrix}, \quad \bar{W}_0 = \begin{pmatrix} 0 & -1 \\ 1 & 0 \end{pmatrix}.$$

(I) L'espace $\mathfrak{X}_1 = SL(2,\mathbb{R})$: Les espaces de Cartan $\mathfrak{a}_1$ et $\mathfrak{b}_1$ définis c-dessous forment un ensemble maximal de sous-espaces de Cartan non-conjugués pour G:

$$\mathfrak{a}_1 = \left\{ \begin{pmatrix} t & 0 \\ 0 & -t \end{pmatrix} ; t \in \mathbb{R} \right\}, \quad \mathfrak{b}_1 = \left\{ \begin{pmatrix} 0 & -\varphi \\ \varphi & 0 \end{pmatrix} ; \varphi \in \mathbb{R} \right\}.$$

Les sous-espaces globaux de Cartan associés sont alors :

$$A_1 = Z_{K_1}(\mathfrak{a}_1) = \left\{ a(\varepsilon e^t) = \begin{pmatrix} \varepsilon e^t & 0 \\ 0 & \varepsilon e^{-t} \end{pmatrix} ; t \in \mathbb{R}, \varepsilon = \pm 1 \right\}, \quad B_1 = Z_{K_1}(\mathfrak{b}_1) = \left\{ b(e^{i\varphi}) = \begin{pmatrix} \cos\varphi & -\sin\varphi \\ \sin\varphi & \cos\varphi \end{pmatrix} ; \varphi \in \mathbb{R} \right\}$$

On a $[\mathfrak{a}_1] \to [\mathfrak{b}_1]$ et $A_1 \cap B_1 = \{ a(\varepsilon) = b(\varepsilon) \}$. En appliquant les définitions du §5 on obtient:

$$\Delta^{\mathfrak{a}_1}(a(\varepsilon e^t)) = (\varepsilon e^t - \varepsilon e^{-t}), \quad \varepsilon_R^{\mathfrak{a}_1}(a(\varepsilon e^t)) = \operatorname{sgn} t,$$
$$(\varepsilon_R^{\mathfrak{a}_1} \Delta^{\mathfrak{a}_1})(a(\varepsilon e^t)) = \varepsilon |e^t - e^{-t}| :$$
$$\Delta^{\mathfrak{b}_1}(b(e^{i\varphi})) = (e^{i\varphi} - e^{-i\varphi}), \quad \varepsilon_R^{\mathfrak{b}_1}(b(e^{i\varphi})) = 1,$$
$$(\varepsilon_R^{\mathfrak{b}_1} \Delta^{\mathfrak{b}_1})(b(e^{i\varphi})) = \Delta^{\mathfrak{b}_1}(b(e^{i\varphi})).$$

Soit ω l'élément de $U(\mathfrak{g}_1)^G$ défini par $\omega = H_0^2 + 2(X_0Y_0 + Y_0X_0) + 1$. Dans $U(\mathfrak{g}_c)^G$, on peut aussi écrire $\omega = W_0^2 + 2(U_0V_0 + V_0U_0) + 1$ Comme ω engende $U(\mathfrak{g}_1)^G$ on peut définir un homomorphisme de $U(\mathfrak{g}_1)^G$ dans par sa valeur sur ω.

(i) Les séries discrétes de DSI: elles sont associées à un homomorphisme λ de $U(\mathfrak{g}_1)^G$ dans $\mathbb{C}$ donné par $\lambda(\omega) = c^2$ où $c = 1, 2, \cdots$

$$\Theta^+_{c+1}(gag^{-1}) = \begin{cases} \dfrac{(\varepsilon e^{-|t|})^c}{\varepsilon(e^t - e^{-t}|} & (a \in A_1), \\[2ex] \dfrac{-(e^{i\varphi})^c}{e^{i\varphi} - e^{-i\varphi}} & (a \in B_1) \qquad (g \in G), \end{cases}$$

$$\Theta^-_{c+1}(gag^{-1}) = \begin{cases} \dfrac{(\varepsilon e^{-|t|})^c}{\varepsilon|e^t - e^{-t}|} & (a \in A_1), \\[2ex] \dfrac{(e^{-i\varphi})^c}{e^{i\varphi} - e^{-i\varphi}} & (a \in B_1) \qquad (g \in G). \end{cases}$$

(ii) Les séries continues de DSI: $\lambda(\omega) = -c^2$ ($c \in \mathbb{R}$), $\epsilon = 0, 1$

$$\Theta_{c,\epsilon}(gag^{-1}) = \begin{cases} \dfrac{\varepsilon^{\epsilon}(e^{ict}+e^{-ict})}{|e^{t}-e^{-t}|} & (a \in A_1), \\ 0 & (a \in B_1) \end{cases}$$

Si on applique le théorème 5.1, on a (cf. [11]):

<u>Proposition 6.1.</u> Soit λ un homomorphisme de $U(\mathfrak{g}_1)^G$ dans $\mathbb{C}$ défini par $\lambda(\omega)=c^2$. Alors la dimension de $\mathcal{D}'_\lambda(G)$ est donnée par:

$$\dim \mathcal{D}'_\lambda(G) = \begin{cases} 2 & (c \in \mathbb{C}\setminus\mathbb{Z}), \\ 4 & (c = 1,2,\cdots), \\ 3 & (c=0) \end{cases}$$

(II) L'espace $X_2 \simeq SL(2,\mathbb{C})/SL(2,\mathbb{R})$: Les espaces de Cartan $\mathfrak{a}_2$ et $\mathfrak{b}_2$ définis ci-dessous forment un ensemble maximal de sous-espace de Cartan non-conjugués pour G :

$$\mathfrak{a}_2 = \left\{ \begin{pmatrix} 0 & it \\ -it & 0 \end{pmatrix} ; t \in \mathbb{R} \right\}, \quad \mathfrak{b}_2 = \left\{ \begin{pmatrix} i\varphi & 0 \\ 0 & -i\varphi \end{pmatrix} ; \varphi \in \mathbb{R} \right\}.$$

Les sous-espaces globaux de Cartan associés sont alors :

$$A_2 = Z_{X_2}(\mathfrak{a}_2) = \left\{ a(\varepsilon e^t) = \begin{pmatrix} \varepsilon \cosh t & i\varepsilon \sinh t \\ -i\varepsilon \sinh t & \varepsilon \cosh t \end{pmatrix} ; t \in \mathbb{R}, \varepsilon = \pm 1 \right\},$$

$$B_2 = Z_{X_2}(\mathfrak{b}_2) = \left\{ b(e^{i\varphi}) = \begin{pmatrix} e^{i\varphi} & 0 \\ 0 & e^{-i\varphi} \end{pmatrix} : \varphi \in \mathbb{R} \right\}.$$

On a $[\mathfrak{a}_2] \to [\mathfrak{b}_2]$ et $A_2 \cap B_2 = \{ a(\varepsilon) = b(\varepsilon) \}$. On obtient:

$$\Delta^{\mathfrak{a}_2}(a(\varepsilon e^t)) = (\varepsilon e^t - \varepsilon e^{-t}), \quad \varepsilon_I^{\mathfrak{a}_2}(a(\varepsilon e^t)) = 1,$$

$$(\varepsilon_I^{\mathfrak{a}_2} \Delta^{\mathfrak{a}_2})(a(\varepsilon e^t)) = \Delta^{\mathfrak{a}_2}(a(\varepsilon e^t));$$

$$\Delta^{\mathfrak{b}_2}(b(e^{i\varphi})) = (e^{i\varphi} - e^{-i\varphi}), \quad \varepsilon_I^{\mathfrak{b}_2}(b(e^{i\varphi})) = \mathrm{sgn}(\sin\varphi)$$

$$(\varepsilon_I^{\mathfrak{b}_2} \Delta^{\mathfrak{b}_2})(b(e^{i\varphi})) = i|e^{i\varphi} - e^{-i\varphi}|$$

Soit $\bar{\omega}$ l'élément de $U(\mathfrak{g}_2)^G$ défini par $\bar{\omega} = \bar{H}_0^2 + 2(\bar{X}_0\bar{Y}_0 + \bar{Y}_0\bar{X}_0) - 1$
Dans $U(\mathfrak{g}_1)^G$, on peut aussi écrire $\bar{\omega} = \bar{W}_0^2 + 2(\bar{U}_0\bar{V}_0 + \bar{V}_0\bar{U}_0) - 1$
Comme $\bar{\omega}$ engende $U(\mathfrak{g}_2)^G$ on peut définir un homomorphisme de $U(\mathfrak{g}_2)^G$

dans $\mathbb{C}$ par sa valeur sur $\bar{\omega}$. Les séries discrétes de DSI et les séries continues de DSI qui contribuent à la formule de Plancherel sont les suivantes:

(i) Les séries discrétes de DSI: elles sont associées à un homomorphisme λ de $U(\mathfrak{g}_2)^G$ dans $\mathbb{C}$ donné par $\lambda(\bar{\omega}) = c^2$ $(c = 1, 2, \cdots)$.

$$\Theta_c(gag^{-1}) = \begin{cases} 0 & (a \in A_2), \\ \dfrac{e^{ic\varphi} + e^{-ic\varphi}}{i|e^{i\varphi} - e^{-i\varphi}|} & (a \in B_2) \end{cases}$$

(ii) Les séries continues de DSI: $\lambda(\bar{\omega}) = -c^2$ $(c \in \mathbb{R})$

$$\Theta_c(gag^{-1}) = \begin{cases} \left.\begin{array}{ll} \dfrac{e^{ict}}{e^t - e^{-t}} & \varepsilon = 1 \\ 0 & \varepsilon = -1 \end{array}\right\} (a \in A_2) \\ \left.\begin{array}{ll} \dfrac{-1}{2\sinh \pi c} \dfrac{e^{c(\theta-\pi)} + e^{-c(\theta-\pi)}}{e^{i(\theta-\pi)} - e^{-i(\theta-\pi)}} & (\varphi \equiv \theta \bmod 2\pi, 0<\theta<\pi) \\ \dfrac{1}{2\sinh \pi c} \dfrac{e^{c(\theta+\pi)} + e^{-c(\theta+\pi)}}{e^{i(\theta+\pi)} - e^{-i(\theta+\pi)}} & (\varphi \equiv \theta \bmod 2\pi, -\pi<\theta<0) \end{array}\right\} (a \in B_2) \end{cases}$$

D'après le théorème 5.1, on a (cf. [22]).

Proposition 6.2. Soit λ un homomorphisme de $U(\mathfrak{g}_2)^G$ dans $\mathbb{C}$ défini par $\lambda(\bar{\omega}) = c^2$. Alors la dimension de $\mathcal{D}'_\lambda(\mathfrak{X}_2)$ est donnée par:

$$\dim \mathcal{D}'_\lambda(\mathfrak{X}_2) = 4 \qquad (c \in \mathbb{C}).$$

Remarque. (i) Dans le cas de groupe semi-simple G, le caractère d'une représentation de classe trace est une DSI. Le support d'une DSI appartenant à la série discréte est égal à G . Par contre dans $SL(2,\mathbb{C})/SL(2,\mathbb{R})$, le support d'une DSI appartenant à la série discréte des DSI (cf. (II)) est égal à $[B_2]$. D'autre part le support d'une DSI appartenant à la série continue des DSI(cf. (II)), contient des éléments des deux sous-espaces globaux de Cartan. On peut donc prévoir que la c-dualité implique une dualité entre série discréte ou DSI (resp.continue) sur un espace (de type I) et série continue de DSI (resp.discréte) sur (un espace de type II) son c-dual.

(ii) En général, par la relation #{ l'orbite fermé de $H^d\backslash G^d/P^d$} =#{l'orbite ouvert de $H^{da}\backslash G^{da}/P^{da}$}, le nombre de la série le plus discréte de G/H égale à le nombre de la série le plus continu de $\overline{G}/H$ (cf. [15] , [16]).

§7. Séries discrétes de DSI sur G_c/G

Nous étudions les distributions sphériques invariantes sur G_c/G dans la notation de §3. Par le G-diffeomorphisme $\varphi_2 : G_c/G \simeq \mathfrak{X}_2$, nous nous resteingnons sur l'espace $\mathfrak{X}_2$. Supposons l'algèbre de Lie symétrique ($\mathfrak{g}_c$, $\mathfrak{g}$, σ_2) ait un sous-espace de Cartan compact $\mathfrak{a}_0$. Dans la condition, il y a des séries discrétes pour $\mathfrak{X}_2$ (cf. [16]). $\Sigma(\mathfrak{a}_0)$ est l'ensemble des racines imaginaires singulières. Soit $\mathfrak{a}_0, \mathfrak{a}_1, \cdots, \mathfrak{a}_n$ un ensemble maximal de Cartan θ-invariant non G-conjués de $\mathfrak{q}_2$. Posons $A_j = Z_{\mathfrak{X}_2}(\mathfrak{a}_j)$ (j = 0,...,n). On définit $\Delta^j(a) = \Delta^{\mathfrak{a}_j}(a)$ et $W_G(A_j)$ (j = 0,...,n)comme §5. Le sous-espace global de Cartan A_0 est une groupe connexe. Soit A_0^* le groupe des caractères de A_0. Soit $\mathfrak{F}$ l'espace des formes linéaires sur $(\mathfrak{a}_0)_c$ qui prennent seulement des valeurs imaginaires pures sur $\mathfrak{a}_0$. Pour $a^* \in A_0^*$, on note par $\langle a^*, a\rangle$ le valeur de le caractère a^* sur le point $a \in A_0$. Il exist un élément unique $\lambda \in \mathfrak{F}$ tel que

$$\langle a^*, \exp H \rangle = e^{\lambda(H)} \qquad (H \in \mathfrak{a}_0).$$

Soit $W = W(\mathfrak{g}, \mathfrak{a}_0)$ le groupe de Weyl. Un élément $a^* \in A_0^*$ est dite réguliers si $s\lambda \neq \lambda$ pour tout $s(\neq 1) \in W$. Soit $(A_0^*)'$ l'ensemble des éléments réguliers de A_0^*. Soit χ_{a^*} un homomorphisme de $U(\mathfrak{g}_2)^G$ dans $\mathbb{C}$ défini par $\chi_{a^*}(D) = \gamma^{\mathfrak{a}_0}(D)(\lambda)$ où $\gamma^{\mathfrak{a}_0}$ est le homomorphisme donné dans §4. Alors on a séries discrétes de DSI qui contribuent à la formule de

Plancherel pour G_c/G par le théorème suivant d'après Théorème 5.1.

Théorème 7.1. Fixons un élément $a^* \in (A_0^*)'$. Soit Θ_{a^*} une distribution sphérique invariante telle que

(i) $D\Theta_{a^*} = \chi_{a^*}(D)\Theta_{a^*}$ pour tout $D \in U(\mathfrak{g}_c)^G$

(ii) $\sup_{x \in X_i'} |D_\ell(x)|^{1/4} |\Theta_{a^*}(x)| < \infty$

Alors on a

$$\Theta_{a^*}(gag^{-1}) = \begin{cases} \dfrac{\sum_{w \in W} e^{w\lambda(\log a)}}{\varepsilon_I^0(a)\,\Delta^0(a)} & (a \in A_0') \\ 0 & (a \in A_j',\ j = 1, \cdots, n) \end{cases} \qquad (g \in G)$$

Démonstration. Si on définit les fonctions G-invariantes $(j = 0, \ldots, n)$ par

$$\begin{cases} \chi_0(a) = \sum_{w \in W} e^{w\lambda(\log a)} & (a \in A_0'), \\ \chi_j(a) = 0 & (a \in A_j',\ j = 1, 2, \cdots, n) \end{cases}$$

la famille des fonctions $\chi_j(a)(j = 0, \ldots, n)$ satisfait les hypothèses de théorème 5.1. D'après ce théorème, la famille des fonctions définit une distribution sphérique invariante Θ_λ qui satisfait les condition (i),(ii) et $\operatorname{supp} \Theta_\lambda \subset Cl(\bigcup_{g \in G} gA_0g^{-1})$. Sous la condition (ii) une distribution sphérique invariante sur vérifiant (i) et (iii) est déterminée uniquement par sa valeur sur le sous-espace de Cartan global A_0' d'après le théorème.

C.Q.F.D.

Remarque. Dans le groupe de Lie semi-simple, la formule du caractère de la série discréte que Harish-chandra a donné [27], a la condition de la racine compacte. Mais dans l'algèbre de Lie symétrique $(\mathfrak{g}_c, \mathfrak{g})$, il n'y a pas de racine compacte. La formule du Théorème 7.1 est plus simple.

BIBLIOGRAPHIE

[1] M.F.Atiyah,, Characters of semi-simple Lie groups (lectures given in Oxford), Mathematical Institute, Oxford, 1976.

[2] J.Faraut, Distributions sphériques sur les espaces hyperboliques, J. Math. Pures Appl., 58(1979), 369-444.

[3] J.Faraut, Analyse harmonique sur les paires de Guelfand et les espaces hyperboliques, Analyse Harmonique, (1983), 315-446.

[4] J.Faraut, Analyse harmonique et fonctions speciales, Ecole d'été d'analyse harmonique de tunis, (1984).

[5] M.Flensted-Jensen, Discrete series for semi-simple symmetric spaces, ANN. of Math., 111(1980), 253-311.

[6] S.Helgason, Differential Geometry and symmetric spaces, Academic Press, 1962.

[7] S.Helgason, Analysis on Lie groups and homogeneous spaces, Regional Conference series in Mathematics, Amer. Math. Soc. 14, 1972.

[8] Harish-Chandra, The characters of semi-simple Lie groups, Trans. Amer. Math. Soc., 83(1956), 98-163.

[9] Harish-Chandra, Some results on an invariant integral on a semi-simple Lie algebra, Ann. of Math., 80(1964), 551-593.

[10] T. Hirai, Invariant eigendistributions of Laplace operators on real simple Lie groups I, Japan. J. Math., 39(1970), 1-68.

[11] T. Hirai, The Plancherel formula for SU(p,q), J. Math. Soc. Japan, 22(1970), 134-179.

[12] T. Hirai, Invariant eigendistributions of Laplace operators on real simple Lie groups II, Japan.J.Math.New series, 2 (1976), 27-89.

[13] M.Kashiwara, The Riemann-Hilbert Problem for Holonomic Systems, Publ.RIMS,Kyoto Univ., 20(1984), 319-365.

[14] O.Loos, Symmetric spaces, Vols.I-II, Benjamin, New York, 1969.

[15] T.Oshima-T.Matsuki, Orbits on affine symmetric spaces under the action of the isotropic subgroup, J.Math.Soc.Japan, 32 (1980),399-414.

[16] T.Oshima-T.Matsuki, A complete description of discrete series for semi-simple symmetric spaces, Advanced studies in Pure Math., 4(1984) 331-390.

[17] T.Oshima-J.Sekiguchi, Eigenspaces of invariant differential Operators on an affine symmetric space, Invent. Math., 57 (1980), 1-81.

[18] T.Oshima-J.Sekiguchi, The Restricted Root System of a Semi-simple Symmetric Pair, Advanced studies in Pure Math., 4 (1984), 433-497.

[19] S.Sano, Invariant spherical distributions and the Fourier inversion formula on $GL(n,\mathbb{C})/GL(n,\mathbb{R})$, J.Math.Soc.Japan, 36 (1984),191-219.

[20] S.Sano-S.Aoki-S.Kato, A Note on Connection Formulas for Invariant Eigendistributions on Certain Semisimple Symmetric Spaces, Bull. of I.V.T., 14-A(1985), 99-108.

[21] S.Sano-N.Bopp, Distributions sphériques invariantes sur l'espace semi-simple symétrique G_c/G_R , preprint.

[22] S.Sano-J.Sekiguchi, The Plancherel formula for SL(2,C)/ SL(2,R), Sci. Papers College Ged.Ed.Tokyo,30(1980), 93-105.

[23] M.Sugiura, Conjugate classes of Cartan subalgebras in real semi-simple Lie algebras, J.Math.Soc.Japan, 11(1959), 374-434.

[24] M.Sugiura, Unitary representations and harmonic analysis Kodansha, Tokyo, 1975.

[25] R.Takahashi, Sur les functions spheriques et la formule de Plancherel dans le groupe hyperbolique, Japan.J.Math.,31 (1961), 55-90.

[26] V.S.Varadarajan, Harmonic Analysis on real reductive groups, Lecture Notes in Math., 576, Springer Verlag, 1977.

[27] G.Warner, Harmonic Analysis on semi-simple Lie groups, Vol. 1, 2, Springer Verlag, 1972.

Département de Mathématique
Université de Shokugyokunren
1960 Aihara Sagamihara,
Kanagawa, 229 Japon

LECTURE NOTES IN MATHEMATICS
Edited by A. Dold and B. Eckmann

Some general remarks on the publication of proceedings of congresses and symposia

Lecture Notes aim to report new developments - quickly, informally and at a high level. The following describes criteria and procedures which apply to proceedings volumes.

1. One (or more) expert participant(s) of the meeting should act as the responsible editor(s) of the proceedings. They select the papers which are suitable (cf. points 2, 3) for inclusion in the proceedings, and have them individually refereed (as for a journal). It should not be assumed that the published proceedings must reflect conference events faithfully and in their entirety. Contributions to the meeting which are not included in the proceedings can be listed by title. The series editors will normally not interfere with the editing of a particular proceedings volume - except in fairly obvious cases, or on technical matters, such as described in points 2, 3. The names of the responsible editors appear on the title page of the volume.

2. The proceedings should be reasonably homogeneous (concerned with a limited area). For instance, the proceedings of a congress on "Analysis" or "Mathematics in Wonderland" would normally not be sufficiently homogeneous.

   One or two longer survey articles on recent developments in the field are often very useful additions to such proceedings - even if they do not correspond to actual lectures at the congress. An extensive introduction on the subject of the congress would be desirable.

3. The contributions should be of a high mathematical standard and of current interest. Research articles should present new material and not duplicate other papers already published or due to be published. They should contain sufficient information and motivation and they should present proofs, or at least outlines of such, in sufficient detail to enable an expert to complete them. Thus resumes and mere announcements of papers appearing elsewhere cannot be included, although more detailed versions of a contribution may well be published in other places later.

   Surveys, if included, should cover a sufficiently broad topic, and should in general not simply review the author's own recent research. In the case of surveys, exceptionally, proofs of results may not be necessary.

   The editors of a volume are strongly advised to inform contributors about these points at an early stage.

.../...

4. Proceedings should appear soon after the meeeting. The publisher should, therefore, receive the complete manuscript within nine months of the date of the meeting at the latest.

5. Plans or proposals for proceedings volumes should be sent to one of the editors of the series or to Springer-Verlag Heidelberg. They should give sufficient information on the conference or symposium, and on the proposed proceedings. In particular, they should contain a list of the expected contributions with their prospective length. Abstracts or early versions (drafts) of some of the contributions are very helpful.

6. Lecture Notes are printed by photo-offset from camera-ready typed copy provided by the editors. For this purpose Springer-Verlag provides editors with technical instructions for the preparation of manuscripts and these should be distributed to all contributing authors. Springer-Verlag can also, on request, supply stationery on which the prescribed typing area is outlined. Some homogeneity in the presentation of the contributions is desirable.

   Careful preparation of manuscripts will help keep production time short and ensure a satisfactory appearance of the finished book. The actual production of a Lecture Notes volume normally takes 6 -8 weeks.

   Manuscripts should be at least 100 pages long. The final version should include a table of contents.

7. Editors receive a total of 50 free copies of their volume for distribution to the contributing authors, but no royalties. (Unfortunately, no reprints of individual contributions can be supplied.) They are entitled to purchase further copies of their book for their personal use at a discount of 33 1/3%, other Springer mathematics books at a discount of 20% directly from Springer-Verlag.

   Commitment to publish is made by letter of intent rather than by signing a formal contract. Springer-Verlag secures the copyright for each volume.

Vol. 1090: Differential Geometry of Submanifolds. Proceedings, 1984. Edited by K. Kenmotsu. VI, 132 pages. 1984.

Vol. 1091: Multifunctions and Integrands. Proceedings, 1983. Edited by G. Salinetti. V, 234 pages. 1984.

Vol. 1092: Complete Intersections. Seminar, 1983. Edited by S. Greco and R. Strano. VII, 299 pages. 1984.

Vol. 1093: A. Prestel, Lectures on Formally Real Fields. XI, 125 pages. 1984.

Vol. 1094: Analyse Complexe. Proceedings, 1983. Edité par E. Amar, R. Gay et Nguyen Thanh Van. IX, 184 pages. 1984.

Vol. 1095: Stochastic Analysis and Applications. Proceedings, 1983. Edited by A. Truman and D. Williams. V, 199 pages. 1984.

Vol. 1096: Théorie du Potentiel. Proceedings, 1983. Edité par G. Mokobodzki et D. Pinchon. IX, 601 pages. 1984.

Vol. 1097: R.M. Dudley, H. Kunita, F. Ledrappier, École d'Éte de Probabilités de Saint-Flour XII – 1982. Edité par P.L. Hennequin. X, 396 pages. 1984.

Vol. 1098: Groups – Korea 1983. Proceedings. Edited by A.C. Kim and B.H. Neumann. VII, 183 pages. 1984.

Vol. 1099: C.M. Ringel, Tame Algebras and Integral Quadratic Forms. XIII, 376 pages. 1984.

Vol. 1100: V. Ivrii, Precise Spectral Asymptotics for Elliptic Operators Acting in Fiberings over Manifolds with Boundary. V, 237 pages. 1984.

Vol. 1101: V. Cossart, J. Giraud, U. Orbanz, Resolution of Surface Singularities. Seminar. VII, 132 pages. 1984.

Vol. 1102: A. Verona, Stratified Mappings – Structure and Triangulability. IX, 160 pages. 1984.

Vol. 1103: Models and Sets. Proceedings, Logic Colloquium, 1983, Part I. Edited by G.H. Müller and M.M. Richter. VIII, 484 pages. 1984.

Vol. 1104: Computation and Proof Theory. Proceedings, Logic Colloquium, 1983, Part II. Edited by M.M. Richter, E. Börger, W. Oberschelp, B. Schinzel and W. Thomas. VIII, 475 pages. 1984.

Vol. 1105: Rational Approximation and Interpolation. Proceedings, 1983. Edited by P.R. Graves-Morris, E.B. Saff and R.S. Varga. XII, 528 pages. 1984.

Vol. 1106: C.T. Chong, Techniques of Admissible Recursion Theory. IX, 214 pages. 1984.

Vol. 1107: Nonlinear Analysis and Optimization. Proceedings, 1982. Edited by C. Vinti. V, 224 pages. 1984.

Vol. 1108: Global Analysis – Studies and Applications I. Edited by Yu.G. Borisovich and Yu.E. Gliklikh. V, 301 pages. 1984.

Vol. 1109: Stochastic Aspects of Classical and Quantum Systems. Proceedings, 1983. Edited by S. Albeverio, P. Combe and M. Sirugue-Collin. IX, 227 pages. 1985.

Vol. 1110: R. Jajte, Strong Limit Theorems in Non-Commutative Probability. VI, 152 pages. 1985.

Vol. 1111: Arbeitstagung Bonn 1984. Proceedings. Edited by F. Hirzebruch, J. Schwermer and S. Suter. V, 481 pages. 1985.

Vol. 1112: Products of Conjugacy Classes in Groups. Edited by Z. Arad and M. Herzog. V, 244 pages. 1985.

Vol. 1113: P. Antosik, C. Swartz, Matrix Methods in Analysis. IV, 114 pages. 1985.

Vol. 1114: Zahlentheoretische Analysis. Seminar. Herausgegeben von E. Hlawka. V, 157 Seiten. 1985.

Vol. 1115: J. Moulin Ollagnier, Ergodic Theory and Statistical Mechanics. VI, 147 pages. 1985.

Vol. 1116: S. Stolz, Hochzusammenhängende Mannigfaltigkeiten und ihre Ränder. XXIII, 134 Seiten. 1985.

Vol. 1117: D.J. Aldous, J.A. Ibragimov, J. Jacod, Ecole d'Été de Probabilités de Saint-Flour XIII – 1983. Édité par P.L. Hennequin. IX, 409 pages. 1985.

Vol. 1118: Grossissements de filtrations: exemples et applications. Seminaire, 1982/83. Edité par Th. Jeulin et M. Yor. V, 315 pages. 1985.

Vol. 1119: Recent Mathematical Methods in Dynamic Programming. Proceedings, 1984. Edited by I. Capuzzo Dolcetta, W.H. Fleming and T. Zolezzi. VI, 202 pages. 1985.

Vol. 1120: K. Jarosz, Perturbations of Banach Algebras. V, 118 pages. 1985.

Vol. 1121: Singularities and Constructive Methods for Their Treatment. Proceedings, 1983. Edited by P. Grisvard, W. Wendland and J.R. Whiteman. IX, 346 pages. 1985.

Vol. 1122: Number Theory. Proceedings, 1984. Edited by K. Alladi. VII, 217 pages. 1985.

Vol. 1123: Séminaire de Probabilités XIX 1983/84. Proceedings. Edité par J. Azéma et M. Yor. IV, 504 pages. 1985.

Vol. 1124: Algebraic Geometry, Sitges (Barcelona) 1983. Proceedings. Edited by E. Casas-Alvero, G.E. Welters and S. Xambó-Descamps. XI, 416 pages. 1985.

Vol. 1125: Dynamical Systems and Bifurcations. Proceedings, 1984. Edited by B.L.J. Braaksma, H.W. Broer and F. Takens. V, 129 pages. 1985.

Vol. 1126: Algebraic and Geometric Topology. Proceedings, 1983. Edited by A. Ranicki, N. Levitt and F. Quinn. V, 423 pages. 1985.

Vol. 1127: Numerical Methods in Fluid Dynamics. Seminar. Edited by F. Brezzi, VII, 333 pages. 1985.

Vol. 1128: J. Elschner, Singular Ordinary Differential Operators and Pseudodifferential Equations. 200 pages. 1985.

Vol. 1129: Numerical Analysis, Lancaster 1984. Proceedings. Edited by P.R. Turner. XIV, 179 pages. 1985.

Vol. 1130: Methods in Mathematical Logic. Proceedings, 1983. Edited by C.A. Di Prisco. VII, 407 pages. 1985.

Vol. 1131: K. Sundaresan, S. Swaminathan, Geometry and Nonlinear Analysis in Banach Spaces. III, 116 pages. 1985.

Vol. 1132: Operator Algebras and their Connections with Topology and Ergodic Theory. Proceedings, 1983. Edited by H. Araki, C.C. Moore, Ş. Strătilă and C. Voiculescu. VI, 594 pages. 1985.

Vol. 1133: K.C. Kiwiel, Methods of Descent for Nondifferentiable Optimization. VI, 362 pages. 1985.

Vol. 1134: G.P. Galdi, S. Rionero, Weighted Energy Methods in Fluid Dynamics and Elasticity. VII, 126 pages. 1985.

Vol. 1135: Number Theory, New York 1983–84. Seminar. Edited by D.V. Chudnovsky, G.V. Chudnovsky, H. Cohn and M.B. Nathanson. V, 283 pages. 1985.

Vol. 1136: Quantum Probability and Applications II. Proceedings, 1984. Edited by L. Accardi and W. von Waldenfels. VI, 534 pages. 1985.

Vol. 1137: Xiao G., Surfaces fibrées en courbes de genre deux. IX, 103 pages. 1985.

Vol. 1138: A. Ocneanu, Actions of Discrete Amenable Groups on von Neumann Algebras. V, 115 pages. 1985.

Vol. 1139: Differential Geometric Methods in Mathematical Physics. Proceedings, 1983. Edited by H. D. Doebner and J. D. Hennig. VI, 337 pages. 1985.

Vol. 1140: S. Donkin, Rational Representations of Algebraic Groups. VII, 254 pages. 1985.

Vol. 1141: Recursion Theory Week. Proceedings, 1984. Edited by H.-D. Ebbinghaus, G.H. Müller and G.E. Sacks. IX, 418 pages. 1985.

Vol. 1142: Orders and their Applications. Proceedings, 1984. Edited by I. Reiner and K. W. Roggenkamp. X, 306 pages. 1985.

Vol. 1143: A. Krieg, Modular Forms on Half-Spaces of Quaternions. XIII, 203 pages. 1985.

Vol. 1144: Knot Theory and Manifolds. Proceedings, 1983. Edited by D. Rolfsen. V, 163 pages. 1985.